De Gruyter Graduate

Tadros • Nanodispersions

Also of interest

Interfacial Phenomena and Colloid Stability:
Volume 1 Basic Principles
Tadros, 2015
ISBN 978-3-11-028344-0, e-ISBN 978-3-11-028343-3

Interfacial Phenomena and Colloid Stability:
Volume 2 Industrial Applications
Tadros, 2015
ISBN 978-3-11-036648-8, e-ISBN 978-3-11-036647-1

An Introduction to Surfactants
Tadros, 2014
ISBN 978-3-11-031212-6, e-ISBN 978-3-11-031213-3

Microencapsulation: Innovative Applications
Giamberini, Fernandez-Prieto, Tylkowski (Eds.), 2015
ISBN 978-3-11-033187-5, e-ISBN 978-3-11-033199-8

Tharwat F. Tadros

Nanodispersions

—

DE GRUYTER

Author
Prof. Tharwat F. Tadros
89 Nash Grove Lane
Workingham RG40 4HE
Berkshire, UK
tharwat@tadros.fsnet.co.uk

ISBN 978-3-11-029033-2
e-ISBN (PDF) 978-3-11-029034-9
e-ISBN (EPUB) 978-3-11-038879-4

Library of Congress Cataloging-in-Publication Data
A CIP catalog record for this book has been applied for at the Library of Congress.

Bibliographic information published by the Deutsche Nationalbibliothek
The Deutsche Nationalbibliothek lists this publication in the Deutsche Nationalbibliografie;
detailed bibliographic data are available on the Internet at http://dnb.dnb.de.

© 2016 Walter de Gruyter GmbH, Berlin/Boston
Cover image: Kesu01/iStock/Thinkstock
Typesetting: PTP-Berlin, Protago-TEX-Production GmbH, Berlin
Printing and binding: CPI books GmbH, Leck
♾ Printed on acid-free paper
Printed in Germany

www.degruyter.com

Preface

It is generally accepted that nanodispersions cover the range 10–200 nm diameter. These systems fall within the colloid range (1 nm–1 μm) and hence one can apply the general theories of colloid stability for these systems. The resulting dispersion can be transparent, translucent or turbid depending on three main factors, namely the particle or droplet radius, the difference in refractive index between the dispersed phase and dispersion medium and the volume fraction of the disperse phase. Several advantages of nanodispersions can be listed: (i) The very small particle or droplet size causes a large reduction in gravity force and Brownian motion may be sufficient for overcoming gravity. This means that no creaming or sedimentation occurs on storage. (ii) The small droplet size also prevents any flocculation of the particles or droplets, due to the small van der Waals attraction between the particles or droplets. The repulsive energy produced by surfactants and/or polymers will give nanodispersions a high kinetic stability, particularly when using polymeric surfactants that provide strong steric repulsion. (iii) The small droplets within nanoemulsions will also prevent coalescence, since these droplets are nondeformable and hence surface fluctuations are prevented. In addition, the significant surfactant film thickness (relative to droplet radius) prevents any thinning or disruption of the liquid film between the droplets. (iv) Nanosuspensions have wide applications in drug delivery systems for poorly insoluble compounds, whereby reduction of particle size to nanoscale dimensions enhances the drug bioavailability. This is due to the increase of solubility of the active ingredient on reduction of particle radius. (v) Nanoemulsions are suitable for efficient delivery of active ingredients through the skin. The large surface area of the emulsion system allows rapid penetration of actives. (vi) Due to their small size, nanoemulsions can penetrate through the "rough" skin surface and this enhances penetration of actives. (vii) The transparent nature of the system, their fluidity (at reasonable oil concentrations) as well as the absence of any thickeners may give them a pleasant aesthetic character and skin feel. (viii) Nanoemulsions can be prepared using reasonable surfactant concentration. For a 20 % O/W nanoemulsion, a surfactant concentration in the region of 5 % may be sufficient. (ix) The small size of the droplets allows them to deposit uniformly on substrates. Wetting, spreading and penetration may also be enhanced as a result of the low surface tension of the whole system and the low interfacial tension of the O/W droplets. (x) Nanoemulsions can be applied for delivery of fragrants which may be incorporated in many personal care products. (xi) Nanoemulsions may be applied as a substitute for liposomes and vesicles (which are much less stable) and in some cases it is possible to build lamellar liquid crystalline phases around the nanoemulsion droplets.

Two methods can be applied for preparation of nanosuspensions: (i) The bottom-up approach where one starts with molecular components and builds up the particles by a process of nucleation and growth. (ii) The top-down process where one starts with the bulk material (which may consist of aggregates and agglomerates), this is dispersed into single particles (using a wetting/dispersing agent) and then the large particles are subdivided into smaller units that fall within the required nanosize. This process requires the application of intense mechanical energy that can be applied using bead milling, high pressure homogenization and/or application of ultrasonics. The preparation of nanopolymer colloids using emulsion and suspension polymerization can be considered a bottom-up process since one starts with the monomer which is then polymerized using an initiator. The preparation of biodegradable nanoparticles by aggregation of A–B block copolymers (such as polylactic polyglycolic block copolymer) can also be considered a bottom-up process. A special case for producing nanodispersions is the formation of liposomes and vesicles. Liposomes are multilamellar structures consisting of several bilayers of lipids (several µm). They are produced by simply shaking an aqueous solution of phospholipids, e.g. egg lecithin. When sonicated, these multilayer structures produce unilamellar structures (with size range of 25–50 nm) that are referred to as liposomes.

There are generally two methods for stabilizing nanodispersions. The first method depends on charge separation and formation of an electrical double layer whose extension (double layer thickness) depends on electrolyte concentration and valency of the counter and co-ions. When the particles approach to a distance h that is smaller than twice the double layer extension, strong repulsion occurs, particularly when the $1:1$ electrolyte (e.g. NaCl) is lower than 10^{-2} mol dm^{-3}. This repulsion can overcome the van der Waals attraction at intermediate particle separation resulting in an energy barrier (maximum) that prevents particle approach and hence flocculation is prevented. The second and more effective mechanism is obtained by using nonionic surfactants or polymers (referred to as polymeric surfactant). When two particles or droplets approach to a distance h that is smaller than 2δ, the stabilizing chains may overlap or become compressed resulting in very strong repulsion as a result of two main effects, namely unfavourable mixing of the stabilizing chains when these are in good solvent conditions and unfavourable loss of configurational entropy of the stabilizing chains on significant overlap. Combination of these two effects is referred to as steric repulsion which, when combined with the van der Waals attraction, results in an energy-distance with only one shallow minimum.

The formation of nanosuspensions by a bottom-up or top-down process seldom results in monodisperse particles. By proper control of the method of preparation, one may at best produce a narrow size distribution which still consists of small and larger particles. The smaller size particles will have a higher solubility than the larger ones. On storage, molecular diffusion will occur from the smaller to the larger particles and this results in a shift in the particle size distribution to larger values. This process is referred to as Ostwald ripening and the driving force is the difference in solubility between the small and the large particles.

Several industrial applications of nanodispersions have emerged in the last few decades. One of the most important applications is in drug delivery of poorly soluble drugs. By reducing the size of the particles to values in the nanosize range, the solubility of the drug is greatly enhanced and this enhances bioavailability. Another important application of nanodispersions is the use of biodegradable polymer nanoparticles for targeted delivery of drugs, e.g. in cancer treatment. Another important area of application of nanodispersions is in the field of cosmetics and personal care. Sunscreen dispersions of semiconductor TiO_2 require particles in the range 30–50 nm which need to be stable against aggregation in the formulation and on application. Nanoemulsions are applied in many hand creams for efficient delivery of anti-wrinkle agents. Liposomes and vesicles are also used in many formulations for efficient delivery of ceramides that protect the skin from ageing. Several other industrial applications of nanoparticles can be mentioned such as preparation of nanosize catalyst particles, nanopolymer colloids, ceramics, etc.

The present text addresses the topic of nanodispersions at a fundamental level and some of their industrial applications. Chapter 1 gives a general introduction of the definition of nanodispersions, their main advantages and general applications in industry. Chapter 2 deals with the colloid stability of nanodispersions, both electrostatic and steric stabilization. The interaction between particles or droplets containing double layers is described and when combined with the van der Waals attraction results in the general theory of colloid stability due to Deryaguin–Landau–Verwey–Overbeek (DLVO theory). This is followed by describing the interaction between particles or droplets containing adsorbed surfactant or polymer layers. Steric interaction is described in terms of the unfavourable mixing of the adsorbed layers (when these are in good solvent conditions) and the unfavourable loss of entropy on significant overlap of the adsorbed layers. Combining these two effects with van der Waals attraction results in the theory of steric stabilization. Chapter 3 describes the Ostwald ripening of nanodispersions and its prevention. It starts with the Kelvin theory that describes the effect of curvature (particle or droplet size) on the solubility of the disperse phases. It shows the rapid increase in solubility when the particle or droplet size is reduced below 500 nm. The kinetics of Ostwald ripening is described showing the change of the cube of the mean radius with time and the effect of oil solubility on the rate. This is followed by the methods that can be applied to reduce Ostwald ripening for both nanosuspensions and nanoemulsions. Chapter 4 describes the methods of preparation of nanosuspensions by the bottom-up process. The theory of nucleation and growth is described with particular reference to the effect of supersaturation and presence of surfactants and polymers. The preparation of nanopolymer colloids (lattices) using miniemulsion polymerization is described. This is followed by a section on the application of microemulsions for the preparation of nanopolymer colloids. Chapter 5 describes the methods of preparation of nanosuspensions using the top-down process. The process of wetting of powder aggregates and agglomerates is described with special reference to the role of surfactants (wetting agents). The different classes of surfactants for enhancing wetting are described. This is followed by a section on dispersion of the aggregates and agglomerates using high speed stirrers. Finally, the methods that can be applied for size reduction are described with reference to microfluidization (using high pressure homogenizers) and bead milling. The main factors responsible for maintenance of colloid stability are described. Chapter 6 describes the industrial applications of nanosuspensions. Two main areas will be considered in some detail, namely the use of nanosuspensions for size reduction of highly insoluble drugs to enhance their bioavailability and application of nanosuspensions of semiconductor titanium dioxide in sunscreens. Chapter 7 deals with the application of nanoparticles as drug carriers for targeted delivery of the active ingredient. It deals with the formation of liposomes (multilamellar lipid bilayers) and vesicles (unilamellar lipid bilayer) and the nature and structure of lipids used in their preparation. The formation of vesicles by sonication of liposomes is described and the origin of their stability by application of thermodynamic principles is analyzed. The use of block copolymers for stabilization of the lipid bilayer is described at a fundamental level. The application of liposomes and vesicles in drug delivery is briefly described. The use of model nanolatex particles for investigating the circulation of the nanoparticles and preventing their engulfment by the Kupffer cells in the liver is described. The influence of particle surface charge, size and presence of adsorbed polymer layers on particle circulation is discussed. The use of biodegradable polymers of polylactic polyglycolic A–B block copolymers for preparation of biodegradable nanoparticles is described followed by the methods of synthesis and characterization of the block copolymer. The characterization of the resulting nanoparticles using light scattering and rheological techniques is briefly described. The prevention of protein adsorption by the hydrophilic chain is explained at a fundamental level. This is followed by an investigation of the circulation of the nanoparticle using radio labelled particles. Chapter 8 describes the preparation of nanoemulsions using high pressure homogenizers. The selection of emulsifiers and description of their role in prevention of coalescence during emulsification are analyzed at a fundamental level. The methods of emulsification and application of high pressure and ultrasonic techniques for formation of the nanoemulsions are described with particular reference to the influence of the formulation variables such as the disperse volume fraction, nature and concentration of the emulsifier on the property of the nanoemulsion. The characterization of the nanoemulsion using dynamic light scattering technique is described. Chapter 9 deals with the preparation of nanoemulsions using low energy techniques. Three methods are described, namely the phase inversion composition, the phase inversion temperature and dilution of microemulsion methods. A comparison of nanoemulsions prepared using low energy emulsification and high energy methods is given. Particular attention is given to the effect of polydispersity and Ostwald ripening rates. Chapter 9 also describes some practical examples of nanoemulsions with particular reference to the effect of oil solubility and nature of the emulsifier. Chapter 10 deals with the topic of solubilized systems and swollen micelles (microemulsions). The thermodynamic definition of microemulsions, their spontaneous formation and stability are described at a fundamental level. This is followed by describing the theories of microemulsion formation and stability and the need for using two surfactants to produce ultra-low interfacial tension. The characterization of microemulsions using scattering, conductivity and NMR techniques is described. Chapter 10 also describes the methods that can be applied for formulation of microemulsions and the importance of the partition of the cosurfactant between the oil and water phases. The various methods that can be applied for selecting emulsifiers for microemulsion formation are described. The application of microemulsions in several industries is briefly described.

The present book provides the reader with the fundamental principles of preparation of nanodispersions and their stabilization. It could be valuable for research workers in academia as well as in industrial research laboratories. It also provides the reader with various practical applications of nanoemulsions in various fields such as pharmaceuticals, cosmetics, agrochemicals, the oil industry, etc. The text is certainly valuable for formulation scientists, chemical engineers and research scientists involved in this important topic.

Tharwat Tadros   November, 2015

Contents

1 Nanodispersions – general introduction

1.1 Definition of colloids

The recognition of the colloidal state of matter started in 1861 with Thomas Graham [1]: "Systems in which a significant proportion of the molecules lie in or are associated with interfacial regions". Simple considerations suggest that the lowest limit for colloids (whereby one can distinguish between molecules in the interfacial region and the bulk) is 1 nm. The upper limit for colloidal dispersions lies in the region of 1000 nm (1 μm), whereby a significant proportion of the total molecules lie at the interface.

Colloids encompass disperse systems of several kinds:
- Solid/liquid suspensions
- Liquid/liquid emulsions
- Liquid/gas aerosols
- Gas/liquid foams
- Liquid/solid gels
- Solid/gas smokes
- Solid/solid composites

Macromolecular solutions may also be regarded as colloids since the coil dimensions of most polymers can exceed 10 nm. Surfactant micelles (aggregates of monomers) are also colloids since the micelle size is in the region of 5–10 nm.

1.2 Definition of nanodispersions

It is generally accepted that nanodispersions of the solid/liquid (nanosuspensions) and liquid/liquid (nanoemulsions) types cover the range 10–200 nm diameter. Some authors [2] consider a smaller range of the order of 10–50 nm diameter. These systems fall within the colloid range described above and hence one can apply the general theories of colloid stability for these systems. The resulting dispersion can be transparent, translucent or turbid depending on three main factors, namely the particle or droplet radius, the difference in refractive index between the dispersed phase and dispersion medium and the volume fraction of the disperse phase. This can be understood if one considered the variation of the light scattering intensity with these parameters.

For small particles, with a radius smaller than $\lambda/20$ (where λ is the wavelength of light, that is $\sim 400–600$ nm), the scattering intensity $I(\theta)$ (where θ is the scattering angle) is related to the incident intensity $I(0)$ by the following equation [3]:

$$\left[\frac{I(\theta)}{I(0)} \right] = \frac{\pi^2 d_p^6}{4r^2\lambda^4} \left(\frac{n_{21}^2 - 1}{n_{21}^2 + 2} \right)^2 , \tag{1.1}$$

where d_p is the particle diameter, r is the distance to the detector and n_{21} is the ratio of the refractive index of the particle or droplet n_2 relative to that of the medium n_1, i.e. $n_{21} = n_2/n_1$.

Equation (1.1) applies to a very dilute dispersion where the separation distance between the particles is so large that the scattering from each particle is not subsequently scattered a second or third time by the neighbouring particles, i.e. there is no multiple scattering.

Equation (1.1) shows that the relative scattering intensity is proportional to the sixth power of the particle or droplet diameter and the square of the ratio of the refractive index of the particle or droplet and the medium. Clearly if $n_{21} = 1$, i.e. the refractive index of the particle and that of the medium are matched, then $I(\theta)/I(0) = 0$ and the whole nanodispersion becomes transparent regardless of the value of the particle diameter. However, this is seldom the case and in most practical systems $n_{21} > 1$. At any given d_p the higher the value of n_{21} the higher the relative scattering intensity. This clearly shows that with nanodispersions with high n_{21} the dispersion may appear translucent or turbid even though d_p is small (say < 30 nm). In contrast, a nanodispersion with low n_{21} may appear transparent even though d_p is large (say > 50 nm). Clearly the relative intensity of scattered light at a given n_{21} is proportional to d_p^6 and hence the larger the value of d_p the less transparent the nanodispersion is.

Another factor that determines the lack of transparency with many nanodispersions is the volume fraction of the disperse phase which determines the number of particles or droplets per unit volume. At high particle or droplet number density multiple scattering occurs and a dilute transparent nanodispersion may become translucent or turbid as the particle or droplet number concentration is increased.

1.3 Main advantages of nanodispersions

(i) The very small particle or droplet size causes a large reduction in gravity force and the Brownian motion may be sufficient for overcoming gravity. Gravity force is given by

$$\text{Gravity Force} = \frac{4}{3}\pi R^3 \Delta\rho gh, \tag{1.2}$$

where R is the particle radius, $\Delta\rho$ is the density difference between the particle or droplet and that of the medium, g is the acceleration due to gravity and h is the height of the container. Brownian motion is given by kT, where k is the Boltzmann constant and T is the absolute temperature.

With nanodispersions

$$kT > \frac{4}{3}R^3 \Delta\rho gh. \tag{1.3}$$

This means that no creaming or sedimentation occurs on storage.

(ii) The small droplet size also prevents any flocculation of the particles or droplets. As will be shown later the driving force for flocculation is van der Waals attraction, which for spherical particles at short distances of separation is proportional to the particle or droplet radius [4]. Since R is much smaller than that of suspensions and emulsions, the van der Waals attraction between the particles or droplets is also much smaller than that of particles of suspensions or emulsions. The repulsive energy produced by surfactants and/or polymers will give nanodispersions a high kinetic stability, particularly when using polymeric surfactants that provide strong steric repulsion. As we will see later weak flocculation is also prevented when the ratio of adsorbed layer thickness to particle radius is large (> 0.1) and this enables the system to remain dispersed with no separation.

(iii) The small droplets in nanoemulsions will also prevent them coalescing, since these droplets are nondeformable and hence surface fluctuations are prevented. In addition, the significant surfactant film thickness (relative to droplet radius) prevents any thinning or disruption of the liquid film between the droplets.

(iv) Nanosuspensions have wide applications in drug delivery systems of poorly insoluble compounds, whereby reduction of particle size to nanoscale dimensions enhances the drug bioavailability. This is due to the increase in solubility of the active ingredient on reduction of particle radius as given by the Kelvin equation [5],

$$S(r) = S(\infty) \exp\left(\frac{2\gamma V_m}{rRT}\right), \tag{1.4}$$

where S(r) is the solubility of a particle with radius r and S(∞) is the solubility of a particle with infinite radius (the bulk solubility), γ is the S/L interfacial tension, R is the gas constant and T is the absolute temperature.

Equation (1.4) shows a significant increase in solubility of the particle with reduction of particle radius, particularly when the latter becomes significantly smaller than 1 μm.

(v) Nanoemulsions are suitable for efficient delivery of active ingredients through the skin. The large surface area of the emulsion system allows rapid penetration of actives.

(vi) Due to their small size, nanoemulsions can penetrate through the "rough" skin surface and this enhances penetration of actives.

(vii) The transparent nature of the system, their fluidity (at reasonable oil concentrations) as well as the absence of any thickeners may give them a pleasant aesthetic character and skin feel.

(viii) Nanoemulsions can be prepared using reasonable surfactant concentration. For a 20 % O/W nanoemulsion, a surfactant concentration in the region of 5 % may be sufficient.

(ix) The small size of the droplets allows them to deposit uniformly on substrates. Wetting, spreading and penetration may also be enhanced as a result of the low surface tension of the whole system and the low interfacial tension of the O/W droplets.

(x) Nanoemulsions can be applied for delivery of fragrants which may be incorporated in many personal care products. This could also be applied in perfumes which are desirable to be formulated alcohol free.

(xi) Nanoemulsions may be applied as a substitute for liposomes and vesicles (which are much less stable) and it is possible in some cases to build lamellar liquid crystalline phases around the nanoemulsion droplets.

1.4 General methods for preparation of nanodispersions

Two methods can be applied for preparation of nanosuspensions: (i) The bottom-up approach where one starts with molecular components and builds up the particles by a process of nucleation and growth. (ii) The top-down process where one starts with the bulk material (which may consist of aggregates and agglomerates), this is dispersed into single particles (using a wetting/dispersing agent) and then the large particles are subdivided into smaller units that fall within the required nanosize. This process requires the application of intense mechanical energy that can be applied using bead milling, high pressure homogenization and/or application of ultrasonics [6].

The preparation of nanopolymer colloids using emulsion and suspension polymerization [7] can be considered a bottom-up process since one starts with the monomer that is polymerized using an initiator. The preparation of biodegradable nanoparticles by aggregation of A–B block copolymers (such as polylactic polyglycolic block copolymer) can also be considered a bottom-up process.

For preparation of nanoemulsions (covering the droplet radius size range 50–200 nm) four methods may be applied: use of high pressure homogenizers (aided by appropriate choice of surfactants and cosurfactants); application of the phase inversion composition method; application of the phase inversion temperature (PIT) concept; dilution of a microemulsion.

Several other methods are applied to prepare nanodispersions, such as solubilization of water insoluble compounds in the hydrophobic core of the micelle. A special case in preparing nanodispersions is referred to as formation of swollen micelles (initially referred to as microemulsions). In this case an oil, water, surfactant and cosurfactant are chosen to produce an ultra-low interfacial tension ($< 10^{-2}$ mN m^{-1}). The swollen micelles are then produced spontaneously and the system becomes thermodynamically stable [7]. A special case in producing nanodispersions is the formation of liposomes and vesicles. Liposomes are multilamellar structures consisting of several bilayers of lipids (several µm). They are produced by simply shaking an aqueous solution of phospholipids, e.g. egg lecithin. When sonicated, these multilayer structures produce unilamellar structures (with size range of 25–50 nm) that are referred to as liposomes [8]. Glycerol-containing phospholipids are used for the preparation of liposomes and vesicles such as phosphatidylcholine, phosphatidylserine, phosphatidylethanolamine, phosphatidylinositol, phosphatidylglycerol, phosphatidic acid and cholesterol. In most preparations, a mixture of lipids is used to obtain the optimum structure.

1.5 General stabilization mechanisms for nanodispersions

There are generally two methods for stabilizing nanodispersions. The first method depends on charge separation and formation of an electrical double layer. Since many nanodispersions contain surface groups that can be dissociated, e.g. oxides that contain OH groups, or latex-containing sulphate groups, a surface charge can be produced as a result of the dissociation of the surface groups. This charge is compensated by unequal distribution of counterions (with opposite charge sign to the surface charge) and co-ions (with the same charge sign as the surface charge). The same mechanism occurs when using ionic surfactants that can adsorb on the particle or droplet surface. As a result an electrical double layer is formed whose extension (double layer thickness) depends on electrolyte concentration and valency of the counter- and co-ions. When the particles approach to a distance h that is smaller than twice the double layer extension, strong repulsion occurs, particularly when the 1 : 1 electrolyte (e.g. NaCl) is lower than 10^{-2} mol dm^{-3}. This repulsion can overcome the van der Waals attraction at intermediate particle separation resulting in an energy barrier (maximum) that prevents particle approach and hence flocculation is prevented [9, 10].

The second and more effective mechanism is using nonionic surfactants or polymers (referred to as polymeric surfactant) [11]. These molecules consist of a hydrophobic chain (such as an alkyl chain B with nonionic surfactants, or polystyrene or polymethylmethacrylate or polypropylene oxide B chain for an A–B, A–B–A or BA$_n$) and a hydrophilic chain A (such as polyethylene oxide). On a hydrophobic particle or an oil droplet in aqueous medium, the insoluble B chain adsorbs strongly on the surface of a particle or droplet, leaving the hydrophilic chain (that is strongly hydrated) in the aqueous medium. The surfactant or polymer layer will have a thickness δ that depends on the number of ethylene oxide units in the A chain and its hydration by water molecules. When two particles or droplets approach to a distance h that is smaller than 2δ, the A chains may overlap or become compressed resulting in very strong repulsion as a result of two main effects [12]: (i) Unfavourable mixing of the A chains when these are in good solvent conditions; (ii) unfavourable loss of configurational entropy of the A chains on significant overlap. Combination of these two effects is referred to as steric repulsion which, when combined with van der Waals attraction, results in an energy-distance with only one shallow minimum. When h < 2δ, the energy increases very sharply with a further decrease in h. This method of stabilization is more effective than electrostatic stabilization in two main respects. Firstly, the repulsion is still maintained at moderate electrolyte concentration (~ 1 mol dm^{-3} NaCl). Secondly, the repulsion can be maintained at high temperatures (exceeding in some cases 50 °C) providing the chains remain hydrated at such high temperatures.

1.6 Ostwald ripening in nanodispersions

The formation of nanosuspensions by bottom-up or top-down processes seldom results in monodisperse particles. By proper control of the method of preparation, one may at best produce a narrow size distribution which still consists of small and larger particles. As shown in equation (1.4) the smaller size particles will have a higher solubility than the larger ones. On storage, molecular diffusion will occur from the smaller to the larger particles and this results in a shift in the particle size distribution to larger values. This process is referred to as Ostwald ripening and the driving force is the difference in solubility between the small and the large particles. This process may continue with increasing time and higher temperatures with the ultimate result that the dispersion reaches sizes outside the nanosize range. This process must be inhibited or eliminated by addition of strongly adsorbed impurities and/or polymeric surfactants.

Nanoemulsions prepared by phase composition or phase inversion temperature techniques usually result in a polydisperse system. The smaller droplets will also have higher solubility than the larger one resulting in Ostwald ripening of the nanoemulsion, which become more turbid on storage with the ultimate formation of an emulsion. The nanoemulsions prepared using high pressure homogenizers are less polydisperse than those prepared using low energy methods. Still these nanoemulsions will show Ostwald ripening but with a lower rate. Ostwald ripening in nanoemulsions can be reduced by addition of a second oil with much lower solubility. During Ostwald ripening partitioning of the more soluble oil will occur with the result that the less soluble oil becomes more concentrated in the smaller droplets. The difference in chemical potential between the small and large droplets due to the curvature effect will be balanced by the difference in chemical potential resulting from the difference in composition of the two droplets. This will show a significant reduction in Ostwald ripening. Further reduction can be produced by addition of an oil soluble polymeric surfactant that strongly adsorbs at the oil/water interface.

1.7 Industrial applications of nanodispersions

Several industrial applications of nanodispersions have emerged in the last few decades. One of the most important applications is in drug delivery of poorly soluble drugs. By reducing the size of the particles to values in the nanosize range, the solubility of the drug is greatly enhanced, as predicted by equation (1.4), and this enhances bioavailability. Another important application of nanodispersions is the use of biodegradable polymer nanoparticles for targeted delivery of drugs, e.g. in cancer treatment. Another important application is the use of nanoemulsions for drug delivery such as anaesthetics. A further application of nanoemulsions is the delivery of lipids (Intralipid formulation) for treating anorexic people.

Another important area of application of nanodispersions is in the field of cosmetics and personal care. Sunscreen dispersions of semiconductor TiO_2 require particles in the range 30–50 nm which need to be stable against aggregation in the formulation and on application. Nanoemulsions are applied in many hand creams for efficient delivery of anti-wrinkle agents. Liposomes and vesicles are also used in many formulations for efficient delivery of ceramides that protect the skin from ageing.

Several other industrial applications of nanoparticles can be mentioned such as preparation of nanosize catalyst particles, nanopolymer colloids, ceramics, etc.

1.8 Outline of the book

Chapter 2 deals with the colloid stability of nanodispersions, both electrostatic and steric stabilization. The origin of charge in nanodispersions and the structure of the electrical double layer are discussed at a fundamental level. The charge and potential distribution at the solid/liquid interface are described with particular reference to the effect of electrolyte concentration and valency on the double layer extension. The interaction between particles or droplets containing double layers is described and a description is given of how the double layer repulsion changes with separation distance between the particle or droplet surfaces. The van der Waals attraction between the particles or droplets is described using the microscopic theory of Hamaker. This results in the description of the variation of van der Waals attraction with separation distance. Electrostatic repulsion is combined with van der Waals attraction at various separation distances in the well-known theory of colloid stability due to Deryaguin–Landau–Verwey–Overbeek (DLVO theory) [8, 9]. Energy-distance curves are presented at various electrolyte concentrations to distinguish between stable and unstable systems. The adsorption and conformation of nonionic surfactants and polymers at various interfaces is described. This is followed by describing the interaction between particles or droplets containing adsorbed surfactant or polymer layers [12]. Steric interaction is described in terms of the unfavourable mixing of the adsorbed layers (when these are in good solvent conditions) and the unfavourable loss of entropy on significant overlap of the adsorbed layers. Combining these two effects with van der Waals attraction results in the theory of steric stabilization. The energy-distance curves of sterically stabilized dispersions are presented with particular reference to the effect of the ratio of adsorbed layer thickness to particle or droplet radius.

Chapter 3 describes the Ostwald ripening of nanodispersions and its prevention. It starts with the Kelvin theory that describes the effect of curvature (particle or droplet size) on the solubility of the disperse phases. It shows the rapid increase in solubility when the particle or droplet size is reduced below 500 nm. The kinetics of Ostwald ripening is described showing the change of the cube of the mean radius with time and the effect of oil solubility on the rate. The effect of Brownian motion, phase volume fraction and surfactant micelles is described. This is followed by the methods that can be applied to reduce Ostwald ripening for both nanosuspensions and nanoemulsions.

Chapter 4 describes the methods of preparation of nanosuspensions by the bottom-up process. The theory of nucleation and growth is described with particular reference to the effect of supersaturation and presence of surfactants and polymers. The preparation of nanopolymer colloids (lattices) using miniemulsion polymerization is described. This is followed by a section on the application of microemulsions for the preparation of nanopolymer colloids.

Chapter 5 describes the methods of preparation of nanosuspensions using the top-down process. The process of wetting of powder aggregates and agglomerates is described with special reference to the role of surfactants (wetting agents). The different classes of surfactants for enhancing wetting are described. This is followed by a section on dispersion of the aggregates and agglomerates using high speed stirrers. Finally, the methods that can be applied for size reduction are described with reference to microfluidization (using high pressure homogenizers) and bead milling. The main factors responsible for maintenance of colloid stability are described.

Chapter 6 describes the industrial applications of nanosuspensions. Two main areas will be considered in some detail: the use of nanosuspensions for size reduction of highly insoluble drugs to enhance their bioavailability and enhancing the rate of dissolution and solubility of drugs using nanosuspensions. The application of nanosuspensions of semiconductor titanium dioxide in sunscreens is described, as are the role of particle size in UV protection; stabilization of nanosuspensions of titanium dioxide by using polymeric surfactants and their subsequent incorporation in creams for application and the importance of transparency of the deposited particles.

Chapter 7 deals with the application of nanoparticles as drug carriers for targeted delivery of the active ingredient. It deals with the formation of liposomes (multilamellar lipid bilayers) and vesicles (unilamellar lipid bilayer) and the nature and structure of lipids used in their preparation. The formation of vesicles by sonication of liposomes is described and the origin of their stability by application of thermodynamic principles is analyzed. The use of block copolymers for stabilization of the lipid bilayer is described at a fundamental level. The application of liposomes and vesicles in drug delivery is briefly described. The use of model nanolatex particles for investigating the circulation of the nanoparticles and preventing their engulfment by the Kupffer cells in the liver is described. The influence of particle surface charge, size and presence of adsorbed polymer layers on particle circulation is discussed. The use of biodegradable polymers of polylactic polyglycolic A–B block copolymers for preparation of biodegradable nanoparticles is described followed by the methods of synthesis and characterization of the block copolymer. The characterization of the resulting nanoparticles using light scattering and rheological techniques is briefly described. The prevention of protein adsorption by the hydrophilic chain is explained at a fundamental level. This is followed by an investigation of the circulation of the nanoparticle using radio labelled particles.

Chapter 8 describes the preparation of nanoemulsions using high pressure homogenizers. The selection of emulsifiers and the description of their role in prevention of coalescence during emulsification are analyzed at a fundamental level. The methods of emulsification and application of high pressure and ultrasonic techniques for formation of the nanoemulsions are described with particular reference to the influence of the formulation variables such as the disperse volume fraction, nature and concentration of the emulsifier on the property of the nanoemulsion. The characterization of the nanoemulsion using dynamic light scattering technique is described.

Chapter 9 deals with the preparation of nanoemulsions using low energy techniques. Three methods are described, namely the phase inversion composition, the phase inversion temperature and dilution of microemulsion methods. A comparison of nanoemulsions prepared using low energy emulsification and high energy methods is given. Particular attention is given to the effect of polydispersity and Ostwald ripening rates. Chapter 9 also describes some practical examples of nanoemulsions with particular reference to the effect of oil solubility and nature of the emulsifier and effect of the hydrophilic-lipophilic-balance (HLB) on nanoemulsion formation.

Chapter 10 deals with the topic of solubilized systems and swollen micelles (microemulsions). The thermodynamic definition of microemulsions, their spontaneous formation and stability are described at a fundamental level. This is followed by describing the theories of microemulsion formation and stability and the need for using two surfactants to produce ultra-low interfacial tension. The characterization of microemulsions using scattering, conductivity and NMR techniques is described. Chapter 10 also describes the methods that can be applied for formulation of microemulsions and the importance of the partition of the cosurfactant between the oil and water phases. The various methods that can be applied for selecting emulsifiers for microemulsion formation are described. The application of microemulsions in several industries is briefly described.

References

[1]   Graham, T., Philos. Mag. Trans. Royal Society, **151**, 183 (1861).
[2]   Capek, I., Nanosuspensions, in "Encyclopedia of Colloid and Interface Science", Th. F. Tadros (ed.), Springer (2013).
[3]   Kerker, M., "The Scattering of Light and Other Electromagnetic Radiations", Academic Press, New York (1969).
[4]   Hamaker, H. C., Physica (Utrecht), **4**, 1058 (1937).
[5]   Thomson, W. (Lord Kelvin), Phil. Mag., **42**, 448 (1871).
[6]   Tadros, Th. F., "Dispersion of Powders in Liquids and Stabilisation of Suspensions", Wiley-VCH, Germany (2012).
[7]   Tadros, Th. F., "Formulation of Disperse Systems", Wiley-VCH, Germany (2014).
[8]   Tadros, Th. F. (ed.), "Colloids in Cosmetics", Wiley-VCH, Germany (2008).
[9]   Deryaguin, B. V. and Landau, L., Acta Physicochem. USSR, **14**, 633 (1941).

[10] Verwey, E. J. W. and Overbeek, J. Th. G. "Theory of Stability of Lyophobic Colloids", Elsevier, Amsterdam (1948).

[11] Tadros, Th. F., in "Polymer Colloids", R. Buscall, T. Corner and J. F. Stageman (eds.), Applied Sciences, Elsevier, London (1985) p. 105.

[12] Napper, D. H., "Polymeric Stabilisation of Colloidal Dispersions", Academic Press, London (1983).

There are generally two mechanisms of stabilization of nanodispersions. The first method depends on charge separation and formation of an electrical double layer. With many nanodispersions containing surface groups that can be dissociated, e.g. oxides that contain OH groups, or latex-containing sulphate groups, a surface charge can be produced as a result of the dissociation of the surface groups. This charge is compensated by an unequal distribution of counterions (with opposite charge sign to the surface charge) and co-ions (with the same charge sign as the surface charge). The same mechanism occurs when using ionic surfactants that can adsorb on the particle or droplet surface. As a result an electrical double layer is formed whose extension (double layer thickness) depends on electrolyte concentration and valency of the counter- and co-ions. When the particles approach to a distance h that is smaller than twice the double layer extension, strong repulsion occurs, particularly when the $1:1$ electrolyte (e.g. NaCl) is lower than 10^{-2} mol dm^{-3}. This repulsion can overcome the van der Waals attraction at intermediate particle separation resulting in an energy barrier (maximum) that prevents particle approach and hence flocculation is prevented [1].

The second and more effective mechanism is obtained by using nonionic surfactants or polymers (referred to as polymeric surfactant) [2, 3]. These molecules consist of a hydrophobic chain (such as an alkyl chain B with nonionic surfactants, or polystyrene or polymethylmethacrylate or polypropylene oxide B chain for an A–B, A–B–A or BA$_n$) and a hydrophilic chain A (such as polyethylene oxide). On a hydrophobic particle or an oil droplet in aqueous medium, the insoluble B chain adsorbs strongly on the surface of a particle or droplet, leaving the hydrophilic chain (that is strongly hydrated) in the aqueous medium. The surfactant or polymer layer will have a thickness δ that depends on the number of ethylene oxide units in the A chain and its hydration by water molecules. When two particles or droplets approach to a distance h that is smaller than 2δ, the A chains may overlap or become compressed resulting in very strong repulsion as a result of two main effects: (i) Unfavourable mixing of the A chains when these are in good solvent conditions; (ii) unfavourable loss of configurational entropy of the A chains on significant overlap. The combination of these two effects is referred to as steric repulsion which, when combined with the van der Waals attraction, results in an energy-distance with only one shallow minimum. When h < 2δ, the energy increases very sharply with a further decrease in h. This method of stabilization is more effective than electrostatic stabilization in two main respects. Firstly, the repulsion is still maintained at moderate electrolyte concentration (~ 1 mol dm^{-3} NaCl). Secondly, the repulsion can be maintained at high temperatures (exceeding in some cases 50 °C) providing the chains remain hydrated at such high temperatures.

These two mechanisms above are discussed in detail below.

2.2 Electrostatic stabilization

A great variety of processes occur to produce a surface charge [1–3], the most important is the ionization of surface groups. Carboxylic groups that are chemically bound to the surface of synthetic latexes provide an example of this. The charge is a function of pH as the degree of dissociation is a function of pH. Although the pKa of an isolated –COOH group is pH ~ 4, this is not the situation with a surface with many –COOH groups in close proximity. The dissociation of one group makes it more difficult for the immediate neighbours to dissociate. This means that the surface has variable pK_a and pH values as high as 9 may be required to ensure dissociation of all surface groups.

Another example where the charge is produced by dissociation is that of oxides which have surface hydroxyl groups. At high pH, these can ionize to give $-O^-$ and at low pH the lone pair of electrons can hold a proton to give $-OH_2^+$. This process can be represented as follows:

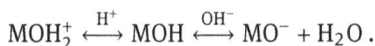

$$MOH_2^+ \xrightarrow{H^+} MOH \xrightarrow{OH^-} MO^- + H_2O .$$

The surface charge follows from

$$\sigma_o = F([MOH_2^+] - [MO^-]) = F(\Gamma_{H^+} - \Gamma_{OH^-}) , \tag{2.1}$$

where Γ refers to the surface concentration in moles per unit area.

A schematic representation of the process of charge formation in an oxide is shown in Fig. 2.1 whereby HNO_3 and KOH are used to provide the H^+ and OH^- ions respectively. In this case one my write equation (2.1) as

$$\sigma_o = F(\Gamma_{HNO_3} - \Gamma_{KOH}) . \tag{2.2}$$

The charge depends on the pH of the solution: Below a certain pH the surface is positive and above a certain pH the surface is negative. At a specific pH ($\Gamma_H = \Gamma_{OH}$) the surface is uncharged; this is referred to as the point of zero charge (pzc). The pzc depends on the type of oxide: For an acidic oxide such as silica the pzc is ~ pH 2–3. For a basic oxide such as alumina pzc is ~ pH 9. For an amphoteric oxide such as titania the pzc ~ pH 6. Some typical values of pzc for various oxides are given in Tab. 2.1.

In some cases, specifically adsorbed ions (that have non-electrostatic affinity to the surface) "enrich" the surface but may not be considered as part of the surface, e.g. bivalent cations on oxides, cationic and anionic surfactants on most surfaces. In particular, ionic surfactants are often added as a component to disperse systems, as wetting and dispersing agents for powders to produce solid/liquid dispersions

H^+

$NO_3^- H^+$ $H^+NO_3^-$ in HNO_3

$NO_3^- H^+$ $H^+NO_3^-$ Counterions   Surface positively charged

H^+

NO_3^-

OH^-

K^+OH^- OH^-K^+ in KOH

K^+OH^- OH^-K^+ Counterions   Surface negatively charged

OH

K^+

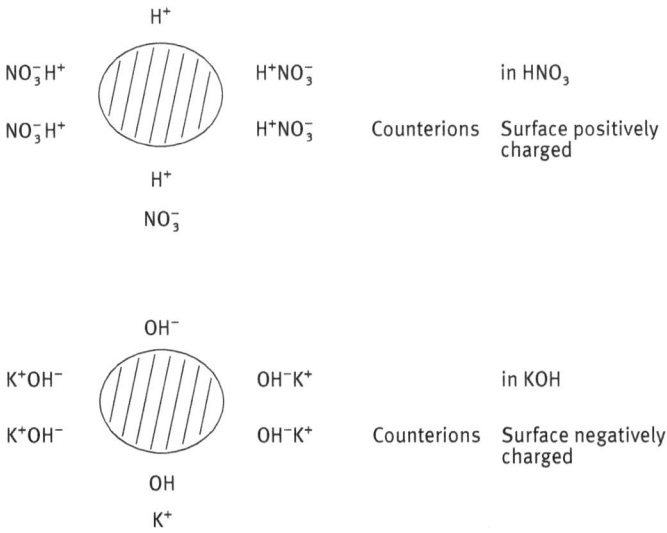

Fig. 2.1: Schematic representation of an oxide surface.

Tab. 2.1: pzc values for some oxides.

Oxide	pzc
SiO_2 (precipitated)	2–3
SiO_2 (quartz)	3.7
SnO_2 (cassiterite)	5–6
TiO_2 (anatase)	6.2
TiO_2 (rutile)	5.7–5.8
RuO_2	5.7
$\alpha\text{-}Fe_2O_3$ (haematite)	8.5–9.5
$\alpha\text{-}FeO\cdot OH$ (goethite)	8.4–9.4
ZnO	8.5–9.5
$\gamma\text{-}Al(OH)_3$ (gibbsite)	8–9

(suspensions). They are also added for emulsification of oils to produce oil/water emulsions. In most cases the surfactant also acts as stabilizer for the final suspension or emulsion. By adsorption at the solid/liquid or liquid/liquid interface the surfactant ions produce a charge (negative for anionic surfactants and positive for cationic surfactants) on the surface of the particle or droplet. This charge is compensated by an unequal distribution of counterions (with opposite charge to the surface) and co-ions (with the same charge of the surface) forming an electrical double layer. In some cases the charge on the particle or droplet is produced by adsorption of anionic or cationic polyelectrolytes. An example is polyacrylates that are used to disperse many pigments such as titania.

The structure of the electrical double layer was first presented by Gouy and Chapman [4, 5] who assumed that the charge is smeared out over a plane surface immersed in an electrolyte solution. This surface has a uniform potential ψ_o and the compensating ions are regarded as point charges immersed in a continuous dielectric medium. The surface charge σ_o is compensated by an unequal distribution of counterions (opposite in charge to the surface) and co-ions (same sign as the surface) which extend to some distance from the surface [4, 5]. The counterion and co-ion concentration n_i near the surface can be related to the value in the bulk n_{io} using the Boltzmann distribution principle,

$$n_i = n_{io} \exp\left[-\frac{Z_i e \psi_x}{kT}\right],$$
(2.3)

where Z_i is the valency of the ion i, e is the electronic charge and ψ_x is the potential at a distance x from the surface, k is the Boltzmann constant and T is the absolute temperature. Since the charge on the counterion is always opposite to that of the surface, the exponent in equation (2.3) will always be negative for the counterion concentration and positive for the co-ion concentration. Equation (2.3) shows that the concentration of counterions increases close to the surface (positively adsorbed) whereas the co-ion concentration is reduced near the surface (negative adsorption).

The number of charges per unit volume, i.e. the space charge density ρ, is given by

$$\rho = \sum_i n_i Z_i e = -2n_o Ze \sinh\left[\frac{Ze\psi_x}{kT}\right].$$
(2.4)

Note that $\sinh x = (\exp x - \exp(-x))/2$.

A schematic picture of the diffuse double layer according to Gouy [4] and Chapman [5] is shown in Fig. 2.2. The potential decays exponentially with distance x.

The potential ψ at a distance x is related to the surface potential ψ_o by the well-known Debye–Huckel approximation:

$$\psi_x = \psi_o \exp -(\kappa x).$$
(2.5)

Note that when $x = 1/\kappa$, $\psi_x = \psi_o/e$; $1/\kappa$ is referred to as the "thickness" of the double layer.

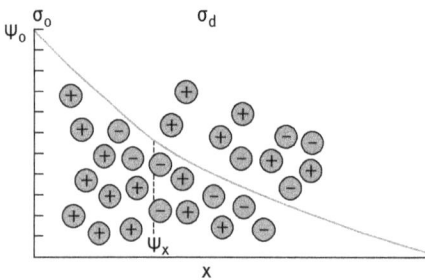

Fig. 2.2: Schematic representation of the diffuse double layer according to Gouy [4] and Chapman [5].

The double layer extension depends on electrolyte concentration and valency of the counterions,

$$\left(\frac{1}{\kappa}\right) = \left(\frac{\varepsilon_r\varepsilon_0 kT}{2n_0 Z_i^2 e^2}\right)^{1/2}.$$ (2.6)

The double layer extension increases with decreasing electrolyte concentration. It also depends on the valency of the ions. An expression for $(1/\kappa)$ in terms of the electrolyte ionic strength I is

$$\left(\frac{1}{\kappa}\right) = \left(\frac{\varepsilon_r\varepsilon_0 RT}{2000F^2}\right)^{1/2} I^{-1/2} \text{ in m}^{-1} = 0.304\, I^{-1/2} \text{ in nm},$$ (2.7)

where

$$I = \sum c_i Z_i^2.$$ (2.8)

c_i is the electrolyte concentration in mol dm^{-3}.

Increasing the ionic strength causes a decrease in $(1/\kappa)$ that is referred to as compression of the double layer. The distance $(1/\kappa)$ is referred to as the thickness of the double layer. For example, for KCl in water at 25 °C, $(1/\kappa) = 96.17\,\text{nm}$ at $I = 10^{-5}\,\text{mol dm}^{-3}$ decreasing to $3.04\,\text{nm}$ at $I = 10^{-2}\,\text{mol dm}^{-3}$. Approximate values of $(1/\kappa)$ for KCl are given in Tab. 2.2.

Tab. 2.2: Approximate values of $(1/\kappa)$ for 1:1 electrolyte (KCl).

C / mol dm^{-3}	10^{-5}	10^{-4}	10^{-3}	10^{-2}	10^{-1}
$(1/\kappa)$ / nm	100	33	10	3.3	1

Equations (2.7) and (2.8) show that $(1/\kappa)$ depends on the valency of the counter- and co-ions.

Stern [6] recognized that the assumption in the Gouy–Chapman theory [4, 5] that the electrolyte ions are regarded as point charges is unsatisfactory. Also the assumption that the solvent can be treated as a structureless dielectric of constant permittivity is unsatisfactory. Stern then introduced the concept of the nondiffuse part of the double layer for specifically adsorbed ions, the rest being diffuse in nature. This is schematically illustrated in Fig. 2.3.

The potential drops linearly in the Stern region and then exponentially. Grahame [7] distinguished two types of ions in the Stern plane, physically adsorbed counterions (outer Helmholtz plane) and chemically adsorbed ions (that lose part of their hydration shell) (inner Helmholtz plane). The outer Helmholtz plane is considered as the plane of closest approach of hydrated counterions, i.e. the Stern plane. The inner Helmholtz plane is that of specifically adsorbed counterions which may have lost part or their complete hydration shell. The number of these specifically adsorbed ions may exceed the number of surface charges causing a reversal of the sign of the potential.

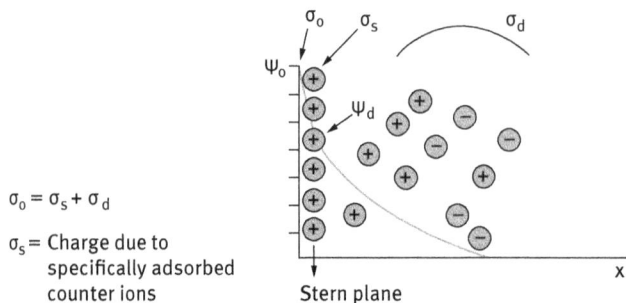

Fig. 2.3: Schematic representation of the double layer according to Stern and Grahame.

When charged colloidal particles in a dispersion approach each other such that the double layers begin to overlap (particle separation becomes less than twice the double layer extension), repulsion occurs. The individual double layers can no longer develop unrestrictedly, since the limited space does not allow complete potential decay [8].

This is illustrated in Fig. 2.4 for two flat plates. The potential $\psi_{H/2}$ half way between the plates is no longer zero (as would be the case for isolated particles at x → ∞). The potential distribution at an interparticle distance H is schematically depicted by the full line in Fig. 2.4. The Stern potential ψ_d is considered to be independent of the particle distance. The dashed curves show the potential as a function of distance x to the Helmholtz plane, had the particles been at infinite distance.

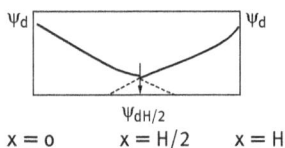

Fig. 2.4: Schematic representation of double layer interaction for two flat plates.

For two spherical particles of radius R and surface potential ψ_0 and condition $\kappa R < 3$, the expression for the electrical double layer repulsive interaction is given by [8]

$$G_{el} = \frac{4\pi\varepsilon_r\varepsilon_0 R^2 \psi_0^2 \exp -(\kappa h)}{2R + h},$$ (2.9)

where h is the closest distance of separation between the surfaces.

Expression (2.9) shows the exponential decay of G_{el} with h. The higher the value of κ (i.e. the higher the electrolyte concentration), the steeper the decay, as schematically shown in Fig. 2.5.

This means that at any given distance h, the double layer repulsion decreases with increasing electrolyte concentration.

An important aspect of the double layer repulsion is the situation during particle approach. If at any stage the assumption is made that the double layers adjust to

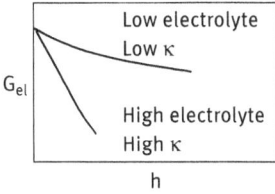

Fig. 2.5: Variation of G_{el} with h at different electrolyte concentrations.

new conditions so that equilibrium is always maintained, then the interaction takes place at constant potential. This would be the case if the relaxation time of the surface charge is much shorter than the time the particles are in each other's interaction sphere as a result of Brownian motion. However, if the relaxation time of the surface charge is appreciably longer than the time particles are in each other's interaction sphere, the charge rather than the potential will be the constant parameter. The constant charge leads to larger repulsion than the constant potential case.

The van der Waals attraction is universal with all disperse systems and it arises from the attractive energies between atoms or molecules in any particle or droplet. As is well known atoms or molecules always attract each other at short distances of separation. The attractive forces are of three different types: Dipole-dipole interaction (Keesom), dipole-induced dipole interaction (Debye) and London dispersion force. The London dispersion force is the most important, since it occurs for polar and nonpolar molecules. It arises from fluctuations in the electron density distribution.

At small distances of separation r in vacuum, the attractive energy between two atoms or molecules is given by

$$G_{aa} = -\frac{\beta_{11}}{r^6} .$$

(2.10)

β_{11} is the London dispersion constant.

For colloidal particles made of atom or molecular assemblies, the attractive energies have to be compounded. In this process, only the London interactions have to be considered, since large assemblies have neither a net dipole moment nor a net polarization. The result relies on the assumption that the interaction energies between all molecules in one particle with all the others are simply additive [9]. The interaction between two identical half-infinite plates at a distance h in vacuum is given by

$$G_A = -\frac{A_{11}}{12\pi h^2} .$$

(2.11)

Whereas for two spheres in vacuum the result is

$$G_A = -\frac{A_{11}}{6} \left(\frac{2}{s^2 - 4} + \frac{2}{s^2} + \ln \frac{s^2 - 4}{s^2} \right) .$$

(2.12)

A_{11} is known as the Hamaker constant and is defined by

$$A = \pi q_{11}^2 \beta_{ii} .$$

(2.13)

q_{11} is number of atoms or molecules of type 1 per unit volume, and $s = (2R + h)/R$. Equation (2.13) shows that A_{11} has the dimension of energy.

For very short distances ($h \ll R$), equation (2.12) may be approximated by

$$G_A = -\frac{A_{11}R}{12h} . \tag{2.14}$$

When the particles are dispersed in a liquid medium, the van der Waals attraction has to be modified to take into account the medium effect. When two particles are brought from infinite distance to h in a medium, an equivalent amount of medium has to be transported the other way round. Hamaker forces in a medium are excess forces.

Consider two identical spheres 1 at a large distance apart in a medium 2 as is illustrated in Fig. 2.6 (a). In this case the attractive energy is zero. Figure 2.6 (b) gives the same situation with arrows indicating the exchange of 1 against 2. Figure 2.6 (c) shows the complete exchange which now shows the attraction between the two particles 1 and 1 and equivalent volumes of the medium 2 and 2.

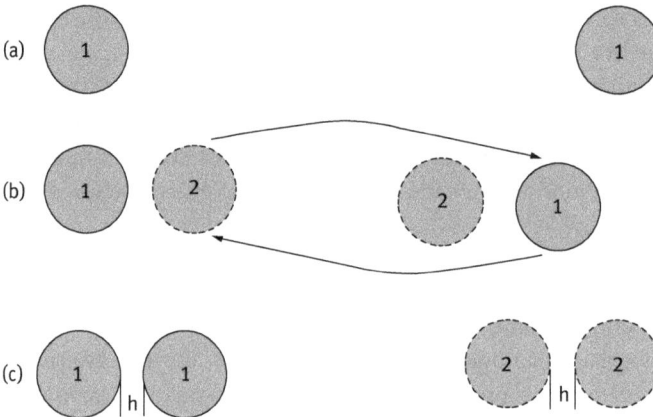

Fig. 2.6: Schematic representation of interaction of two particles in a medium.

The effective Hamaker constant for two identical particles 1 and 1 in a medium 2 is given by

$$A_{11(2)} = A_{11} + A_{22} - 2A_{12} = (A_{11}^{1/2} - A_{22}^{1/2})^2 . \tag{2.15}$$

Equation (2.11) shows that two particles of the same material attract each other unless their Hamaker constants exactly match each other. Equation (2.10) now becomes

$$G_A = -\frac{A_{11(2)}R}{12h} , \tag{2.16}$$

where $A_{11(2)}$ is the effective Hamaker constant of two identical particles with Hamaker constant A_{11} in a medium with Hamaker constant A_{22}.

In most cases the Hamaker constant of the particles is higher than that of the medium. Examples of Hamaker constant for some liquids are given in Tab. 2.3. Table 2.4 gives values of the effective Hamaker constant for some particles in some liquids. Generally speaking, the effect of the liquid medium is to reduce the Hamaker constant of the particles below its value in vacuum (air).

G_A decreases with increasing h as schematically shown in Fig. 2.7.

Tab. 2.3: Effective Hamaker constant $A_{11(2)}$ of some particles of water.

Liquid	$A_{22} \times 10^{20}$ J
Water	3.7
Ethanol	4.2
Decane	4.8
Hexadecane	5.2
Cyclohexane	5.2

Tab. 2.4: Hamaker constant of some liquids.

System	$A_{11(2)} \times 10^{20}$ J
Fused Quartz/Water	0.83
Al_2O_3/Water	5.32
Copper/Water	30.00
Poly(methylmethacrylate/Water	1.05
Poly(vinylchloride)/Water	1.03
Poly(tetrafluoroethylene)/Water	0.33

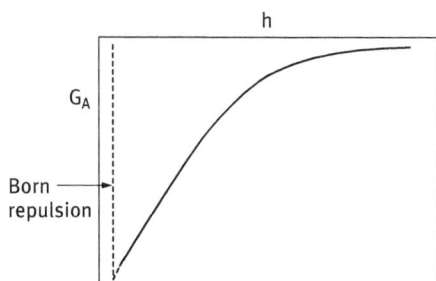

Fig. 2.7: Variation of G_A with h.

As shown in Fig. 2.7 G_A increases very sharply with h at small h values. A capture distance can be defined at which all the particles become strongly attracted to each other (coagulation). At very short distances, the Born repulsion appears.

The Hamaker approach, referred to as a "microscopic" theory is based on the interactions between pairs of atoms or molecules. The more accurate "macroscopic" approach originally suggested by Lifshitz and described in detail by Mahanty and Ninham [10] is based on the principle that the spontaneous electromagnetic fluctuations in two particles become correlated when the latter approach each other, causing a decrease in the free energy of the system. The elaboration of this theory is rather complex and its application requires extensive data on the electromagnetic interaction energies. Nevertheless, the theory allows the important conclusion that the mostly qualitative aspects of the "microscopic" theory given by equations (2.11)–(2.16) are fully confirmed. The only exception concerns the decay of G_A with h at large separations. Owing to the time required for electromagnetic waves to cover the distance between the particles, the h^{-2} dependence in equation (2.16) gradually changes to h^{-3} dependence at large separations, a phenomenon known as retardation.

Combining G_{el} and G_A results in the well-known theory of stability of colloids due to Deryaguin–Landau–Verwey–Overbeek (DLVO theory) [11, 12]:

$$G_T = G_{el} + G_A . \tag{2.17}$$

A plot of G_T versus h is shown in Fig. 2.8, which represents the case at low electrolyte concentrations, i.e. strong electrostatic repulsion between the particles. G_{el} decays exponentially with h, i.e. $G_{el} \to 0$ as h becomes large. G_A is $\propto 1/h$, i.e. G_A does not decay to 0 at large h.

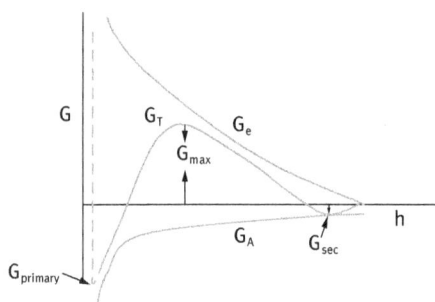

Fig. 2.8: Schematic representation of the variation of G_T with h according to the DLVO theory.

At long distances of separation, $G_A > G_{el}$ resulting in a shallow minimum (secondary minimum) particularly for nanodispersions. At very short distances, $G_A \gg G_{el}$ resulting in a deep primary minimum. At intermediate distances, $G_{el} > G_A$ resulting in energy maximum, G_{max}, whose height depends on ψ_o (or ψ_d) and the electrolyte concentration and valency.

At low electrolyte concentrations ($< 10^{-2}$ mol dm^{-3} for a 1 : 1 electrolyte), G_{max} is high (> 25kT) and this prevents particle aggregation into the primary minimum. The higher the electrolyte concentration (and the higher the valency of the ions), the lower the energy maximum. This is illustrated in Fig. 2.9 which shows the variation of G_T with h at various electrolyte concentrations.

Since approximate formulae are available for G_{el} and G_A, quantitative expressions for $G_T(h)$ can also be formulated. These can be used to derive expressions for the

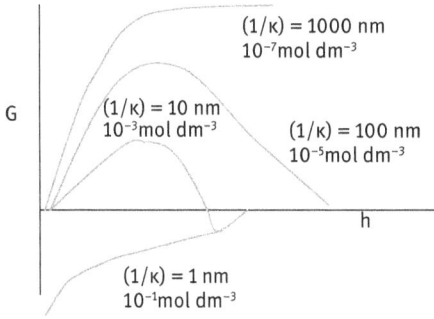

$(1/\kappa) = 1000$ nm
10^{-7} mol dm^{-3}

G

$(1/\kappa) = 10$ nm
10^{-3} mol dm^{-3}

$(1/\kappa) = 100$ nm
10^{-5} mol dm^{-3}

h

$(1/\kappa) = 1$ nm
10^{-1} mol dm^{-3}

Fig. 2.9: Variation of G with h at various electrolyte concentrations.

coagulation concentration, which is that concentration that causes every encounter between two colloidal particles to lead to destabilization. Verwey and Overbeek [12] introduced the following criteria for transition between stability and instability:

$$G_T(= G_{el} + G_A) = 0 , \tag{2.18}$$

$$\frac{dG_T}{dh} = 0 , \tag{2.19}$$

$$\frac{dG_{el}}{dh} = -\frac{dG_A}{dh} . \tag{2.20}$$

Using the equations for G_{el} and G_A, the critical coagulation concentration, ccc, could be calculated. The theory predicts that the ccc is directly proportional to the surface potential ψ_0 and inversely proportional to the Hamaker constant A and the electrolyte valency Z. The ccc is inversely proportional to Z^6 at high surface potential and inversely proportional to Z^2 at low surface potential.

As shown in equation (2.9) G_{el} is directly proportional to ψ_0^2 and at a given electrolyte concentration the energy maximum G_{max} increases with increasing ψ_0 and this leads to an enhanced electrostatic stabilization. In most cases ψ_0 cannot be directly measured and this can be replaced by the measurable electrokinetic or zeta potential [13]. The latter can be obtained by measuring particle mobility on application of an electric field (electrophoresis). This results in charge separation at the interface between two phases and on application of an electric field one of the phases is caused to move tangentially past the second phase. One measures the particle or droplet velocity v (m s^{-1}) from which the particle or droplet mobility u (m^2 V^{-1} s^{-1}) is calculated by dividing v by the field strength (E/l, where E is the applied potential in volts and l is the distance between the two electrodes used).

The main problem in any analysis of electrophoresis is defining the plane at which the liquid begins to move past the surface of the particle or droplet. This is defined as the "shear plane", which at some distance from the surface. One usually defines an "imaginary" surface close to the particle surface within which the fluid is stationary. The point just outside this imaginary surface is described as the surface of shear and the potential at this point is described as the zeta potential (ζ). A schematic representation of the surface of shear, the surface and zeta potential is shown in Fig. 2.10.

$\eta' \gg \eta_b$　　　　Permittivity $\epsilon' < \epsilon_b$

Bulk
Viscosity
η_b

Particle

Liquid Immobile
Viscosity ?'

Particle
surface

Surface of Shear

Potential Ψ_0

Electrokinetic
Potencial
ζ potential

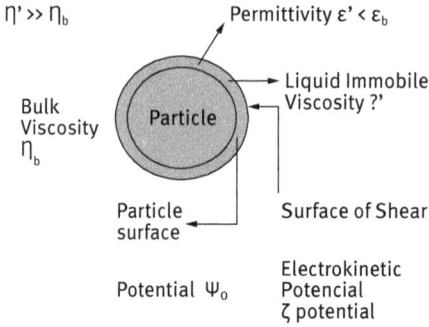

Fig. 2.10: Surface (plane) of shear.

The exact position of the plane of shear is not known; it is usually in the region of few A; in some cases one may equate the shear plane with the Stern plane (the centre of specifically adsorbed ions) although this may be an underestimate of its location. Several layers of liquid may be immobilized at the particle surface (which means that the shear plane is farther apart from the Stern plane). The particle or droplet plus its immobile liquid layer form the kinetic unit that moves under the influence of the electric field. The viscosity of the liquid in the immobile sheath around the particles (η') is much larger than the bulk viscosity η. The permittivity of the liquid in this liquid sheath ϵ' is also lower than the bulk permittivity (due to dielectric saturation in this layer). In the absence of specific adsorption, the assumption is usually made that $\zeta \sim \psi_0$. The latter potential is the value that is commonly used to calculate the repulsive energy between two particles.

The zeta potential ζ is calculated from the mobility u using the Huckel equation [14]

$$u = \frac{2}{3} \frac{\epsilon_r \epsilon_0 \zeta}{\eta} .$$
(2.21)

Equation (2.21) applies for small particles or droplets (< 100 nm) and thick double layers (low electrolyte concentration), i.e. for the case $\kappa R < 1$.

The most suitable method for measurement of zeta potential is the laser velocimetry method that is suitable for small particles that undergo Brownian motion [15]. The light scattered by small particles will show intensity fluctuation as a result of Brownian diffusion (Doppler shift). When a light beam passes through a colloidal dispersion, an oscillating dipole movement is induced in the particles, thereby radiating the light. Due to the random position of the particles, the intensity of scattered light, at any instant, appears as random diffraction ("speckle" pattern). As the particles undergo Brownian motion, the random configuration of the pattern will fluctuate, such that the time taken for an intensity maximum to become a minimum (the coherence time), corresponds approximately to the time required for a particle to move one wavelength λ. Using a photomultiplier of active area about the diffraction maximum (i.e. one coherent area) this intensity fluctuation can be measured. The analogue output is digitized (using a digital correlator) that measures the photocount (or intensity) correlation

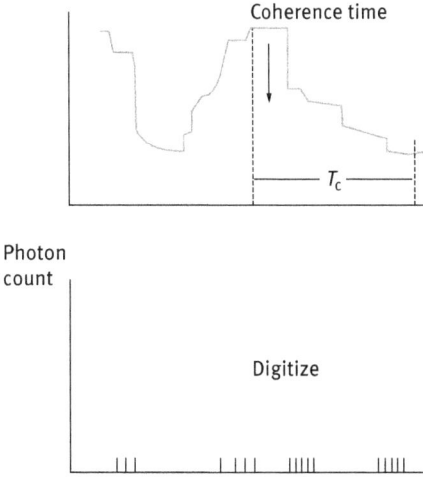

Fig. 2.11: Schematic representation of intensity fluctuation of scattered light.

function of scattered light. The intensity fluctuation is schematically illustrated in Fig. 2.11.

The photocount correlation function $g^{(2)}(\tau)$ is given by

$$g^{(2)} = B[1 + \gamma^2 g^{(1)}(\tau)]^2 , \qquad (2.22)$$

where τ is the correlation delay time.

The correlator compares $g^{(2)}(\tau)$ for many values of τ. B is the background value to which $g^{(2)}(\tau)$ decays at long delay times. $g^{(1)}(\tau)$ is the normalized correlation function of the scattered electric field and γ is a constant (~ 1).

For monodispersed noninteracting particles or droplets,

$$g^{(1)}(\tau) = \exp(-\Gamma\gamma) . \qquad (2.23)$$

Γ is the decay rate or inverse coherence time, which is related to the translational diffusion coefficient D,

$$\Gamma = DK^2 , \qquad (2.24)$$

where K is the scattering vector,

$$K = \left(\frac{4\pi n}{\lambda_o}\right) \sin\left(\frac{\theta}{2}\right) . \qquad (2.25)$$

The particle radius R can be calculated from D using the Stokes–Einstein equation,

$$D = \frac{kT}{6\pi\eta_o R} , \qquad (2.26)$$

where η_o is the viscosity of the medium.

If an electric field is placed at right angles to the incident light and in the plane defined by the incident and observation beam, the line broadening is unaffected but the centre frequency of the scattered light is shifted to an extent determined by the electrophoretic mobility. The shift is very small compared to the incident frequency (~100 Hz for an incident frequency of ~6 × 10^{14} Hz) but with a laser source it can be detected by heterodyning (i.e. mixing) the scattered light with the incident beam and detecting the output of the difference frequency. A homodyne method may be applied in which case a modulator to generate an apparent Doppler shift at the modulated frequency is used. To increase the sensitivity of the laser Doppler method, the electric fields are much higher than those used in conventional electrophoresis. The Joule heating is minimized by pulsing the electric field in opposite directions. The Brownian motion of the particles also contributes to the Doppler shift and an approximate correction can be made by subtracting the peak width obtained in the absence of an electric field from the electrophoretic spectrum. An He-Ne laser is used as the light source and the output of the laser is split into two coherent beams which are cross-focused in the cell to illuminate the sample. The light scattered by the particle, together with the reference beam is detected by a photomultiplier. The output is amplified and analyzed to transform the signals to a frequency distribution spectrum. At the intersection of the beam interferences of known spacing are formed.

The magnitude of the Doppler shift Δv is used to calculate the electrophoretic mobility u using the following expression:

$$\Delta v = \left(\frac{2n}{\lambda_0} \right) \sin \left(\frac{\theta}{2} \right) uE, \qquad (2.27)$$

where n is the refractive index of the medium, λ_0 is the incident wavelength in vacuum, θ is the scattering angle and E is the field strength.

Several commercial instruments are available for measuring electrophoretic light scattering: (i) The Coulter DELSA 440SX (Coulter Corporation, USA) is a multiangle laser Doppler system employing heterodyning and autocorrelation signal processing. Measurements are made at four scattering angles (8, 17, 25, and 34°) and the temperature of the cell is controlled by a Peltier device. The instrument reports the electrophoretic mobility, zeta potential, conductivity and particle size distribution. (ii) Malvern (Malvern Instruments, UK) has two instruments: The ZetaSizer 3000 and the ZetaSizer 5000: The ZetaSizer 3000 is a laser Doppler system using crossed beam optical configuration and homodyne detection with photon correlation signal processing. The zeta potential is measured using laser Doppler velocimetry and the particle size is measured using photon correlation spectroscopy (PCS). The ZetaSizer 5000 uses PCS to measure both (a) movement of the particles in an electric field for zeta potential determination and (b) random diffusion of particles at different measuring angles for size measurement on the same sample. In both instruments, a Peltier device is used for temperature control.

2.3 Steric stabilization

A more effective stabilization mechanism is produced by the use of polymeric surfactants of the A–B, A–B–A block and BA_n graft copolymers. The "anchor" chain B is chosen to be highly insoluble in the medium and strongly adsorbed to the particle or droplet surface. The stabilizing chain A is chosen to be highly soluble in the medium and strongly solvated by the molecules of the solvent. Examples of B chains for hydrophobic nanoparticles or oil droplets in aqueous medium are polystyrene, polymethylmethacrylate or polypropylene oxide. Examples of A chains in aqueous media are polyethylene oxide, polyvinylpyrrolidone or polysaccharide. Figure 2.12 shows a schematic representation of the adsorption and conformation of the block and graft copolymers on a flat surface.

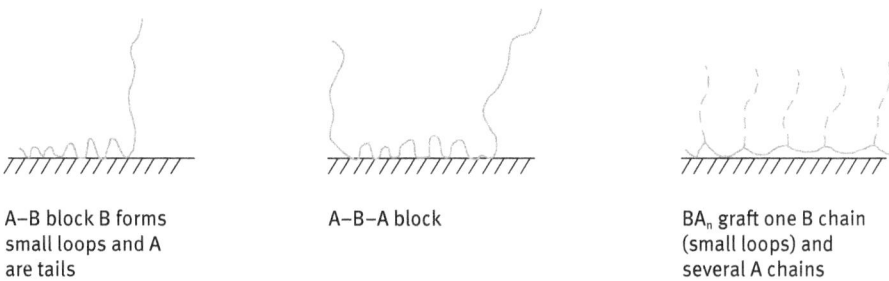

A–B block B forms
small loops and A
are tails

A–B–A block

BA_n graft one B chain
(small loops) and
several A chains

Fig. 2.12: Adsorption and conformation of A–B, A–B–A block and BA_n graft copolymer on a flat surface.

For a full description of polymer adsorption one needs to obtain information on the following: (i) The amount of polymer adsorbed Γ (in mg or mol) per unit area of the particles. It is essential to know the surface area of the particles in the nanodispersion. (ii) The fraction of segments in direct contact with the surface, i.e. the fraction of segments in trains, p (p = (number of segments in direct contact with the surface)/total number). (iii) The adsorption energy per segment of the B chain, χ^s. (iv) The distribution of segments in loops and tails, $\rho(z)$, which extend in several layers from the surface. $\rho(z)$ is usually difficult to obtain experimentally although recently application of small angle neutron scattering could obtain such information. An alternative and useful parameter for assessing "steric stabilization" is the hydrodynamic thickness, δ_h (thickness of the adsorbed or grafted polymer layer plus any contribution from the hydration layer). (v) The solvency of the medium for the chains as determined by the Flory–Huggins interaction parameter χ. For a good solvent for the A chain $\chi < 0.5$.

Several experimental techniques are available to obtain the above parameters. The amount of polymer adsorbed Γ can be determined by measuring the adsorption isotherm. One measures the polymeric surfactant concentration before ($C_{initial}$, C_1)

and after ($C_{equilibrium}$, C_2)

$$\Gamma = \frac{(C_1 - C_2)V}{A} , \qquad (2.28)$$

where V is the total volume of the solution and A is the specific surface area ($m^2 \, g^{-1}$). It is necessary in this case to separate the particles from the polymer solution after adsorption. This could be carried out by centrifugation and/or filtration. One should make sure that all particles are removed. To obtain this isotherm, one must develop a sensitive analytical technique for determination of the polymeric surfactant concentration in the ppm range. It is essential to follow the adsorption as a function of time to determine the time required to reach equilibrium. For some polymer molecules such as polyvinyl alcohol, PVA, and polyethylene oxide, PEO, (or blocks containing PEO), analytical methods based on complexation with iodine/potassium iodide or iodine/ boric acid potassium iodide have been established. For some polymers with specific functional groups spectroscopic methods may be applied, e.g. UV, IR or fluorescence spectroscopy. A possible method is to measure the change in refractive index of the polymer solution before and after adsorption. This requires very sensitive refractometers. High resolution NMR has been recently applied since the polymer molecules in the adsorbed state are in a different environment to those in the bulk. The chemical shift of functional groups within the chain is different in these two environments. This has the attraction of measuring the amount of adsorption without separating the particles.

The fraction of segments in direct contact with the surface can be directly measured using spectroscopic techniques: (i) IR if there is specific interaction between the segments in trains and the surface, e.g. polyethylene oxide on silica from nonaqueous solutions [16]. (ii) Electron spin resonance (ESR); this requires labelling of the molecule. (iii) NMR, pulse gradient or spin ECO NMR. This method is based on the fact that the segments in trains are "immobilized" and hence they have lower mobility than those in loops and tails [17, 18].

An indirect method of determining p is to measure the heat of adsorption ΔH using microcalorimetry [19]. One should then determine the heat of adsorption of a monomer H_m (or molecule representing the monomer, e.g. ethylene glycol for PEO); p is then given by the equation

$$p = \frac{\Delta H}{H_m n} , \qquad (2.29)$$

where n is the total number of segments in the molecule.

The above indirect method is not very accurate and can only be used in a qualitative sense. It also requires very sensitive enthalpy measurements (e.g. using an LKB microcalorimeter).

The segment density distribution $\rho(z)$ is given by the number of segments parallel to the surface in the z-direction. Three direct methods can be applied for determination of adsorbed layer thickness: ellipsometry, attenuated total reflection (ATR) and neutron scattering. Both ellipsometry and ATR [20] depend on the difference between refractive indices between the substrate, the adsorbed layer and bulk solution and they require a flat reflecting surface. Ellipsometry [20] is based on the principle that light undergoes a change in polarizability when it is reflected at a flat surface whether covered or uncovered with a polymer layer.

The above limitations when using ellipsometry or ATR are overcome by the application technique of neutron scattering, which can be applied to both flat surfaces as well as particulate dispersions. The basic principle of neutron scattering is to measure the scattering due to the adsorbed layer, when the scattering length density of the particle is matched to that of the medium (the so-called "contrast-matching" method). Contrast matching of particles and medium can be achieved by changing the isotopic composition of the system (using deuterated particles and a mixture of D_2O and H_2O). It was used for measuring the adsorbed layer thickness of polymers, e.g. PVA or poly(ethylene oxide) (PEO) on polystyrene latex [21]. Apart from obtaining δ, one can also determine the segment density distribution $\rho(z)$.

The above technique of neutron scattering clearly gives a quantitative picture of the adsorbed polymer layer. However, its application in practice is limited since one needs to prepare deuterated particles or polymers for the contrast matching procedure. The practical methods for determination of the adsorbed layer thickness are mostly based on hydrodynamic methods. Several methods may be applied to determine the hydrodynamic thickness of adsorbed polymer layers of which viscosity, sedimentation coefficient (using an ultracentrifuge) and dynamic light scattering measurements are the most convenient. A less accurate method is from zeta potential measurements.

The most rapid technique for measuring δ_h is photon correlation spectroscopy (PCS) (sometime referred to as quasi-elastic light scattering) which allows one to obtain the diffusion coefficient of the particles with and without the adsorbed layer (D_δ and D respectively). This is obtained from measuring the intensity fluctuation of scattered light as the particles undergo Brownian diffusion as described above. Thus, by measuring D_δ and D, one can obtain δ_h. It should be mentioned that the accuracy of the PCS method depends on the ratio of δ_h/R, since δ_h is determined by difference. Since the accuracy of the measurement is ±1 %, δ_h should be at least 10 % of the particle radius. This method can be used with small particles of nanodispersions and reasonably thick adsorbed layers.

When two particles each with a radius R and containing an adsorbed polymer layer with a hydrodynamic thickness δ_h, approach each other to a surface-surface separation distance h that is smaller than $2\delta_h$, the polymer layers interact with each other resulting in two main situations [22]: (i) The polymer chains may overlap with each other. (ii) The polymer layer may undergo some compression. In both cases, there will be an increase in the local segment density of the polymer chains in the interaction region. This is schematically illustrated in Fig. 2.13. The real situation is perhaps in between the above two cases, i.e. the polymer chains may undergo some interpenetration and some compression.

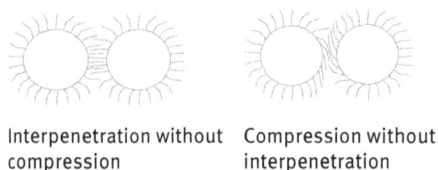

Interpenetration without compression   Compression without interpenetration

Fig. 2.13: Schematic representation of the interaction between particles containing adsorbed polymer layers.

Provided that the dangling chains (the A chains in A–B, A–B–A block or BA_n graft copolymers) are in a good solvent, this local increase in segment density in the interaction zone will result in strong repulsion as a result of two main effects [22]: (i) An increase in the osmotic pressure in the overlap region as a result of the unfavourable mixing of the polymer chains when these are in good solvent conditions. This is referred to as osmotic repulsion or mixing interaction and it is described by a free energy of interaction G_{mix}. (ii) Reduction of the configurational entropy of the chains in the interaction zone; this entropy reduction results from the decrease in the volume available for the chains when these are either overlapped or compressed. This is referred to as volume restriction interaction, entropic or elastic interaction and it is described by a free energy of interaction G_{elas}.

The combination of G_{mix} and G_{elas} is usually referred to as the steric interaction free energy, G_s, i.e.

$$G_s = G_{mix} + G_{elas} . \tag{2.30}$$

The sign of G_{mix} depends on the solvency of the medium for the chains. If in a good solvent, i.e. the Flory–Huggins interaction parameter χ is less than 0.5, then G_{mix} is positive and the mixing interaction leads to repulsion (see below). In contrast, if $\chi > 0.5$ (i.e. the chains are in a poor solvent condition), G_{mix} is negative and the mixing interaction becomes attractive. G_{elas} is always positive and hence in some cases one can produce stable dispersions in a relatively poor solvent (enhanced steric stabilization).

As mentioned above, G_{mix} results from the unfavourable mixing of the polymer chains when these are in good solvent conditions. This is schematically shown in Fig. 2.14. Consider two spherical particles with the same radius and each containing an adsorbed polymer layer with thickness δ. Before overlap, one can define in each polymer layer a chemical potential for the solvent μ_i^α and a volume fraction for the polymer in the layer ϕ_2^α. In the overlap region (volume element dV), the chemical potential of the solvent is reduced to μ_i^β. This results from the increase in polymer segment concentration in this overlap region.

In the overlap region, the chemical potential of the polymer chains is now higher than in the rest of the layer (with no overlap). This amounts to an increase in the osmotic pressure in the overlap region; as a result solvent will diffuse from the bulk to the overlap region, thus separating the particles and hence a strong repulsive energy arises from this effect. The above repulsive energy can be calculated by considering the free energy of mixing two polymer solutions, as for example treated by Flory and Krigbaum [23]. The free energy of mixing is given by two terms: (i) An entropy term that depends on the volume fraction of polymer and solvent. (ii) An energy term that is determined by the Flory–Huggins interaction parameter:

$$\delta(G_{mix}) = kT(n_1 \ln \phi_1 + n_2 \ln \phi_2 + \chi n_1 \phi_2), \tag{2.31}$$

where n_1 and n_2 are the number of moles of solvent and polymer with volume fractions ϕ_1 and ϕ_2, k is the Boltzmann constant and T is the absolute temperature.

The total change in free energy of mixing for the whole interaction zone, V, is obtained by summing over all the elements in V,

$$G_{mix} = \frac{2kTV_2^2}{V_1} v_2 \left(\frac{1}{2} - \chi\right) R_{mix}(h), \tag{2.32}$$

where V_1 and V_2 are the molar volumes of solvent and polymer respectively, v_2 is the number of chains per unit area and $R_{mix}(h)$ is a geometric function which depends on the form of the segment density distribution of the chain normal to the surface, $\rho(z)$. k is the Boltzmann constant and T is the absolute temperature.

Using the above theory one can derive an expression for the free energy of mixing of two polymer layers (assuming a uniform segment density distribution in each layer) surrounding two spherical particles as a function of the separation distance h between the particles [24].

The expression for G_{mix} is

$$G_{mix} = \left(\frac{2V_2^2}{V_1}\right) v_2 \left(\frac{1}{2} - \chi\right) \left(3R + 2\delta + \frac{h}{2}\right). \tag{2.33}$$

The sign of G_{mix} depends on the value of the Flory–Huggins interaction parameter χ: if $\chi < 0.5$, G_{mix} is positive and the interaction is repulsive; if $\chi > 0.5$, G_{mix} is negative and the interaction is attractive. The condition $\chi = 0.5$ and $G_{mix} = 0$ is termed the θ-condition. The latter corresponds to the case where the polymer mixing behaves as ideal, i.e. mixing of the chains does not lead to an increase or decrease of the free energy.

The elastic interaction results from the loss in configurational entropy of the chains on the approach of a second particle. As a result of this approach, the volume available for the chains becomes restricted, resulting in loss of the number of configurations. This can be illustrated by considering a simple molecule, represented by a rod that rotates freely in a hemisphere across a surface (Fig. 2.15). When the two surfaces are separated by an infinite distance ∞ the number of configurations of the rod is $\Omega(\infty)$, which is proportional to the volume of the hemisphere. When a second particle approaches to a distance h such that it cuts the hemisphere (losing some volume), the volume available to the chains is reduced and the number of configurations becomes $\Omega(h)$ which is less than $\Omega(\infty)$. For two flat plates, G_{elas} is given by the following expression [25]:

$$\frac{G_{elas}}{kT} = -2v_2 \ln \left[\frac{\Omega(h)}{\Omega(\infty)}\right] = -2v_2 R_{elas}(h), \tag{2.34}$$

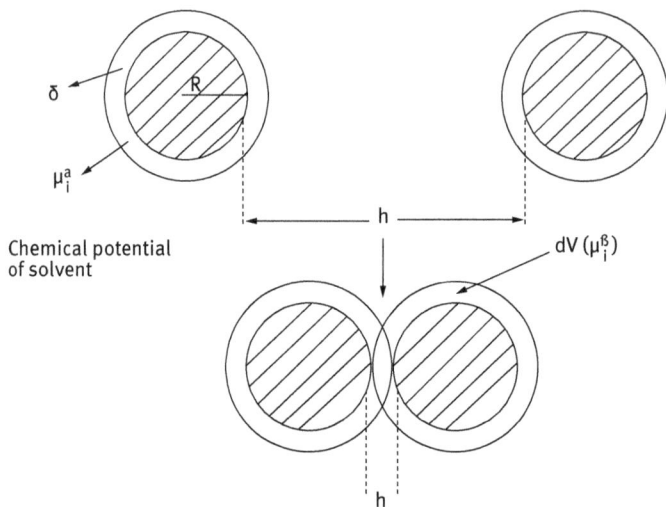

Fig. 2.14: Schematic representation of polymer layer overlap.

where $R_{elas}(h)$ is a geometric function whose form depends on the segment density distribution. It should be stressed that G_{elas} is always positive and could play a major role in steric stabilization. It becomes very strong when the separation distance between the particles becomes comparable to the adsorbed layer thickness δ.

Combining G_{mix} and G_{elas} with G_A gives the total energy of interaction G_T (assuming there is no contribution from any residual electrostatic interaction), i.e. [26]

$$G_T = G_{mix} + G_{elas} + G_A . \tag{2.35}$$

A schematic representation of the variation of G_{mix}, G_{elas}, G_A and G_T with surface-surface separation distance h is shown in Fig. 2.16. G_{mix} increases very sharply with decreasing h, when $h < 2\delta$. G_{elas} increases very sharply with decreasing h, when $h < \delta$. G_T versus h shows a minimum, G_{min}, at separation distances comparable to 2δ. When $h < 2\delta$, G_T shows a rapid increase with decreasing h. The depth of the minimum depends on the Hamaker constant A, the particle radius R and adsorbed layer thickness δ. G_{min} increases with increasing A and R. At a given A and R, G_{min} decreases with increasing δ (i.e. with increasing molecular weight, M_w, of the stabilizer). This is illustrated in Fig. 2.17 which shows the energy-distance curves as a function of δ/R. The larger the value of δ/R, the smaller the value of G_{min}. In this case the system may approach thermodynamic stability as is the case with nanodispersions.

For effective steric stabilization, the following criteria must be satisfied: (i) The particles should be completely covered by the polymer (the amount of polymer should correspond to the plateau value). Any bare patches may cause flocculation either by van der Waals attraction (between the bare patches) or by bridging flocculation (whereby a polymer molecule will become simultaneously adsorbed on two or more

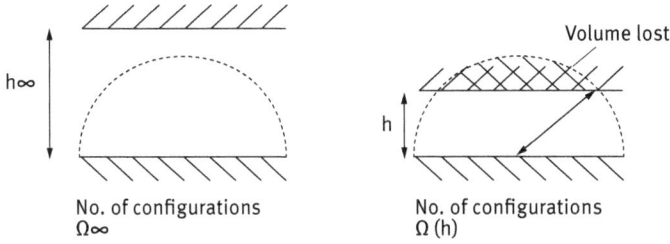

Fig. 2.15: Schematic representation of configurational entropy loss on approach of a second particle.

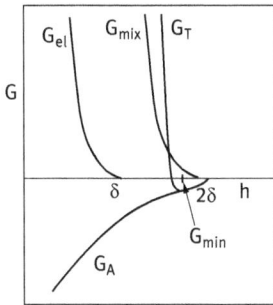

Fig. 2.16: Energy-distance curves for sterically stabilized systems.

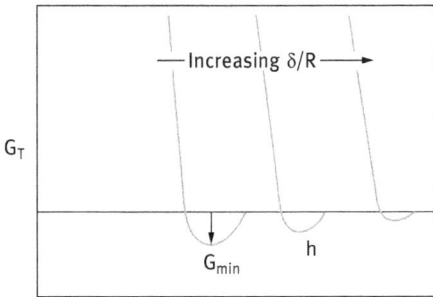

Fig. 2.17: Variation of G_{min} with δ/R.

particles). (ii) The polymer should be strongly "anchored" to the particle surfaces, to prevent any displacement during particle approach. For this purpose A–B, A–B–A block and BA_n graft copolymers are the most suitable where the chain B is chosen to be highly insoluble in the medium and has a strong affinity to the surface. Examples of B groups for hydrophobic particles in aqueous media are polystyrene and polymethylmethacrylate. (iii) The stabilizing chain A should be highly soluble in the medium and strongly solvated by its molecules. Examples of A chains in aqueous media are poly(ethylene oxide) and poly(vinyl alcohol). (iv) δ should be sufficiently large (> 5 nm) to prevent weak flocculation.

References

[1]   Kruyt, H. R. (ed.), "Colloid Science", Vol. I, Elsevier, Amsterdam (1952).
[2]   Tadros, Th. F., "Applied Surfactants", Wiley-VCH, Germany (2005).
[3]   Tadros, Th. F., "Dispersions of Powders in Liquids and Stabilisation of Suspensions", Wiley-VCH, Germany (2012).
[4]   Gouy, G., J. Phys., **9**, 457 (1910); Ann. Phys., **7**, 129 (1917).
[5]   Chapman, D. L., Phil. Mag., **25**, 475 (1913)
[6]   Stern, O., Z. Electrochem., **30**, 508 (1924).
[7]   Grahame, D. C., Chem. Rev., **41**, 44 (1947).
[8]   Bijesterbosch, B. H., "Stability of Solid/Liquid Dispersions", in "Solid/Liquid Dispersions", Th. F. Tadros (ed.), Academic Press, London (1987).
[9]   Hamaker, H. C., Physica (Utrecht), **4**, 1058 (1937).
[10]   Mahanty, J. and Ninham, B. W., "Dispersion Forces", Academic Press, London (1976).
[11]   Deryaguin, B. V. and Landau, L., Acta Physicochem. USSR, **14**, 633 (1941).
[12]   Verwey, E. J. W. and Overbeek, J. Th. G., "Theory of Stability of Lyophobic Colloids", Elsevier, Amsterdam (1948).
[13]   Hunter, R. J., "Zeta Potential in Colloid Science; Principles and Applications", Academic Press, London (1981).
[14]   Huckel, E., Phys. Z., **25**, 204 (1924).
[15]   Pusey, P. N. in "Industrial Polymers: Characterisation by Molecular Weights", J. H. S. Green and R. Dietz (eds.), Transcripta Books, London (1973).
[16]   Killmann, E., Eisenlauer, E. and Horn, M. J., Polymer Sci. Polymer Symposium, **61**, 413 (1977).
[17]   Fontana, B. J., and Thomas, J. R., J. Phys. Chem., **65**, 480 (1961).
[18]   Robb, I. D. and Smith, R., Eur. Polym. J., **10**, 1005 (1974).
[19]   Cohen-Stuart, M. A., Fleer, G. J. and Bijesterbosch, B., J. Colloid Interface Sci., **90**, 321 (1982).
[20]   Abeles, F., in "Ellipsometry in the Measurement of Surfaces and Thin Films", E. Passaglia, R. R. Stromberg and J. Kruger, Nat. Bur. Stand. Misc. Publ., **256**, 41 (1964).
[21]   Barnett, K. G., Cosgrove, T., Vincent, B., Burgess, A., Crowley, T. L., Kims, J., Turner J. D. and Tadros, Th. F., Disc. Faraday Soc., **22** , 283 (1981).
[22]   Napper, D. H., "Polymeric Stabilisation of Colloidal Dispersions", Academic Press, London (1981).
[23]   Flory, P. J. and Krigbaum, W. R., J. Chem. Phys., **18**, 1086 (1950).
[24]   Fischer, E. W., Kolloid Z., **160**, 120 (1958).
[25]   Mackor, E. L. and van der Waals, J. H., J. Colloid Sci., **7**, 535 (1951).
[26]   Hesselink, F. Th., Vrij, A. and Overbeek, J. Th. G., J. Phys. Chem., **75**, 2094 (1971).

3 Ostwald ripening in nanodispersions

3.1 Driving force for Ostwald ripening

The driving force of Ostwald ripening is the difference in solubility between the smaller and larger droplets [1]. The small particles or droplets with radius r_1 will have higher solubility than the larger particles or droplets with radius r_2. This can be easily recognized from the Kelvin equation [2] which relates the solubility of a particle or droplet $S(r)$ to that of a particle or droplet with infinite radius $S(\infty)$,

$$S(r) = S(\infty) \exp\left(\frac{2\gamma V_m}{rRT}\right), \tag{3.1}$$

where γ is the solid/liquid or liquid/liquid interfacial tension, V_m is the molar volume of the disperse phase, R is the gas constant and T is the absolute temperature. The quantity $(2\gamma V_m/RT)$ has the dimension of length and is termed the characteristic length with an order of $\sim 1\,nm$.

A schematic representation of the enhancement of the solubility $c(r)/c(0)$ with decreasing particle or droplet size according to the Kelvin equation is shown in Fig. 3.1.

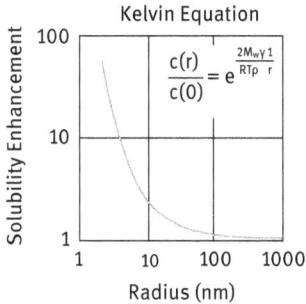

Fig. 3.1: Solubility enhancement with decreasing particle or droplet radius.

It can be seen from Fig. 3.1 that the solubility of nanodispersion particles or droplets increases very rapidly with decreasing radius, particularly when $r < 100\,nm$. This means that a particle with a radius of say 4 nm will have its solubility enhanced about 10 times compared say with a particle or droplet with a 10 nm radius whose solubility is only enhanced 2 times. Thus with time, molecular diffusion will occur between the smaller and larger particles or droplets, with the ultimate disappearance of most of the small particles or droplets. This results in a shift in the particle or droplet size distribution to larger values on storage of the nanodispersion. This could lead to the formation of a dispersion with average particle or droplet size in the μm range. This instability can cause severe problems, such as creaming or sedimentation, flocculation and even coalescence of the nanoemulsion.

For two particles with radii r_1 and r_2 $(r_1 < r_2)$,

$$\frac{RT}{V_m} \ln \left[\frac{S(r_1)}{S(r_2)} \right] = 2\gamma \left[\frac{1}{r_1} - \frac{1}{r_2} \right] . \tag{3.2}$$

Equation (3.2) is sometimes referred to as the Ostwald equation and it shows that the larger the difference between r_1 and r_2, the higher the rate of Ostwald ripening. That is why when preparing nanodispersions, one aims at producing a narrow size distribution.

A second driving force for Ostwald ripening in nanosuspensions is due to polymorphic changes. If a drug has two polymorphs A and B, the more soluble polymorph, say A (which may be more amorphous) will have higher solubility than the less soluble (more stable) polymorph B. During storage, polymorph A will dissolve and recrystallize as polymorph B. This can have a detrimental effect on bioefficacy, since the more soluble polymorph may be more active.

3.2 Kinetics of Ostwald ripening

The kinetics of Ostwald ripening is described in terms of the theory developed by Lifshitz and Slesov [3] and by Wagner [4] (referred to as LSW theory). The LSW theory assumes that: (i) Mass transport is due to molecular diffusion through the continuous phase. (ii) The dispersed phase particles are spherical and fixed in space. (iii) There are no interactions between neighbouring droplets (the droplets are separated by a distance much larger than the diameter of the droplets). (iv) The concentration of the molecularly dissolved species is constant except adjacent to the droplet boundaries.

The rate of Ostwald ripening ω is given by:

$$\omega = \frac{d}{dr}\left(r_c^3 \right) = \left(\frac{8\gamma DS(\infty)V_m}{9RT} \right) f(\phi) = \left(\frac{4DS(\infty)\alpha}{9} \right) f(\phi) , \tag{3.3}$$

where r_c is the radius of a particle or droplet that is neither growing nor decreasing in size, D is the diffusion coefficient of the disperse phase in the continuous phase, $f(\phi)$ is a factor that reflects the dependence of ω on the disperse volume fraction and α is the characteristic length scale $(= 2\gamma V_m/RT)$.

Droplets with $r > r_c$ grow at the expense of smaller ones, while droplets with $r < r_c$ tend to disappear. The validity of the LSW theory was tested by Kabalnov et al. [5] who used 1,2 dichloroethane-in-water emulsions and fixed the droplets to the surface of a microscope slide to prevent their coalescence. The evolution of the droplet size distribution as a function of time was followed by microscopic investigations.

LSW theory predicts that the droplet growth over time will be proportional to r_c^3. This is illustrated in Fig. 3.2 for dichloroethane-in-water emulsions.

Another consequence of LSW theory is the prediction that the size distribution function $g(u)$ for the normalized droplet radius $u = r/r_c$ adopts a time independent

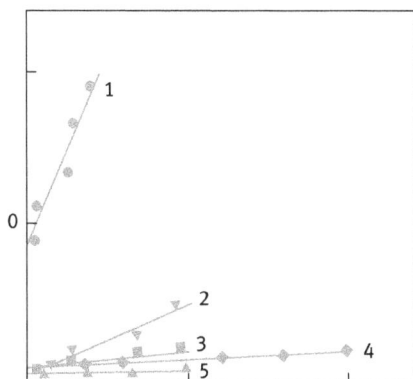

Fig. 3.2: Variation of average cube radius with time during Ostwald ripening in emulsions of: (1) 1,2 dichloroethane; (2) benzene; (3) nitrobenzene; (4) toluene; (5) p-xylene.

form given by:

$$g(u) = \frac{81eu^2 \exp[1/(2u/3 - 1)]}{32^{1/3}(u + 3)^{7/3}(1.5 - u)^{11/3}} \quad \text{for } 0 < u \leq 1.5 \tag{3.4}$$

and

$$g(u) = 0 \quad \text{for } u > 1.5. \tag{3.5}$$

A characteristic feature of the size distribution is the cut-off at $u > 1.5$.

A comparison of the experimentally determined size distribution (dichloroethane-in-water emulsions) with the theoretical calculations based on the LSW theory is shown in Fig. 3.3.

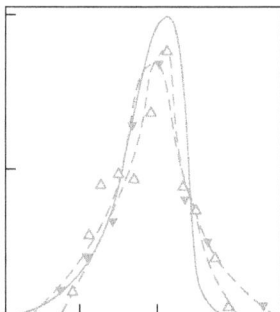

Fig. 3.3: Comparison between theoretical function g(u) (full line) and experimentally determined functions obtained for 1,2 dichloroethane droplets at time 0 (open triangles) and 300 s (inverted solid triangles).

The influence of the alkyl chain length of the hydrocarbon on the Ostwald ripening rate of nanoemulsions was systematically investigated by Kabalanov et al. [6]. Increasing the alkyl chain length of the hydrocarbon used for the emulsion results in a decrease in the oil solubility. According to LSW theory this reduction in solubility should result in a decrease in the Ostwald ripening rate. This was confirmed by the results of Kabalnov et al. [6] who showed that the Ostwald ripening rate decreases with increasing the alkyl chain length from C_9-C_{16}. Table 3.1 shows the solubility of the hydrocarbon, the experimentally determined rate ω_e and the theoretical values ω_t and the ratio of ω_e/ω_t.

Although the results showed the linear dependence of the cube of the droplet radius with time in accordance with LSW theory, the experimental rates were ~ 2–3 times higher than the theoretical values. The deviation between theory and experiment has been ascribed to the effect of Brownian motion [6]. LSW theory assumes that the droplets are fixed in space and that molecular diffusion is the only mechanism of mass transfer. For droplets undergoing Brownian motion, one must take into account the contributions of molecular and convective diffusion as predicted by the Peclet number,

$$Pe = \frac{rv}{D},\tag{3.6}$$

where v is the velocity of the droplets that is approximately given by

$$v = \left(\frac{3kT}{M}\right)^{1/2},\tag{3.7}$$

where k is the Boltzmann constant, T is the absolute temperature and M is the mass of the droplet. For r = 100 nm, Pe = 8, indicating that mass transfer will be accelerated with respect to that predicted by LSW theory.

Tab. 3.1: Influence of the alkyl chain length on the Ostwald ripening rate.

Hydrocarbon	$c(\infty)$ / ml ml^{-1a}	ω_e / cm^{-3} s^{-1}	ω_t / cm^{-3} s^{-1b}	$\omega_r = \omega_e/\omega_t$
C_9H_{20}	3.1×10^{-7}	6.8×10^{-19}	2.9×10^{-19}	2.3
$C_{10}H_{22}$	7.1×10^{-8}	2.3×10^{-19}	0.7×10^{-19}	3.3
$C_{11}H_{24}$	2.0×10^{-8}	5.6×10^{-20}	2.2×10^{-20}	2.5
$C_{12}H_{26}$	5.2×10^{-9}	1.7×10^{-20}	0.5×10^{-20}	3.4
$C_{13}H_{28}$	1.4×10^{-9}	4.1×10^{-21}	1.6×10^{-21}	2.6
$C_{14}H_{30}$	3.7×10^{-10}	1.0×10^{-21}	0.4×10^{-21}	2.5
$C_{15}H_{32}$	9.8×10^{-11}	2.3×10^{-22}	1.4×10^{-22}	1.6
$C_{16}H_{34}$	2.7×10^{-11}	8.7×10^{-23}	2.2×10^{-23}	4.0

a Molecular solubilities of hydrocarbons in water taken from: C. McAuliffe, *J. Phys. Chem.*, 1966, 1267.
b For theoretical calculations, the diffusion coefficients were estimated according to the Hayduk–Laudie equation (W. Hayduk and H. Laudie, *AIChE J.*, 1974, **20**, 611) and the correction coefficient f(ϕ) assumed to be equal to 1.75 for ϕ = 0.1 (P. W. Voorhees. *J. Stat. Phys.*, 1985, **38**, 231).

LSW theory assumes that there are no interactions between the droplets and is limited to low oil volume fractions. At higher volume fractions the rate of ripening depends on the interaction between diffusion spheres of neighbouring droplets. It is expected that emulsions with higher volume fractions of oil will have broader droplet size distribution and faster absolute growth rates than those predicted by LSW theory. However, experimental results using high surfactant concentrations (5 %) showed the rate to be independent of the volume fraction in the range $0.01 \leq \phi \leq 0.3$. It has been suggested that the emulsion droplets may have been screened from one another by surfactant micelles [7]. A strong dependence on volume fraction has been observed for fluorocarbon-in-water emulsions [8]. A threefold increase in ω was found when ϕ was increased from 0.08 to 0.52.

It has been suggested that micelles play a role in facilitating mass transfer between emulsion droplets by acting as carriers of oil molecules [9]. Three mechanisms were suggested: (i) Oil molecules are transferred via direct droplet/micelle collisions. (ii) Oil molecules exit the oil droplet and are trapped by micelles in the immediate vicinity of the droplet. (iii) Oil molecules exit the oil droplet collectively with a large number of surfactant molecules to form a micelle.

In mechanism (i) the micellar contribution to the rate of mass transfer is directly proportional to the number of droplet/micelle collisions, i.e. to the volume fraction of micelles in solution. In this case the molecular solubility of the oil in the LSW equation is replaced by the micellar solubility which is much higher. Large increases in the rate of mass transfer would be expected with increasing micelle concentration. Numerous studies indicate, however, that the presence of micelles affects mass transfer to only a small extent [10]. Results were obtained for decane-in-water emulsions using sodium dodecyl sulphate (SDS) as emulsifier at concentrations above the critical micelle concentration (cmc). This is illustrated in Fig. 3.4 which shows plots of $(d_{inst}/d_{inst}^o)^3$ (where d_{inst} is the diameter at time t and d_{inst}^o is the diameter at time 0) as a function of time. The results showed only a twofold increase in ω above the cmc. This result is consistent with many other studies which showed an increase in mass transfer of only 2–5 times with increasing micelle concentration. The lack of strong dependence of mass transfer on micelle concentration for ionic surfactants may result from electrostatic repulsion between the emulsion droplets and micelles, which provide a high energy barrier preventing droplet/micelle collision.

In mechanism (ii), a micelle in the vicinity of an emulsion droplet rapidly takes up dissolved oil from the continuous phase. This "swollen" micelle diffuses to another droplet, where the oil is redeposited. Such a mechanism would be expected to result in an increase in mass transfer over and above that expected from LSW theory by a factor φ given by equation (3.8):

$$\varphi = 1 + \frac{\phi_s \Gamma D_m}{D} = 1 + \frac{\chi^{eq} D_m}{c^{eq} D} , \tag{3.8}$$

where ϕ_s is the volume fraction of micelles in solution, $\chi^{eq} = \phi_s c_m^{eq}$ is the net oil solubility in the micelle per unit volume of micellar solution reduced by the density of the solute, $\Gamma = c_m^{eq}/c^{eq} \sim 10^6$–$10^{11}$ is the partition coefficient for the oil between the micelle and bulk aqueous phase at the saturation point, D_m is the micellar diffusivity ($\sim 10^{-6}$–10^{-7} cm^2 s^{-1}. For a decane-water nanoemulsion in the presence of

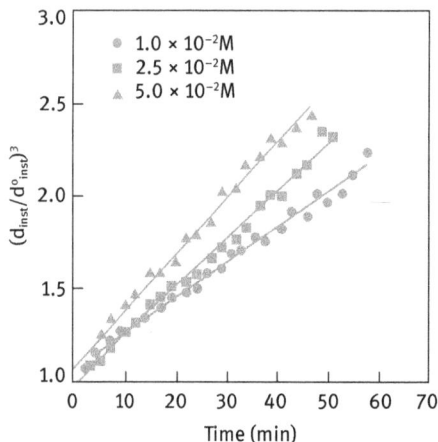

Fig. 3.4: Variation of $(d_{inst}/d^o_{inst})^3$ with time for decane-in-water emulsions for different SDS concentrations above the cmc.

0.1 mol dm^{-3} SDS, equation (3.8) predicts an increase in the rate of ripening by three orders of magnitude, in sharp contrast to the experimental results.

To account for the discrepancy between theory and experiment in the presence of surfactant micelles, Kabalanov [11] considered the kinetics of micellar solubilization and he proposed that the rate of oil monomer exchange between the oil droplets and the micelles is slow, and rate-determining. Thus at low micellar concentration, only a small proportion of the micelles are able to rapidly solubilize the oil. This leads to a small, but measurable increase in the Ostwald ripening rate with micellar concentration. Taylor and Ottewill [12] proposed that micellar dynamics may also be important. According to Aniansson et al. [13], micellar growth occurs in a stepwise fashion and is characterized by two relaxation times τ_1 and τ_2. The short relaxation time τ_1 is related to the transfer of monomers in and out of the micelles, while the long relaxation time τ_2 is the time required for break-up and reformation of the micelle. At low SDS (0.05 mol dm^{-3}) concentration $\tau_2 \approx 0.01$ s, whereas at higher SDS concentration (0.2 mol dm^{-3}) $\tau_2 \approx 6$ s. Taylor and Ottewill [12] suggested that, at low SDS concentration, τ_2 may be fast enough to have an effect on the Ostwald ripening rate, but at 5 % SDS τ_2 may be as long as 1000 s (taking into account the effect of solubilization on τ_2), which is too long to have a significant effect on the Ostwald ripening rate.

When using nonionic surfactant micelles, larger increases in the Ostwald ripening rate might be expected due to the larger solubilization capacities of the nonionic surfactant micelles and absence of electrostatic repulsion between the nanoemulsion droplets and the uncharged micelles. This was confirmed by Weiss et al. [14] who found a large increase in the Ostwald ripening rate in tetradecane-in-water emulsions in the presence of Tween 20 micelles.

3.3 Reduction of Ostwald ripening

3.3.1 Reduction of Ostwald ripening in nanosuspensions

Crystal growth in nanosuspensions may be inhibited by the addition of additives, usually referred to as crystal growth inhibitors. Trace concentrations of certain additives (one part in 10^4) can have profound effects on crystal growth and habit modification. It is generally accepted that the additive must adsorb on the surface of the crystals. Surfactants and polymers (interfacially active) are expected to affect crystal size and habit. They may affect the diffusion process (transport to and from the boundary layer) and they may also affect the surface controlled process (by adsorption at the surface, edge or specific sites). A good example of a polymer that inhibits crystal growth of sulphathiazole is polyvinyl pyrrolidone (PVP). The molecular weight of the polymer can be important. Many block A–B–A and graft BA_n copolymers (with B being the "anchor" part and A the stabilizing chain) are very effective in inhibiting crystal growth. The B chain adsorbs very strongly on the surface of the crystal and sites become unavailable for deposition. This has the effect of reducing the rate of crystal growth. Apart from their influence on crystal growth, the above copolymers also provide excellent steric stabilization, providing the A chain is chosen to be strongly solvated by the molecules of the medium. This was discussed in detail in Chapter 2.

3.3.2 Reduction of Ostwald ripening in nanoemulsions

3.3.2.1 Addition of a small proportion of highly insoluble oil

Huguchi and Misra [15] suggested that by addition of a second disperse phase that is virtually insoluble in the continuous phase, such as squalane, can significantly reduce the Ostwald ripening rate. In this case, significant partitioning between different droplets is predicted, with the component having the low solubility in the continuous phase (e.g. squalane) being expected to be concentrated in the smaller droplets. During Ostwald ripening in a two-component disperse system, equilibrium is established when the difference in chemical potential between different sized droplets, which results from curvature effects, is balanced by the difference in chemical potential resulting from partitioning of the two components. Huguchi and Misra [15] derived expression (3.9) for the equilibrium condition, wherein the excess chemical potential of the medium soluble component, $\Delta\mu_1$, is equal for all of the droplets in a polydisperse medium.

$$\frac{\Delta\mu_i}{RT} = \left(\frac{a_1}{r_{eq}}\right) + \ln(1 - X_{eq2}) = \left(\frac{a_1}{r_{eq}}\right) - X_{02}\left(\frac{r_0}{r_{eq}}\right)^3 = \text{const.}, \qquad (3.9)$$

where $\Delta\mu_1 = \mu_1 - \mu_1^*$ is the excess chemical potential of the first component with respect to the state μ_1^* when the radius $r = \infty$ and $X_{02} = 0$, r_0 and r_{eq} are the radii of an arbitrary drop under initial and equilibrium conditions respectively, X_{02} and X_{eq2} are the initial and equilibrium mole fractions of the medium insoluble component 2, a_1 is the characteristic length scale of the medium soluble component 1.

The equilibrium determined by equation (3.9) is stable if the derivative $\partial \Delta \mu_1 / \partial r_{eq}$ is greater than zero for all the droplets in a polydisperse system. Based on this analysis, Kabalanov et al. [16] derived the criterion (3.10):

$$X_{02} > \frac{2a_1}{3d_o},$$ (3.10)

where d_o is the initial droplet diameter. If the stability criterion is met for all droplets, two patterns of growth will result, depending on the solubility characteristic of the secondary component. If the secondary component has zero solubility in the continuous phase, then the size distribution will not deviate significantly from the initial one, and the growth rate will be equal to zero. In the case of limited solubility of the secondary component, the distribution is governed by rules similar to LSW theory, i.e. the distribution function is time-variant. In this case, the Ostwald ripening rate ω_{mix} will be a mixture growth rate that is approximately given by equation (3.11) [16]:

$$\omega_{mix} = \left(\frac{\phi_1}{\omega_1} + \frac{\phi_2}{\omega_2} \right)^{-1},$$ (3.11)

where ϕ_1 is the volume fraction of the medium soluble component and ϕ_2 is the volume fraction of the medium insoluble component respectively.

If the stability criterion is not met, a bimodal size distribution is predicted to emerge from the initially monomodal one. Since the chemical potential of the soluble component is predicted to be constant for all the droplets, it is also possible to derive equation (3.12) for the quasi-equilibrium component 1,

$$X_{02} + \frac{2a_1}{d} = const.,$$ (3.12)

where d is the diameter at time t.

Kabalanov et al. [17] studied the effect of addition of hexadecane to a hexane-in-water nanoemulsion. Hexadecane, which is less soluble than hexane, was studied at three levels X_{02} = 0.001, 0.01, and 0.1. For the higher mole fraction of hexadecane, namely 0.01 and 0.1, the emulsion had a physical appearance similar to that of an emulsion containing only hexadecane and the Ostwald ripening rate was reliably predicted by equation (3.11). However, the emulsion with X_{02} = 0.001 quickly separated into two layers, a sedimented layer with a droplet size of ca. 5 µm and a dispersed population of submicron droplets (i.e. a bimodal distribution). Since the stability criterion was not met for this low volume fraction of hexadecane, the observed bimodal distribution of droplets is predictable.

3.3.2.2 Modification of the interfacial layer for reduction of Ostwald ripening
According to LSW theory, the Ostwald ripening rate ω is directly proportional to the interfacial tension γ. Thus by reducing γ, ω is reduced. This could be confirmed by measuring ω as a function of SDS concentration for decane-in-water emulsion [10] below the critical micelle concentration (cmc). Below the cmc, γ shows a linear decrease with an increase in log [SDS] concentration. The results are summarized in Tab. 3.2.

Tab. 3.2: Variation of Ostwald ripening rate with SDS concentration for decane-in-water emulsions.

[SDS] Concentration / mol dm^{-3}	ω / cm^3 s^{-1}
0.0	2.50×10^{-18}
1.0×10^{-4}	4.62×10^{-19}
5.0×10^{-4}	4.17×10^{-19}
1.0×10^{-3}	3.68×10^{-19}
5.0×10^{-3}	2.13×10^{-19}

cmc of SDS = 8.0×10^{-3}

Several other mechanisms have been suggested to account for a reduction of the Ostwald ripening rate by modification of the interfacial layer. For example, Walstra [18] suggested that emulsions could be effectively stabilized against Ostwald ripening by the use of surfactants that are strongly adsorbed at the interface and which do not desorb during the Ostwald ripening process. In this case, an increase in interfacial dilational modulus ε and a decrease in interfacial tension γ would be observed for the shrinking droplets. Eventually the difference in ε and γ between droplets would balance the difference in capillary pressure (i.e. curvature effects) leading to a quasi-equilibrium state. In this case, emulsifiers with low solubilities in the continuous phase such as proteins would be preferred. Long-chain phospholipids with a very low solubility (cmc ~ 10^{-10} mol dm^{-3}) are also effective in reducing Ostwald ripening of some emulsions. The phospholipid would have to have a solubility in water about three orders of magnitude lower than the oil [19].

3.4 Influence of initial droplet size of nanoemulsions on the Ostwald ripening rate

The influence of initial droplet size on Ostwald ripening can be realized by considering the droplet size dependence of the characteristic time, τ_{OR},

$$\tau_{OR} \approx \frac{r^3}{\alpha S(\infty)D} \approx \frac{r^3}{\omega}. \tag{3.13}$$

Values of τ_{OR} when r = 100 nm are given in Tab. 3.3 for a series of hydrocarbons with increasing chain length which clearly show the reduction in Ostwald ripening

Tab. 3.3: Characteristic time for Ostwald ripening in hydrocarbon-in-water nanoemulsions stabilized by 0.1 mol dm^{-3} SDS.

Hydrocarbon	ω_e / cm^3 s^{-1}	$\tau_{OR} \sim (r^3/\omega_e)$
C_9H_{20}	6.8×10^{-19}	25 min
$C_{10}H_{22}$	2.3×10^{-19}	73 min
$C_{11}H_{24}$	5.6×10^{-20}	5 h
$C_{12}H_{26}$	1.7×10^{-20}	16 h
$C_{13}H_{28}$	4.1×10^{-21}	3 d
$C_{14}H_{30}$	1.0×10^{-21}	12 d
$C_{15}H_{32}$	2.3×10^{-22}	50 d
$C_{16}H_{34}$	8.7×10^{-23}	133 d

rate with increasing chain length (due to reduction in solubility S(∞)). The dramatic dependence of τ_{OR} on increasing chain length is apparent. The characteristic time shows a large dependence on the initial average radius as is illustrated in Fig. 3.5 for a series of emulsions. It can be seen that the Ostwald ripening rate can be extremely rapid for small droplet sizes, thereby providing a key component in determining initial droplet size. For example, it is not likely that droplets with radii less than 100 nm will be observed for decane-in-water nanoemulsions since the droplets will ripen to this size on the timescale of few minutes. This was confirmed by Kabalanov et al. [5] who noted large differences in initial droplet size for hydrocarbon-in-water emulsions as the chain length of the hydrocarbon was decreased. For example, nonane-in-water

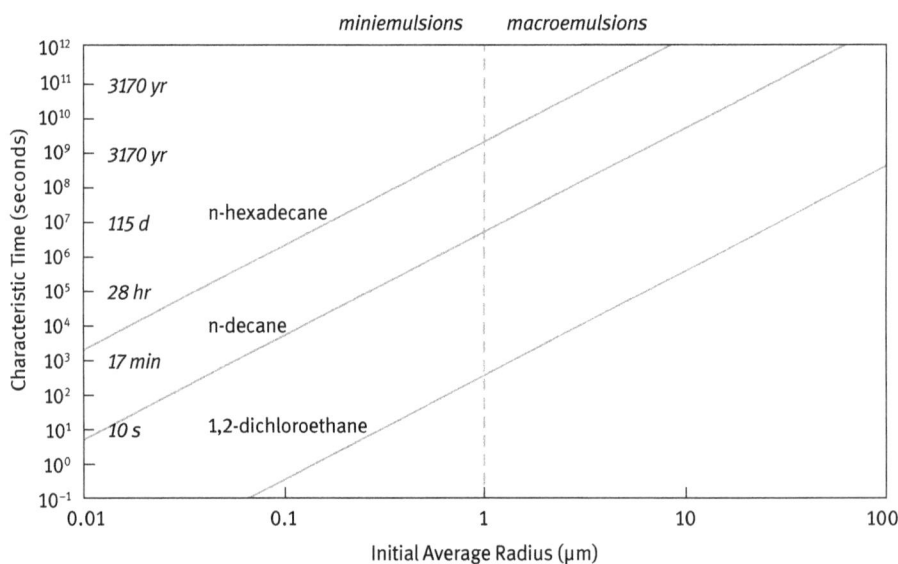

Fig. 3.5: Characteristic time for Ostwald ripening versus droplet size.

nanoemulsions had an initial droplet size of 178 nm, decane-in-water nanoemulsions had a size of 124 nm, and undecane-in-water nanoemulsions a size of 88 nm. It is clear from Fig. 3.5 that the driving force for Ostwald ripening decreases dramatically with increasing droplet size.

References

[1]   Weers, J. G., "Molecular Diffusion in Emulsions and Emulsion Mixtures", in "Modern Aspects of Emulsion Science", B. P. Binks (ed.), Royal Society of Chemistry Publication, Cambridge, UK (1999).
[2]   Thomson, W. (Lord Kelvin), Phil. Mag., **42**, 448 (1871).
[3]   Lifshitz, I. M. and Slesov, V. V., Sov. Phys. JETP, **35**, 331 (1959).
[4]   Wagner, C., Z. Electrochem., **35**, 581 (1961).
[5]   Kabalanov, A. S. and Shchukin, E. D., Adv. Colloid Interface Sci., **38**, 69 (1992).
[6]   Kabalanov, A. S., Makarov, K. N., Pertsov, A. V. and Shchukin, E. D., J. Colloid Interface Sci, **138**, 98 (1990).
[7]   Taylor, P., Colloids and Surfaces A, **99**, 175 (1995).
[8]   Ni, Y., Pelura, T. J., Sklenar, T. A., Kinner, R. A. and Song, D., Art. Cells Blood Subs. Immob. Biorech., **22**, 1307 (1994).
[9]   Karaboni, S., van Os, N. M., Esselink, K. and Hilbers, P. A. J., Langmuir, **9**, 1175 (1993).
[10] Soma, J. and Papadadopoulos, K. D., J. Colloid Interface Sci., **181**, 225 (1996).
[11] Kabalanov, A. S., Langmuir, **10**, 680 (1994).
[12] Taylor, P. and Ottewill, R. H., Colloids and Surfaces A, **88**, 303 (1994).
[13] Aniansson, E. A. G., Wall, S. N., Almegren, M., Hoffmann, H., Kielmann, I., Ulbricht, W., Zana, R., Lang, J. and Tondre, C., J. Phys. Chem., **80**, 905 (1976).
[14] Weiss, J., Coupland, J. N., Brathwaite, D. and McClemments, D. J., Colloids and Surfaces A, **121**, 53 (1997).
[15] Higuchi, W. I. and Misra, J., J. Pharm. Sci., **51**, 459 (1962).
[16] Kabalanov, A. S., Pertsov, A. V. and Shchukin, E. D., Colloids and Surfaces, **24**, 19 (1987).
[17] Kabalanov, A. S., Pertsov, A. V., Aprosin, Yu. D. and Shchukin, E. D., Kolloid Zh., **47**, 1048 (1095).
[18] Walstra, P., in "Encyclopedia of Emulsion Technology", Vol. 4, P. Becher (ed.), Marcel Dekker, NY (1996).
[19] Kabalanov, A. S., Weers, J., Arlauskas, P. and Tarara, T., Langmuir, **11**, 2966 (1995).

4.1 Introduction

As mentioned in Chapter 1, the bottom-up process for preparing nanosuspensions involves forming particles from molecular units. A good example is the preparation of nanoparticles of inorganic materials, such as silica, titania, ZnO, etc., i.e. a process of precipitation, nucleation and growth. This will form the first part of this chapter. Another example is the preparation of nanopolymer colloids by emulsion or dispersion polymerization and this will form the second part of this chapter.

One of the main advantages of the bottom-up process over the top-down process is the possibility of controlling particle size and shape distribution as well as the morphology of the resulting particles [1]. By controlling the process of nucleation and growth it is possible to obtain nanosuspensions with a narrow size distribution. This is particularly important for many practical applications such as photonic materials and semiconductor colloids. However, for other processes such as in paints and ceramic processes, a modest polydispersity can be beneficial to enhance the random packing density of the spheres and consequently the viscosity of the mixtures is generally below that for monodisperse spheres at the same volume fraction [2].

A very important aspect of nanosuspensions is the maintenance of their colloid stability, i.e. absence of any flocculation. This can be achieved by three different mechanisms which have been discussed in detail in Chapter 2 and only a brief summary is given here: (i) Electrostatic stabilization by creation of a surface charge as a result of ionization of surface groups or adsorption of ionic surfactants and formation of electrical double layer. Stabilization is obtained by double layer overlap on the approach of the nanoparticles. This repulsion will overcome van der Waals attraction, particularly at intermediate separation distances. As a result an energy barrier is produced that prevents any flocculation. (ii) Steric stabilization produced by adsorption of nonionic surfactants or polymeric surfactants. These molecules consist of an "anchor" chain B that strongly adsorbs on the particle surface and a stabilizing chain A that is strongly solvated by the molecules of the medium. When two particles approach to a separation distance h that is smaller than twice the adsorbed layer thickness (2δ), overlap of the layers occurs resulting in strong repulsion as a result of two main effects: unfavourable mixing of the A chains when these are in good solvent conditions and reduction of configurational entropy of the A chains on considerable overlap. As a result of these two effects, the energy-distance curve between the particles shows a very rapid increase in repulsion when $h < 2\delta$ and this prevents any flocculation. (iii) Electrosteric repulsion, which is a combination of electrostatic and steric repulsion. This can be produced when using a mixture of ionic and nonionic surfactant or polymer or when using a polyelectrolyte.

4.2 Preparation of nanosuspensions by precipitation

This method is usually applied for preparation of inorganic particles such as metal oxides and nanoparticles of metals. As an illustration, ferric oxide and aluminium oxide or hydroxide are prepared by hydrolysis of metal salts [1],

$$2FeCl_3 + 3H_2O \rightarrow F_2O_3 \downarrow + 6HCl, \tag{4.1}$$

$$AlCl_3 + 3H_2O \rightarrow AlOOOH + 3HCl. \tag{4.2}$$

Another example of the preparation of nanoparticles by precipitation is that of silica sols. These can be prepared by acidification of water glass, a strongly alkaline solution of glass (that consists of essentially amorphous silica). Acidification is necessary to achieve a highly supersaturated solution of dissolved silica. Another method for obtaining high supersaturation of silica is obtained by a change of solvent, instead of a change in pH. In this method a stock solution of sodium-silica solution ($Na_2O \cdot SiO_2$, 27 wt % SiO_2) is diluted with double distilled water to 0.22 wt % SiO_2. Under vigorous stirring 0.2 ml of this water glass solution is rapidly pipetted into 10 ml of absolute ethanol. A sudden turbidity increase manifests the formation of small, smooth silica spheres with a diameter around 30 nm and a typical polydispersity of 20–30 % [1]. A third method for preparation of silica spheres is the well-known Stober method [3]. The precursor tetraethyl silicate (TES) is dissolved in an ethanol-ammonia mixture which is gently stirred in a closed vessel. Silica spheres with a radius of about 60 nm and typical polydispersity of 10–15 % are produced.

An example of nanometal particles is the reduction of metal salt shown in equation (4.3) [1],

$$H_2PtCl_6 + BH_4^- + 3H_2O \rightarrow Pt \downarrow + 2H_2 \uparrow + 6HCl + H_2BO_3^- . \tag{4.3}$$

To understand the process of formation of nanoparticles by the bottom-up process we must consider the process of homogeneous precipitation at a fundamental level. If a substance becomes less soluble by a change of some parameter, such as a temperature decrease, the solution may enter a metastable state by crossing the bimodal as illustrated in the phase diagram (Fig. 4.1) of a solution which becomes supersaturated upon cooling [1].

In the metastable region, the formation of small nuclei initially increases the Gibbs free energy. Thus, demixing by nucleation is an activation process, occurring at a rate which is extremely sensitive to the precipitation in this metastable region. In contrast, when the solution is quenched into the unstable region on crossing the spinodal (see Fig. 4.1) there is no activation barrier to form a new phase.

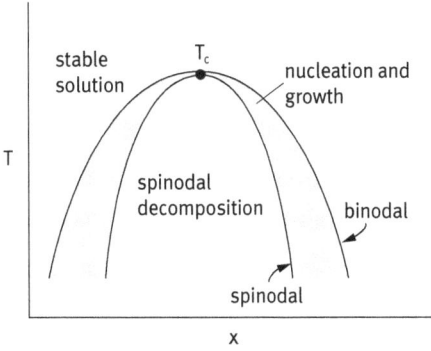

Fig. 4.1: Phase diagram of a solution which becomes supersaturated upon cooling; x is the solute mole fraction and T is the temperature.

4.2.1 Nucleation and growth

The classical nucleation theory considers a precipitating particle (referred to as a nucleus or cluster) to consist of a bulk phase containing N_i^s molecules and a shell with N_i^σ molecules which have a higher free energy per molecule than the bulk. The particle is embedded in a solution containing dissolved molecules i. This is schematically represented in Fig. 4.2. The Gibbs free energy of the nucleus G^s is made of a bulk part and a surface part [4],

$$G^s = \mu_i^s N_i^s + \sigma A, \tag{4.4}$$

where μ_i^s is the chemical potential per molecule, σ is the solid/liquid interfacial tension and A is the surface area of the nucleus.

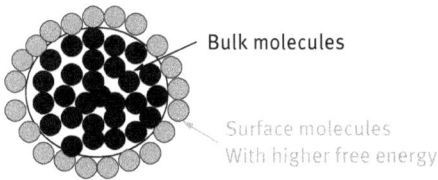

Bulk molecules

Surface molecules
With higher free energy

Fig. 4.2: Schematic representation of a nucleus.

In a supersaturated solution the activity a_i is higher than that of a saturated solution $a_i(\text{sat})$. As a result molecules are transferred from the solution to the nucleus's surface. The free energy change ΔG^s upon the transfer of a small number N_i from the solution to the particle is made up of two contributions from the bulk and the surface:

$$\Delta G^s = \Delta G^s(\text{bulk}) + \Delta G^s(\text{surface}). \tag{4.5}$$

The first term on the right-hand side of equation (4.5) is negative (it is the driving force) whereas the second term is positive (work has to be carried out in expanding the interface). $\Delta G^s(\text{bulk})$ is determined by the relative supersaturation, whereas $\Delta G^s(\text{surface})$ is determined by the solid/liquid interfacial tension σ and the interfacial area A which is proportional to $(N_i^s)^{2/3}$. ΔG^s is given by expression (4.6),

$$\Delta G^s = -N_i kT \ln S + \beta \sigma N_i^{2/3} , \tag{4.6}$$

where k is the Boltzmann constant, T is the absolute temperature and β is a proportionality constant that depends on the shape of the nucleus. S is the relative supersaturation that is equal to $a_i/a_i(sat)$.

For small clusters, the surface area term dominates whereas ΔG^s only starts to decrease due to the bulk term being beyond a critical value N^*.

N^* can be obtained by differentiating equation (4.6) with respect to N and equating the result to 0 ($dG^s/dN = 0$)

$$(N^*)^{1/3} = \frac{2\sigma\beta}{3kT \ln S} . \tag{4.7}$$

The maximum in the Gibbs energy is given by

$$\Delta G^* = \frac{1}{3}\beta(N^*)^{2/3} . \tag{4.8}$$

Equation (4.7) shows that the critical cluster size decreases with an increase in the relative supersaturation S or reduction of σ by addition of surfactants. This explains why a high supersaturation and/or addition of surfactants favour the formation of small particles. A large S pushes the critical cluster size N^* to smaller values and simultaneously lowers the activation barrier as illustrated in Fig. 4.3, which shows the variation of ΔG with radius at increasing S.

Assuming the nuclei to be spherical, equation (4.6) can be given in terms of the nucleus radius r

$$\Delta G = 4\pi r^2 \sigma - \left(\frac{4}{3}\right)\pi r^3 \left(\frac{kT}{V_m}\right) \ln S , \tag{4.9}$$

where V_m is the molecular volume.

ΔG^* and r^* are given by

$$\Delta G^* = \frac{4}{3}\pi(r^*)^2\sigma , \tag{4.10}$$

$$r^* = \frac{2V_m\sigma}{kT \ln S} . \tag{4.11}$$

When no precautions are taken, precipitation from a supersaturated solution produces polydisperse particles. This is because nucleation of new particles and further particle growth overlap in time. This overlap is the consequence of the statistical nature of the nucleation process; near the critical size particles may grow as well as dissolve. To narrow down the particle size distribution as much as possible, nucleation should take place in a short time, followed by equal growth of a constant number of particles. This can be achieved by rapidly creating the critical supersaturation required to initiate homogeneous nucleation after which particle growth lowers the saturation sufficiently to suppress new nucleation events. Another option is to add nuclei (seeds) to a solution with subcritical supersaturation. A fortunate consequence of particle growth is that in many cases the size distribution is self-sharpening.

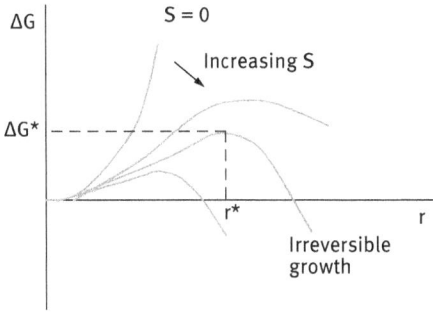

Fig. 4.3: Schematic representation of the effect of supersaturation on particle growth.

4.2.2 Precipitation kinetics

The kinetics of precipitation in a metastable solution can be considered to follow two regimes, according to Fig. 4.3. The initial regime is that where small particles struggle with their own solubility to pass the Gibbs energy barrier ΔG^*. This passage is called a nucleation event, which is simply defined as the capture of one molecule by a critical cluster [1], assuming that after this capture the cluster enters the irreversible growth regime from which a new colloid is born. In this case the number I of colloids that exist per second is proportional to c_m (the concentration of single unassociated molecules) and c^* (the concentration of critical clusters),

$$I = kc_m c^* ,$$ (4.12)

where k is the rate constant. Equation (4.12) predicts second-order reaction kinetics.

To quantify I, one must evaluate the frequency at which molecules encounter a spherical nucleus of radius a by diffusion. This can be evaluated using Smoluchowski's diffusion model for coagulation kinetics [5]. The diffusion flux J of molecules through any spherical envelope with radius a is given by Fick's first law,

$$J = 4\pi r^2 D \frac{dc(a)}{da} ,$$ (4.13)

where D is the molecular diffusion coefficient relative to the sphere positioned at the origin a = 0. Each molecule that reaches the sphere surface irreversibly attaches to the insoluble sphere and it is assumed that the concentration c_m of molecules in the liquid far away from the sphere radius remains constant [1],

$$c(a - r) = 0; \quad c(a \to \infty) = c_m .$$ (4.14)

For these boundary conditions equation (4.13) becomes,

$$J = 4\pi D r^* C_n$$ (4.15)

if it assumed that J is independent of a, i.e. if the diffusion of the molecules towards the sphere has reached a steady state. Such a state is approached by the concentration gradient around a sphere in a time of the order of r^2/D needed by the molecules to diffuse over a sphere diameter. Assuming that the sphere's growth is a consequence of stationary states, one can identify the nucleation rate I as the flux J multiplied by the concentration c^* of sphere with critical radius r^*,

$$I = 4\pi Dr^* c_m c^* \quad [m^{-3} s^{-1}].$$

(4.16)

The concentration c^* can be evaluated by considering the reversible work to form a cluster out of N molecules using the Boltzmann distribution principle,

$$c(N) = c_m \exp\left(\frac{-\Delta G}{kT}\right).$$

(4.17)

$c(N)$ represents the equilibrium concentration of clusters composed of N molecules and ΔG is the free energy of formation of a cluster. Applying this result to clusters with a critical size r^*, we obtain on substitution in equation (4.16),

$$I = 4\pi Dr^* c_m^2 \exp\left(\frac{-\Delta G^*}{kT}\right),$$

(4.18)

$$\Delta G^* = \left(\frac{4\pi}{3}\right)(r^*)^2 \sigma,$$

(4.19)

where ΔG^* is the height of the nucleation barrier. The exponent may be identified as the probability (per particle) that a spontaneous fluctuation will produce a critical cluster. Equation (4.18) shows that the nucleation rate is very sensitive to the value of r^* and hence to supersaturation as given by equation (4.11). The maximum nucleation rate at very large supersaturation, i.e. the pre-exponential term in equation (4.18), can be obtained by substitution of D using the Stokes–Einstein equation,

$$D = \frac{kT}{6\pi\eta r^*},$$

(4.20)

where η is the viscosity of the medium. This maximum nucleation rate is given by

$$I \approx \frac{kT}{\eta} = c_m^2.$$

(4.21)

For an aqueous solution at room temperature with a molar concentration c_m, the maximum nucleation rate is of the order of 10^{29} m^{-3} s^{-1}. A decrease in supersaturation to values around S = 5 is sufficient to reduce this very high rate to practically zero. For silica precipitation in dilute, acidified water glass solutions, $S \sim 5$ and nucleation may take hours or days.

As mentioned above, when no precautions are taken, precipitation from a supersaturated solution produces polydisperse particles. Fortunately in many cases, the size distribution is self-sharpening. This can be illustrated by considering the colloidal spheres with radius r, which irreversibly grow by the uptake of molecules from a solution according to the following rate law [1],

$$\frac{dr}{dt} = k_0 r^n,$$

(4.22)

where k_o and n are constants. This growth equation leads either to spreading or sharpening of the relative distribution, depending on the value of n. Consider at a given time t any pair of spheres with arbitrary size from the population of independently growing particles. Let $1 + \varepsilon$ be their size ratio such that $r(1 + \varepsilon)$ and r are the radius of the larger and smaller spheres, respectively. The former grow according to equation (4.23),

$$\frac{d}{dt}r(1 + \varepsilon) = kr^n(1 + \varepsilon)^n ,$$
(4.23)

which can be combined with the growth equation (4.22) for the smaller spheres to obtain the time evolution of the size ratio,

$$\frac{d\varepsilon}{dt} = k_o r^{n-1} \left[(1 + \varepsilon)^n - (1 + \varepsilon) \right]; \quad \varepsilon \geq 0 .$$
(4.24)

The relative size difference ε increases with time for $n > 1$, in which case particle growth broadens the distribution. For $n = 1$, the size ratio between the spheres remains constant, whereas for $n < 1$ it monotonically decreases with time. Since this decrease holds for any pair of particles in the growing population, it follows that for $n < 1$ the relative size distribution is self-sharpening [1]. This condition is practically realistic. For example, when the growth rate is completely determined by a slow reaction of molecules at the sphere radius,

$$\frac{dr^3}{dt} = k_o r^2 .$$
(4.25)

Implying that dr/dt is a constant, so $n = 0$. The opposite limiting case is growth governed by the rate at which molecules reach a colloid by diffusion. The diffusion flux for molecules with a diffusion coefficient D, relative to a sphere centred at the origin at $a = 0$, is given by equation (4.13). The saturation concentration is assumed to be maintained at the particle surface, neglecting the influence of particle size on c(sat), and keeping the bulk concentration of molecules constant [1],

$$c(a = r) = c(sat); \quad c(a \to \infty) = c(\infty) .$$
(4.26)

For these boundary conditions, the stationary flux towards the sphere equals

$$J = 4\pi Dr \left[c(\infty) - c(sat) \right] ,$$
(4.27)

showing that the rate at which the colloid intercepts the diffusing molecules is proportional to its radius and not to its surface area. If every molecule contributes a volume v_m to the growing colloid, then for a homogeneous sphere, the volume increases at a rate given by

$$\frac{d}{dt} \frac{4}{3}\pi r^3 = Jv_m ,$$
(4.28)

which on substitution in equation (4.27) leads to

$$\frac{dr}{dt} = Dv_m \left[c(\infty) - c(sat) \right] r^{-1} ,$$
(4.29)

with the typical scaling $r^2 \sim t$, as expected for a diffusion-controlled process. Thus the exponent in equation (4.22) for diffusion controlled growth is $n = -1$, and consequently the relative width of the size distribution decreases with time.

For charged species, an electrostatic interaction may be present between the growing colloids and the molecules they consume. This will either enhance or retard growth, depending on whether colloids and monomers attract or repel each other. In this case the diffusion coefficient D of the monomers in the diffusion flux J has to be replaced by an effective coefficient of the form,

$$D_{eff} = \frac{D}{r \int_r^\infty \exp\left(-\frac{u(a)}{kT}\right) a^{-2}\, da}, \tag{4.30}$$

where $u(a)$ is the interaction energy between molecule and colloid.

If the molecules are ions with charge ze and the colloid sphere has a surface potential ψ^o, then for low electrolyte where the interaction is unscreened (upper estimate of the ion-colloid interaction), $u(a)$ is given by

$$\frac{u(a)}{kT} = u_o\frac{r}{a}, \tag{4.31}$$

$$u_o = \frac{ze\psi^o}{kT} = zy^o, \tag{4.32}$$

where u_o is the colloid-ion interaction energy and $y^o = (e\psi^o/kT)$.

Thus the Coulombic interaction, equation (4.30), gives,

$$D_{eff} = D\frac{zy^o}{\exp(zy^o) - 1}. \tag{4.33}$$

Thus, the growth rate is slowed down exponentially by Coulombic interaction. For example when $\psi^o = 75$ mV, the effective diffusion coefficient for divalent ions is about $0.01\,D$. Addition of electrolyte screens the colloid-ion interaction and this moderates the effect of y^o on the growth kinetics.

The interaction between charged monomers and the growing colloid within the approximation given by equation (4.33) does not change the growth equation (4.29) and, therefore, does not affect the conclusion that diffusional growth sharpens the size distribution.

The effect of Ostwald ripening on the kinetics of particle growth can be analyzed using the Gibbs–Kelvin [1, 4] equation that relates the solubility $c(r)$ of a particle with radius r to that of an infinitely large particle $c(sat)$, i.e. a flat surface, by the equation,

$$\ln\left[\frac{c(r)}{c(sat)}\right] = \frac{2\sigma v_m}{r^* kT}. \tag{4.34}$$

The increased solubility according to equation (4.34) is referred to as the Gibbs–Kelvin effect. By considering the Gibbs energy maximum of Fig. 4.3, it is clear that it represents an unstable equilibrium, which can only be maintained for particles of exactly the same size. For polydisperse particles (with the same interfacial tension), there is no single, common equilibrium solubility; particles either grow or dissolve. Clearly, the largest particles have the strongest tendency to grow owing to their low solubility. This coarsening of colloids is referred to as Ostwald ripening and it is an important ageing effect which occurs with most polydisperse systems with small particles.

In a polydisperse system, the bulk concentration c(bulk) is not constant, but slowly decreasing in time due to the gradual disappearance of small particles. At any moment in time there is one sphere with radius r^o which is in metastable equilibrium with the bulk concentration,

$$c(bulk) = c(sat) \exp \left[\frac{2\sigma v_m}{r^o kT} \right] , \tag{4.35}$$

where c(sat) is the solubility of a flat surface. If the local solute concentration near a sphere with radius r_i is also fixed by the Gibbs–Kelvin equation, the steady state diffusion flux for sphere I is given by

$$J_i = 4\pi D r_i c(sat) \left\{ \exp \left[\frac{2\sigma v_m}{r^o kT} \right] - \exp \left[\frac{2\sigma v_m}{r_i kT} \right] \right\} . \tag{4.36}$$

It is clear that particles with radii $r_i < r^o$ dissolve because $J < 0$, whereas for $r_i > r^o$ the particles grow. The average particle radius and the critical radius r^o increase in time, so that the exponents in the diffusion flux can be linearized at a later stage of the ripening process. In this case, one can write for the growth or dissolution rate of sphere i the following approximate equation,

$$\frac{d}{dt} r_i^3 = 6 D r_i c(sat) \frac{\sigma v_m^2}{kT} \left[\frac{1}{r^o} - \frac{1}{r_i} \right] . \tag{4.37}$$

One limiting case of Ostwald ripening allows for a simple analytical solution, namely a monodisperse sphere with radius r, from which dissolved matter is deposited on very large particles, or a flat substrate. If that substrate controls the bulk concentration, r^o is infinitely large and consequently,

$$\frac{dr^3}{dt} = -6 D c(sat) \frac{\sigma v_m^2}{kT} . \tag{4.38}$$

Thus, for this case the particle volume decreases at a constant rate.

The time evolution of a continuous size distribution was analyzed by Lifshitz and Slesov [6] and Wagner [7] (referred to as the LSW theory) which predicts for large times the asymptotic result,

$$\frac{d\langle r \rangle^3}{dt} = \frac{8}{9} D c(sat) \frac{\sigma v_m^2}{kT} . \tag{4.39}$$

Equation (4.39) predicts that at a late stage of the ripening process, the average particle radius increases as $t^{1/3}$. The supersaturation falls as $t^{-1/3}$ and the number of spheres as t^{-1}. A remarkable finding of LSW theory is that due to Ostwald ripening the size distribution approaches a certain universal, time independent shape, irrespective of the initial distribution.

4.2.3 Seeded nucleation and growth

In the above analysis it is assumed that particle nucleation and growth occur in a so-lution of one solute. In practice this process of homogeneous nucleation is difficult to realize due to the presence of contaminants, dust, motes and irregularities on the ves-sel wall. This process of heterogeneous nucleation may have a dramatic effect on the kinetics. This process may be advantageous resulting in particle size polydispersity. The process of seed nucleation was first exploited for preparation of quite monodis-perse gold colloids by using a finely divided Faraday gold sol as the seed. The latter can also differ chemically from the precipitating material, leading to the formation of core-shell colloids [1]. Good examples are the growth of silica on gold cores, and other inorganic particles for the preparation of core-shell semiconductor particles [1]. Such well-defined composite colloids are increasingly important in materials science, in addition to their use in fundamental studies.

The efficiency of seeds or the container wall to catalyze nucleation is due to the reduction of the interfacial Gibbs energy of a precipitating particle. Steps and kinks on the seed substrate may act as active sites because they enable more of the surface of the nucleus to be in contact with the seed which lowers its surface excess Gibbs energy.

4.2.4 Surface modification

Surface modification is the deliberate attachment of a polymeric surfactant to the sur-face of the colloid to change its physical properties or chemical functionality. This modification is permanent if the attached polymer is not desorbed by thermal motion. Such surface modification occurs either via a chemical bond or significant adsorption energy (lack of desorption). The polymeric surfactant provides steric repulsion for the particles as discussed in Chapter 2. Surface modification is generally straightforward by choosing a molecule with suitable chemical linker. For example for metal hydroxide particles, such as silica, one can use a linkage between the -OH group on the surface of the particles and a carboxylic group or alcohol. For example the surface silanol groups on silica react, under mild conditions, with silage coupling agents (SCAs) and these materials are suitable for in situ modification of the colloid in a sol. The SCAs hydrolyze to reactive silanes, which graft themselves onto silica via the formation of a siloxane linkage.

Once reactive oligomers or polymers attach to a colloidal core, the core-shell par-ticle behaves as one kinetic unit with an average kinetic energy of $(3/2)kT$ (where k is the Boltzmann constant and T is the absolute temperature). This energy has to be weighed against the replacement of a large number of solvent molecules by the ad-sorbed species. Even a very small Gibbs energy penalty per replacement may suffice to produce aggregates that do not break apart by thermal motion. Such aggregation can also be induced by minute changes in the nature or composition of the solvent, a subtle effect that is difficult to predict or explain. Any small change in the composition involves a large number of low-molecular species, with a net enthalpy change that easily compensates the entropy loss due to aggregation of large colloids. One obvious counterexample is any solvent adsorption on modified or unmodified colloids. Water adsorption on silica is well known, but polar organic solvents such as dimethylformamide or triethylphosphate also adsorb in significant amounts on bare silica particles, often sufficient to prevent this aggregation. It should be noted that small particles also have a disadvantage since the coagulation rate is proportional to the square of the number density. For modified, stable colloids, the small particle size becomes a benefit in view of the many functional groups per gram. One attractive option is the simultaneous synthesis and modification of inorganic colloids by nucleation and growth in the presence of the modifying agent, which also influences and controls the particle size [1].

4.2.5 Other methods for preparation of nanosuspensions by the bottom-up process

Several other methods can be applied for preparing of nanosuspensions using the bottom-up processes of which the following are worth mentioning: (i) Precipitation of nanoparticles by addition of a nonsolvent (containing a stabilizer for the particles formed) to a solution of the compound in question; (ii) preparation of a nanoemulsion of the substance by using a solvent in which it is soluble following emulsification of the solvent in another immiscible solvent. This is then followed by removal of the solvent making the emulsion droplets by evaporation; (iii) preparation of the particles by mixing two microemulsions containing two chemicals that react together when the microemulsion droplets collide with each other; (iv) sol-gel processes particularly used for preparation of silica nanoparticles; (v) production of polymer nanosuspensions by miniemulsion or suspension polymerization; (vi) preparation of polymer nanosuspensions by polymerization of microemulsions. A brief description of each process is given below.

(i) Solvent-antisolvent method [8]. In this method, the substance (e.g. a hydrophobic drug) is dissolved in a suitable solvent such as acetone. The resulting solution is carefully added to another miscible solvent in which the resulting compound is insoluble. This results in precipitation of the compound by nucleation and growth. The particle size distribution is controlled by using a polymeric surfactant that is strongly adsorbed on the particle surface and providing an effective repulsive barrier to prevent aggregation of the particles. The polymeric surfactant is chosen to have specific adsorption on the particle surface to prevent Ostwald ripening. This method can be adapted for preparation of low water solubility drug nanosuspensions. In this case the drug is dissolved in acetone and the resulting solution is added to an aqueous solution of Poloxamer (an A–B–A block copolymer consisting of two A polyethylene oxide

(PEO) chains and a B polypropylene oxide (PPO) chain, i.e. PEO–PPO–PEO. After precipitation of the particles the acetone is removed by evaporation. The main problem with this method is the possibility of formation of several unstable polymorphs that will undergo crystal growth. In addition, the resulting particles may be of needle shape structure. However, by proper choice of the polymeric surfactant one can control the particle morphology and shape. Another problem may be the lack of removal of the solvent after precipitation of the particles.

(ii) Use of a nanoemulsion. In this case the compound is dissolved in a volatile organic solvent that is immiscible with water, such as methylene dichloride. The oil solution is emulsified in water using a high speed stirrer followed by high pressure homogenization [9]. A suitable emulsifier for the oil phase is used which has the same HLB number as the oil. The volatile oil in the resulting nanoemulsion is removed by evaporation and the formed nanosuspension particles are stabilized against aggregation by the use of an effective polymeric surfactant that could be dissolved in the aqueous phase. The main problem with this technique is the possible interaction between the emulsifier which may result in destabilization of the resulting nanosuspension. However, by careful selection of the emulsifier/stabilizing system one can form a colloidally stable nanosuspension.

(iii) Preparation of nanosuspensions by mixing two microemulsions [10]. Reverse microemulsions lend themselves as suitable "nonreactors" for the synthesis of nanoparticles. Inorganic salts can be dissolved in the water pools of a W/O microemulsion. Another W/O microemulsion with reducing agent dissolved in the water pools is then prepared. The two microemulsions are then mixed and the reaction between the inorganic salt and the reducing agent starts at the interface and proceeds towards the centre of the droplet. The rate limiting step appears to be droplet diffusion. Control of the exchange can be achieved by tuning the film rigidity. This procedure has been applied for the preparation of noble metal particles that could be applied in electronics, catalysis and in potential medical application.

(iv) Sol-gel process. This method is particularly applicable for preparation of silica nanoparticles [11]. This involves the development of networks through an arrangement of colloidal suspension (sol) and gelation to form a system in continuous liquid phase (gel). A sol is basically a dispersion of colloidal particles (1–100 nm) in a liquid and a gel is an interconnected rigid network with pores of submicron dimensions and polymeric chains. The sol-gel process, depending on the nature of the precursors, may be divided into two classes; namely inorganic precursors (chlorides, nitrates, sulphides, etc.) and alkoxide precursors. Extensively used precursors are tetramethyl silane and tetraethoxysilane. In this process, the reaction of metal alkoxides and water, in the presence of acid or base, forms a one phase solution that goes through a solution-to-gel transition to form a rigid, two phase system comprised of metal oxides and solvent filled pores. The physical and electrochemical properties of the resultant materials largely depend on the type of catalyst used in the reaction. In the case of silica alkoxides, the acid catalyzed reaction results in weakly crosslinked linear polymers. These polymers entangle and form additional branches leading to gelation. In the base-catalyzed reaction, due to rapid hydrolysis and condensation of the alkoxide silanes, the system forms highly branched clusters. The difference in cluster formation is due to the solubility of the resulting metal oxide in the reaction medium. The solubility of the silicon oxide is more in alkaline medium which favours the interlinking of the silica clusters than in acidic medium. A general procedure of sol-gel includes four stages; namely hydrolysis, condensation, growth and aggregation. The complete hydrolysis to form $M(OH)_4$ is very difficult to achieve. Instead condensation may occur between two $-OH$ or $M-OH$ groups and an alkoxy group to form bridging oxygen and a water or alcohol molecule. The hydrolysis and polycondensation reactions are initiated at numerous sites and the kinetics of the reaction can be very complex. When a sufficient number of interconnected $M-O-M$ bonds are formed in a particular region, they interact cooperatively to form colloidal particles or a sol. With time the colloidal particles link together to form three-dimensional networks. The size, shape and morphological features of the silica nanoparticles can be controlled by reaction kinetics, use of templates such as cationic, nonionic surfactants, polymers, electrolytes, etc.

(v) Preparation of polymer nanoparticles by miniemulsion or minisuspension polymerization [12]. In emulsion polymerization, the monomer, e.g. styrene or methyl methacrylate that is insoluble in the continuous phase, is emulsified using a surfactant that adsorbs at the monomer/water interface [12]. The surfactant micelles in bulk solution solubilize some of the monomer. A water soluble initiator such as potassium persulphate $K_2S_2O_8$ is added and this decomposes in the aqueous phase forming free radicals that interact with the monomers forming oligomeric chains. It has long been assumed that nucleation occurs in the "monomer swollen micelles". The reasoning behind this mechanism was the sharp increase in the rate of reaction above the critical micelle concentration and that the number of particles formed and their size depend to a large extent on the nature of the surfactant and its concentration (which determines the number of micelles formed). However, later this mechanism was disputed and it was suggested that the presence of micelles means that excess surfactant is available and molecules will readily diffuse to any interface.

The most accepted theory of emulsion polymerization is referred to as the coagulative nucleation theory [13, 14]. A two-step coagulative nucleation model has been proposed by Napper and co-workers [13, 14]. In this process the oligomers grow by propagation and this is followed by a termination process in the continuous phase. A random coil is produced which is insoluble in the medium and this produces a precursor oligomer at the θ-point. The precursor particles subsequently grow primarily by coagulation to form true latex particles. Some growth may also occur by further polymerization. The colloidal instability of the precursor particles may arise from their small size, and the slow rate of polymerization can be due to reduced swelling of the particles by the hydrophilic monomer [13, 14]. The role of surfactants in these processes is crucial since they determine the stabilizing efficiency and the effectiveness of the surface active agent ultimately determines the number of particles formed. This was confirmed by using surface active agents of different nature. The effectiveness of any surface active agent in stabilizing the particles was the dominant factor and the number of micelles formed was relatively unimportant.

A typical emulsion polymerization formulation contains water, 50 % monomer blended for the required glass transition temperature, T_g, surfactant (and often colloid), initiator, pH buffer and fungicide. Hard monomers with a high T_g used in emulsion polymerization may be vinyl acetate, methyl methacrylate and styrene. Soft monomers with a low T_g include butyl acrylate, 2-ethylhexyl acrylate, vinyl versatate and maleate esters. Most suitable monomers are those with low, but not too low, water solubility. Other monomers such as acrylic acid, methacrylic acid, or adhesion promoting monomers may be included in the formulation. It is important that the latex particles coalesce as the diluents evaporate. The minimum film forming temperature (MFFT) of the latex is a characteristic of say a paint system and is closely related to the T_g of the polymer. However, the latter can be affected by materials present such as surfactant and the inhomogeneity of the polymer composition at the surface. High T_g polymers will not coalesce at room temperature and in this case a plasticizer ("coalescing agent") such as benzyl alcohol is incorporated in the formulation to reduce the T_g of the polymer, thus reducing the MFFT of the paint. Several types of surfactants can be used in emulsion polymerization such as anionic surfactants (sulphates, sulphonates and phosphates), cationic surfactants (alkyl ammonium salts), amphoteric surfactants (such as alkyl betaine) and nonionic surfactants (such as alcohol ethoxylates). The role of surfactants is twofold, firstly to provide a locus for the monomer to polymerize and secondly to stabilize the polymer particles as they form. In addition, surfactants aggregate to form micelles (above the critical micelle concentration) and these can solubilize the monomers. In most cases, a mixture of anionic and nonionic surfactant is used for optimum preparation of polymer latexes. Cationic surfactants are seldom used, except for some specific applications where a positive charge is required on the surface of the polymer particles.

In addition to surfactants, most latex preparations require the addition of a polymer (sometimes referred to as "protective colloid") such as partially hydrolyzed polyvinyl acetate (commercially referred to as polyvinyl alcohol, PVA), hydroxyethyl cellulose or a block copolymer of polyethylene oxide (PEO) and polypropylene oxide (PPO). These polymers can be supplied with various molecular weights or proportions of PEO and PPO. When used in emulsion polymerization they can be grafted by the growing chain of the polymer being formed. They assist in controlling the particle size of the latex, enhancing the stability of the polymer dispersion and controlling the rheology of the final paint.

A typical emulsion polymerization process involves two stages known as the seed stage and the feed stage. In the feed stage, an aqueous charge of water, surfactant, and colloid is raised to the reaction temperature (85–90 °C) and 5–10 % of the monomer mixture is added along with a proportion of the initiator (a water soluble persulphate). In this seed stage, the formulation contains monomer droplets stabilized by surfactant, a small amount of monomer in solution as well as surfactant monomers and micelles. Radicals are formed in solution from the breakdown of the initiator and these radicals polymerize the small amount of monomer in solution. These oligomeric chains will grow to some critical size, the length of which depends on the solubility of the monomer in water. The oligomers build up to a limiting concentration and this is followed by a precipitous formation of aggregates (seeds), a process similar to micelle formation, except in this case the aggregation process is irreversible (unlike surfactant micelles which are in dynamic equilibrium with monomers).

In the feed stage, the remaining monomer and initiator are fed together and the monomer droplets become emulsified by the surfactant remaining in solution (or by extra addition of surfactant). Polymerization proceeds as the monomer diffuses from the droplets, through the water phase, into the already forming growing particles. At the same time radicals enter the monomer-swollen particles causing both termination and re-initiation of polymerization. As the particles grow, the remaining surfactant from the water phase is adsorbed onto the surface of particles to stabilize the polymer particles. The stabilization mechanism involves both electrostatic and steric repulsion. The final stage of polymerization may include a further shot of initiator to complete the conversion.

Most aqueous emulsion and dispersion polymerization reported in the literature is based on a few commercial products with a broad molecular weight distribution and varying block composition. The results obtained from these studies could not establish what effect the structural features of the block copolymer has on their stabilizing ability and effectiveness in polymerization. Fortunately, model block copolymers with well-defined structures could be synthesized and their role in emulsion polymerization has been carried out using model polymers and model latexes.

A series of well-defined A–B block copolymers of polystyrene-block-polyethylene oxide (PS-PEO) were synthesized [13] and used for emulsion polymerization of styrene. These molecules are "ideal" since the polystyrene block is compatible with the polystyrene formed and thus it forms the best anchor chain. The PEO chain (the stabilizing chain) is strongly hydrated with water molecules and it extends into the aqueous phase forming the steric layer necessary for stabilization. However, the PEO chain can become dehydrated at high temperature (due to breaking hydrogen bonds) thus reducing the effective steric stabilization. Thus, the emulsion polymerization should be carried out at temperatures well below the theta (θ)-temperature of PEO.

The above method of emulsion polymerization was adapted for the preparation of nanopolymer colloids. The main advantage of using miniemulsions (with diameters in the range 20–200 nm) in place of macroemulsions (with diameters > 500 nm) is the inherent greater surface area which may allow them to compete far more effectively for radicals than macroemulsions. The main problem with using miniemulsions is Ostwald ripening since the monomers with low molecular weight have finite solubility in water. This problem can be overcome by addition of a secondary disperse phase that is highly insoluble in water such as hexadecane, hexadecanol, dodecanethiol

and other monomer soluble polymers (see Chapter 3). In miniemulsion polymerization, nucleation occurs predominantly by radical entry into monomer in the interior
of the miniemulsion droplet. Due to the improved physical stability of the miniemulsion droplets, it is possible to adjust the total surfactant concentration so as to limit
the total number of micelles in solution, thereby limiting aqueous phase nucleation.
Relative to micelles, the droplets can have higher radical numbers (due to their larger
size) and can, therefore, significantly enhance the early stages of the polymerization
process. The rate of polymerization per particle is faster, and the systems are converted
faster. Following nucleation, the reaction proceeds by polymerization of the monomer
in the miniemulsion droplets.

A novel graft copolymer of hydrophobically modified inulin (INUTEC® SP1) has
been used to produce nanolatex particles in emulsion polymerization of styrene,
methyl methacrylate, butyl acrylate and several other monomers [15]. All latexes were
prepared by emulsion polymerization using potassium persulphate as initiator. The
z-average particle size was determined by photon correlation spectroscopy (PCS) and
electron micrographs were also taken.

Emulsion polymerization of styrene or methylmethacrylate showed an optimum
weight ratio of (INUTEC)/monomer of 0.0033 for PS and 0.001 for PMMA particles. The
(initiator)/(monomer) ratio was kept constant at 0.00125. The monomer conversion
was higher than 85 % in all cases. Latex dispersions of PS reaching 50 % and of PMMA
reaching 40 % could be obtained using such low concentrations of INUTEC® SP1. Figure 4.4 shows the variation of particle diameter with monomer concentration.

Nanolatex particles could be prepared up to 10 % styrene monomer, whereas for
PMMA, the nanosize could be maintained up to 2 % monomer.

The suspension polymerization process can be adapted to prepare polymer
nanoparticles. This is divided into three stages for both polymer soluble (A) and
insoluble (B) in its monomer. In the first stage for the A system, when the viscosity
of the disperse phase remains low, the bulk monomer phase is dispersed into small
droplets due to the shear stress imposed by the stirring conditions. Simultaneously,
through the reverse process of coalescence, the drops tend to reverse to the original monomer mass. The droplet size distribution results from break-up-coalescence
dynamic equilibrium. The adsorption of polymeric stabilizers at the monomer droplet-
water interface decreases the interfacial tension and the adsorbed layer prevents
coalescence. During the second stage, the viscosity within the droplets increases
with increasing conversion causing coalescence to overcome break-up. However, if
the stabilizer is present in sufficient amount and gives strong repulsion between the
droplets, this coalescence process is delayed resulting in a small increase in particle
size. Towards the end of this stage, coalescence is stopped due to the elastic nature
of the particle collisions. After this point, the particle size remains virtually constant.
The degree of agitation and design of the stirrer/reactor system has a big influence on
the dispersion of monomer droplets as well as on the overall process. An increase in
agitation improves the mixing and the heat transfer, and promotes the break-up of the

(a) PS latexes

(b) PMMA Latexes

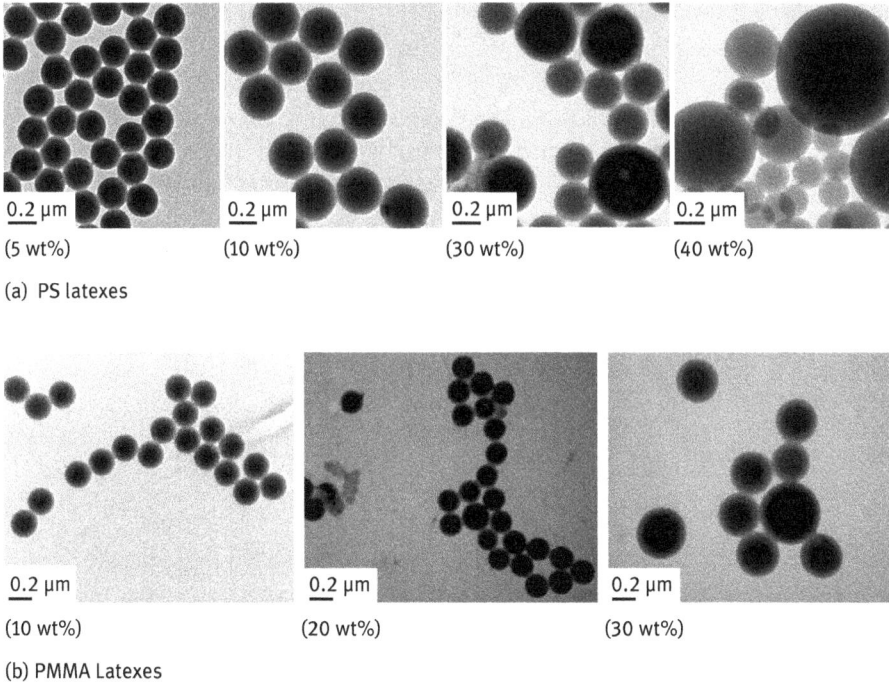

Fig. 4.4: Electron micrographs of the latexes.

droplets. However, increasing agitation increases the frequency of collisions thus increasing coalescence. These conflicting mechanisms show a reduction in droplet size with an increase in speed of agitation reaching a minimum at an optimum agitation speed followed by an increase in droplet size with further increases in stirrer speed due to coalescence. The formation of nanoparticles is also determined by the concentration and nature of the stabilizer. In most cases a mixture of polymeric stabilizer such as poly(vinyl alcohol) or Pluronic (an A–B–A block copolymer of polyethylene oxide, A, and polypropylene oxide, PPO) with an anionic surfactant such as sodium dodecyl sulphate is used. In this case the stabilizing mechanism is the combination of electrostatic and steric mechanisms, referred to as electrosteric.

(v) Preparation of nanopolymer latexes by polymerization of microemulsions. This method has attracted considerable attention and several comprehensive investigations have been carried out by Candau and co-workers [16, 17]. The polymerization of water-in-oil microemulsions containing acrylamide monomer in the aqueous droplets was investigated to prepare polyacrylamide with high molecular weight. The size of the nanolatex produced was larger than that of the water droplets of the microemulsion. This was explained by considering the possible fusion by coalescence of the droplets during polymerization. A systematic study of the polymerization of styrene-in-water and methylmethacrylate-in-water microemulsion was carried out by Larpent and Tadros [18] and Girard et al. [19]. The oil-in-water (o/w) microemulsion was prepared by the inversion method by addition of water containing a high HLB surfactant (nonyl phenol with 20 mol ethylene oxide) to an oil solution containing a low HLB nonionic surfactant (nonyl phenol with 4 mol ethylene oxide) and a small amount of an anionic surfactant (Aerosol OT). The oil soluble initiator (azobisisobutylynitrile; AIBN) was introduced in the oil phase before the formation of the microemulsion. Various polymerization methods were investigated, namely thermally induced, chemically induced and photochemically induced. The latter method gave the best results with latex size (~ 10 nm diameter) similar to that of the microemulsion droplet size.

4.3 Characterization of nanoparticles [1, 11]

4.3.1 Visual observations and microscopy

A great deal of information can already be obtained from visual inspection of a sol, aided by a torch or small laser. For colloids which do not absorb light at visible wavelengths, the turbidity is only due to light scattering. A bluish appearance in this case is due to Rayleigh scattering of particles (see below) with a typical diameter of 100 nm or smaller. This bluish Tyndall effect can be observed for dilute dispersions of latex particles and several metal hydroxide colloids such as boehmite and silica. Milky white appearance may be due to anything that shortens the mean free path of photons in the dispersion, namely large particle size, high refractive index and high colloid concentration. Multiple scattering is easy to demonstrate as it spreads an incoming narrow beam of laser light. A white appearance sometimes manifests aggregation; the bluish Tyndall effect for small aluminium hydroxide or silica colloids changes to white turbidity when the particles coagulate. Inspection of a (either stirred or shaken) sol with a light beam between cross polarizers reveals optical birefringence when the particles have an anisometric, plate or rod-like shape.

When the particles settle significantly within a few days, it is worthwhile to estimate the effective Stokes radius, which would produce the observed settling rate. If this radius is much larger than the expected colloid size, the colloids may be aggregated. The sediment on the bottom should also be observed when the vessel is tilted; stable colloids tend to flow like a liquid, but this flow can be slow if the particles are densely packed. Aggregated particles for sediments or gels with a yield stress which can be measured using rheology.

Instability of colloidal dispersions with respect to aggregation or phase separation is easy to detect. Shaking a dilute, unstable sol usually produces visible specs of aggregated particles, which stick to the glass surface. The onset of coagulation or phase separation is illustrated by a strong increase in the light scattering on approach of a critical point due to the occurrence of large fluctuations in density and hence in refractive index. When such fluctuations can be observed in a gently shaken sol, one can be sure that the sol will gel or phase separate soon after.

4.3.2 Electron microscopy

Electron microscopy is by far the most quantitative technique for determination of particle size in the nanosize range. It utilizes an electron beam to illuminate the sample. The electrons behave as charged particles which can be focused by annular electrostatic or electromagnetic fields surrounding the electron beam. Due to the very short wavelength of electrons, the resolving power of an electron microscope exceeds that of an optical microscope by ~ 200 times. The resolution depends on the accelerating voltage which determines the wavelength of the electron beam and magnifications as high as 200 000 can be reached with intense beams, but this could damage the sample. Mostly the accelerating voltage is kept below 100–200 kV and the maximum magnification obtained is below 100 000. The main advantage of electron microscopy is the high resolution, sufficient for resolving details separated by only a fraction of a nanometer. The increased depth of field, usually by about 10 μm or about 10 times that of an optical microscope, is another important advantage of electron microscopy. Nevertheless, electron microscopy also has some disadvantages such as sample preparation, selection of the area viewed and interpretation of the data. The main drawback of electron microscopy is the potential risk of altering or damaging the sample that may introduce artefacts and possible aggregation of the particles during sample preparation. The suspension has to be dried or frozen and the removal of the dispersion medium may alter the distribution of the particles. If the particles do not conduct electricity, the sample has to be coated with a conducting layer, such as gold, carbon or platinum to avoid negative charging by the electron beam. Two main types of electron microscopes are used: transmission and scanning.

4.3.2.1 Transmission electron microscopy (TEM)

TEM displays an image of the specimen on a fluorescent screen and the image can be recorded on a photographic plate or film. TEM can be used to examine particles in the range 0.001–5 μm. The sample is deposited on a Formvar (polyvinyl formal) film resting on a grid to prevent charging of the simple. The sample is usually observed as a replica by coating with an electron transparent material (such as gold or graphite). The preparation of the sample for TEM may alter the state of dispersion and cause aggregation. Freeze fracturing techniques have been developed to avoid some of the alterations of the sample during sample preparation. Freeze fracturing allows the dispersions to be examined without dilution and replicas can be made of dispersions containing water. It is necessary to have a high cooling rate to avoid the formation of ice crystals.

4.3.2.2 Scanning electron microscopy (SEM)

SEM can show particle topography by scanning a very narrowly focused beam across the particle surface. The electron beam is directed normally or obliquely at the surface. The back-scattered or secondary electrons are detected in a raster pattern and displayed on a monitor screen. The image provided by secondary electrons exhibits good three-dimensional detail. The back-scattered electrons, reflected from the incoming electron beam, indicate regions of high electron density. Most SEMs are equipped with both types of detectors. The resolution of the SEM depends on the energy of the electron beam which does not exceed 30 kV and hence the resolution is lower than that obtained by TEM. A very important advantage of SEM is elemental analysis by energy dispersive X-ray analysis (EDX). If the electron beam impinging on the specimen has sufficient energy to excite atoms on the surface, the sample will emit X-rays. The energy required for X-ray emission is characteristic of a given element and since the emission is related to the number of atoms present, quantitative determination is possible.

Scanning transmission electron microscopy (STEM) coupled with EDX has been used for the determination of metal particle sizes. Specimens for STEM were prepared by ultrasonically dispersing the sample in methanol and one drop of the suspension was placed onto a Formvar film supported on a copper grid.

4.3.2.3 Confocal laser scanning microscopy (CLSM)

CLSM is a very useful technique for identification of nanosuspensions. It uses a variable pinhole aperture or variable width slit to illuminate only the focal plane by the apex of a cone of laser light. Out-of-focus items are dark and do not distract from the contrast of the image. As a result of extreme depth discrimination (optical sectioning) the resolution is considerably improved (up to 40 % when compared with optical microscopy). The CLSM technique acquires images by laser scanning or uses computer software to subtract out-of-focus details from the in-focus image. Images are stored as the sample is advanced through the focal plane in elements as small as 50 nm. Three-dimensional images can be constructed to show the shape of the particles.

4.3.2.4 Scanning probe microscopy (SPM)

SPM can measure physical, chemical and electrical properties of the sample by scanning the particle surface with a tiny sensor of high resolution. Scanning probe microscopes do not measure a force directly; they measure the deflection of a cantilever which is equipped with a tiny stylus (the tip) functioning as the probe. The deflection of the cantilever is monitored by (i) a tunnelling current, (ii) laser deflection beam from the back side of the cantilever, (iii) optical interferometry, (iv) laser output controlled by the cantilever used as a mirror in the laser cavity, and (v) change in capacitance. SPM generates a three-dimensional image and allows calibrated measurements in three (x, y, z) coordinates. SPM not only produces a highly magnified image, but provides valuable information on sample characteristics. Unlike EM which requires vacuum for its operation, SPM can be operated under ambient conditions and, with some limitation, in liquid media.

4.3.2.5 Scanning tunnelling microscopy (STM)

STM measures an electric current that flows through a thin insulating layer (vacuum or air) separating two conductive surfaces. The electrons are visualized to "tunnel" through the dielectric and generate a current, I, that depends exponentially on the distance, s, between the tiny tip of the sensor and the electrically conductive surface of the sample. The STM tips are usually prepared by etching a tungsten wire in an NaOH solution until the wire forms a conical tip. Pt/Ir wire has also been used. In the contrast current imaging mode, the probe tip is raster-scanned across the surface and a feedback loop adjusts the height of the tip in order to maintain a constant tunnel current. When the energy of the tunnelling current is sufficient to excite luminescence, the tip-surface region emits light and functions as an excitation source of subnanometer dimensions. In situ STM has revealed a two-dimensional molecular lamellar arrangement of long chain alkanes adsorbed on the basal plane of graphite. Thermally induced disordering of adsorbed alkanes was studied by variable temperature STM and atomic scale resolution of the disordered phase was claimed by studying the quenched high-temperature phase.

4.3.2.6 Atomic force microscopy (AFM)

AFM allows one to scan the topography of a sample using a very small tip made of silicon nitride. The tip is attached to a cantilever that is characterized by its spring constant, resonance frequency and a quality factor. The sample rests on a piezoceramic tube which can move the sample horizontally (x, y motion) and vertically (z motion). The displacement of the cantilever is measured by the position of a laser beam reflected from the mirrored surface on the top side of the cantilever. The reflected laser beam is detected by a photodetector. AFM can be operated in either a contact or a noncontact mode. In the contact mode the tip travels in close contact with the surface, whereas in the noncontact mode the tip hovers 5–10 nm above the surface.

4.3.3 Scattering techniques

These are by far the most useful methods for characterization of nanosuspensions and in principle they can give quantitative information on the particle size distribution, floc size and shape. The only limitation of the methods is the need to use sufficiently dilute samples to avoid interference such as multiple scattering which makes interpretation of the results difficult. However, recently back scattering methods have been designed to allow one to measure the sample without dilution. In principle, one can use any electromagnetic radiation such as light, X-ray or neutrons but in most industrial labs only light scattering is applied (using lasers).

4.3.3.1 Light scattering techniques

These can be divided into three main classes: (i) Time-average light scattering, static or elastic scattering. (ii) Dynamic (quasi-elastic) light scattering that is usually referred as Photon Correlation Spectroscopy. This is a rapid technique that is very suitable for measuring submicron particles (nanosize range). (iii) Back scattering techniques that are suitable for measuring concentrated samples. Application of any of these methods depends on the information required and availability of the instruments.

(i) Time average light scattering

In this method the dispersion that is sufficiently diluted to avoid multiple scattering is illuminated by a collimated light (usually laser) beam and the time-average intensity of scattered light is measured as a function of scattering angle θ. Static light scattering is termed elastic scattering. Three regimes can be identified:

Rayleigh regime. Whereby the particle radius R is smaller than $\lambda/20$ (where λ is the wave length of incident light). The scattering intensity is given by equation (4.40),

$$I(Q) = [\text{Instrument constant}][\text{Material constant}]NV_p^2 . \tag{4.40}$$

Q is the scattering vector that depends on the wavelength of light λ used and is given by

$$Q = \left(\frac{4\pi n}{\lambda}\right)\sin\left(\frac{\theta}{2}\right), \tag{4.41}$$

where n is the refractive index of the medium.

The material constant depends on the difference between the refractive index of the particle and that of the medium. N is the number of particles and V_p is the volume of each particle. Assuming that the particles are spherical, one can obtain the average size using equation (4.40).

The Rayleigh equation reveals two important relationships: (i) The intensity of scattered light increases with the square of the particle volume and consequently with the sixth power of the radius R. Hence the scattering from larger particles may dominate the scattering from smaller particles. (ii) The intensity of scattering is inversely proportional to λ^4. Hence a decrease in the wavelength will substantially increase the scattering intensity.

Rayleigh–Gans–Debye regime (RGD), $\lambda/20 < R < \lambda$. The RGD regime is more complicated than the Rayleigh regime and the scattering pattern is no longer symmetrical about the line corresponding to the 90° angle but favours forward scattering ($\theta < 90°$) or back scattering ($180° > \theta > 90°$). Since the preference for forward scattering increases with increasing particle size, the ratio $I_{45°}/I_{135°}$ can indicate the particle size.

Mie regime, $R > \lambda$. The scattering behaviour is more complex than the RGD regime and the intensity exhibits maxima and minima at various scattering angles depending on particle size and refractive index. The Mie theory for light scattering can be used to obtain the particle size distribution using numerical solutions. One can also obtain information on particle shape.

(ii) Dynamic light scattering – photon correlation spectroscopy (PCS)

Dynamic light scattering (DLS) is a method that measures the time-dependent fluctuation of scattered intensity. It is also referred to as quasi-elastic light scattering (QELS) or photon correlation spectroscopy (PCS). The latter is the most commonly used term for describing the process since most dynamic scattering techniques employ autocorrelation.

PCS is a technique that utilizes Brownian motion to measure particle size. As a result of Brownian motion of dispersed particles the intensity of scattered light undergoes fluctuations that are related to the velocity of the particles. Since larger particles move less rapidly than smaller ones, the intensity fluctuation (intensity versus time) pattern depends on particle size as illustrated in Fig. 4.5. The velocity of the scatterer is measured in order to obtain the diffusion coefficient.

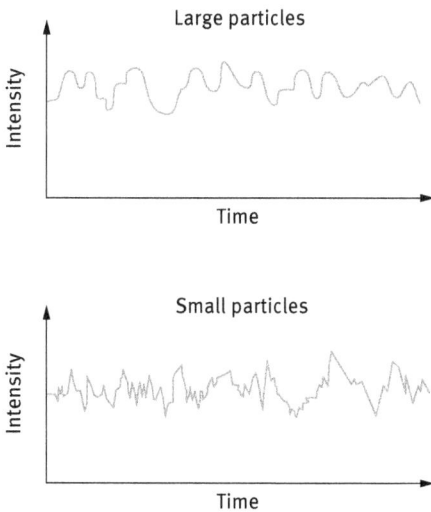

Fig. 4.5: Schematic representation of the intensity fluctuation for large and small particles.

In a system where Brownian motion is not interrupted by sedimentation or particle-particle interaction, the movement of particles is random. Hence, the intensity fluctuations observed after a large time interval do not resemble those fluctuations observed initially, but represent a random distribution of particles. Consequently, the fluctuations observed at large time delay are not correlated with the initial fluctuation pattern. However, when the time differential between the observations is very small (a nanosecond or a microsecond) both positions of particles are similar and the scattered intensities are correlated. When the time interval is increased, the correlation decreases. The decay of correlation is particle size dependent. The smaller the particles are, the faster is the decay.

The fluctuations in scattered light are detected by a photomultiplier and are recorded. The data containing information on the particles' motion are processed by a digital correlator. The latter compares the intensity of scattered light at time t, $I(t)$, to the intensity at a very small time interval τ later, $I(t + \tau)$, and it constructs the second-order autocorrelation function $G_2(\tau)$ of the scattered intensity,

$$G_2(\tau) = \langle I(t) \, I(t + \tau) \rangle \, . \tag{4.42}$$

The experimentally measured intensity autocorrelation function $G_2(\tau)$ depends only on the time interval τ, and is independent of t, the time when the measurement started.

PCS can be measured in a homodyne where only scattered light is directed to the detector. It can also be measured in heterodyne mode where a reference beam split from the incident beam is superimposed on scattered light. The diverted light beam functions as a reference for the scattered light from each particle.

In the homodyne mode, $G_2(\tau)$ can be related to the normalized field autocorrelation function $g_1(\tau)$ by

$$G_2(\tau) = A + Bg_1^2(\tau) \, , \tag{4.43}$$

where A is the background term designated as the baseline value and B is an instrument-dependent factor. The ratio B/A is regarded as a quality factor of the measurement or the signal-to-noise ratio and expressed sometimes as the % merit.

The field autocorrelation function $g_1(\tau)$ for a monodisperse suspension decays exponentially with τ,

$$g_1(\tau) = \exp(-\Gamma\tau) \, , \tag{4.44}$$

where Γ is the decay constant (s^{-1}).

Substitution of equation (4.44) into equation (4.43) yields the measured autocorrelation function,

$$G_2(\tau) = A + B \exp(-2\Gamma\tau) \, . \tag{4.45}$$

The decay constant Γ is linearly related to the translational diffusion coefficient D_T of the particle,

$$\Gamma = D_T q^2 \tag{4.46}$$

The modulus q of the scattering vector is given by

$$q = \frac{4\pi n}{\lambda_o} \sin\left(\frac{\theta}{2}\right),$$ (4.47)

where n is the refractive index of the dispersion medium, θ is the scattering angle and λ_o is the wavelength of the incident light in vacuum.

PCS determines the diffusion coefficient and the particle radius R is obtained using the Stokes–Einstein equation,

$$D = \frac{kT}{6\pi\eta R},$$ (4.48)

where k is the Boltzmann constant, T is the absolute temperature and η is the viscosity of the medium.

The Stokes–Einstein equation is limited to noninteracting, spherical and rigid spheres. The effect of particle interaction at relatively low particle concentration c can be taken into account by expanding the diffusion coefficient into a power series of concentration:

$$D = D_0(1 + k_D c),$$ (4.49)

where D_0 is the diffusion coefficient at infinite dilution and k_D is the virial coefficient that is related to particle interaction. D_0 can be obtained by measuring D at several particle number concentrations and extrapolating to zero concentration.

For polydisperse suspension the first-order autocorrelation function is an intensity-weighted sum of an autocorrelation function of particles contributing to the scattering:

$$g_1(\tau) = \int_0^\infty C(\Gamma) \exp(-\Gamma\tau)\, d\Gamma.$$ (4.50)

C(Γ) represents the distribution of decay rates.

For a narrow particle size distribution the cumulant analysis is usually satisfactory The cumulant method is based on the assumption that for monodisperse suspensions $g_1(\tau)$ is monoexponential. Hence, the log of $g_1(\tau)$ versus τ yields a straight line with a slope equal to Γ,

$$\ln g_1(\tau) = 0.5 \ln(B) - \Gamma\tau,$$ (4.51)

where B is the signal-to-noise ratio.

The cumulant method expands the Laplace transform about an average decay rate,

$$\langle\Gamma\rangle = \int_0^\infty \Gamma C(\Gamma)\, d\Gamma.$$ (4.52)

The exponential in equation (4.52) is expanded about an average and integrated term,

$$\ln g_1(\tau) = \langle\Gamma\rangle\tau + (\mu_2\tau^2)/2! - (\mu_3\tau^3)/3! + \dots$$ (4.53)

An average diffusion coefficient is calculated from $\langle\Gamma\rangle$ and the polydispersity (termed the polydispersity index) is indicated by the relative second moment, $\mu_2/\langle\Gamma\rangle^2$. A constrained regulation method (CONTIN) yields several numerical solutions to the particle size distribution and this is normally included in the software of the PCS machine.

PCS is a rapid, absolute and nondestructive method for particle size measurements. It has some limitations. The main disadvantage is the poor resolution of particle size distribution. Also it suffers from the limited size range (absence of any sedimentation) that can be accurately measured. Several instruments are commercially available, e.g. by Malvern, Brookhaven, Coulters, etc. The most recent instrument that is convenient to use is supplied by Malvern (UK) and this allows one to measure the particle size distribution without the need for too much dilution (which may cause some particle dissolution).

(iii) Back scattering techniques

This method is based on the use of fibre optics, sometimes referred to as fibre optic dynamic light scattering (FODLS) and it allows one to measure at high particle number concentrations. FODLS employs either one or two optical fibres. Alternatively, fibre bundles may be used. The exit port of the optical fibre (optode) is immersed in the sample and the scattered light in the same fibre is detected at a scattering angle of $180°$ (i.e. back scattering).

The above technique is suitable for on-line measurements during manufacture of a nanosuspension or nanoemulsion. Several commercial instruments are available, e.g. Lesentech (USA).

4.3.4 Measurement of charge and zeta potential

Many nanoparticles acquire a charge either by dissociation of surface groups (such as metal oxides or latexes) or by adsorption of ionic surfactants. A surface potential can be ascribed to the particle surface which, as discussed in Chapter 2, determines the electrostatic repulsion between the particles. In most cases it is not easy to determine the surface potential which is replaced by the measurable zeta potential. Two methods can be applied for measurement of the zeta potential. The first is based on laser velocimetry which was described in detail in Chapter 2. In this method one has to use a dilute dispersion in order to prevent multiple scattering by the particles. An alternative method that can be applied on more concentrated dispersions is the electroacoustic method which is described below.

The mobility of a particle in an alternating field is termed dynamic mobility, to distinguish it from the electrophoretic mobility in a static electric field [21]. The principle of the technique is based on the creation of an electric potential by a sound wave transmitted through an electrolyte solution, as described by Debye [22]. The potential, termed the ionic vibration potential (IVP), arises from the difference in the frictional forces and the inertia of hydrated ions subjected to ultrasound waves. The effect of the ultrasonic compression is different for ions of different masses and the displacement amplitudes are different for anions and cations. Hence the sound waves create period- ically changing electric charge densities. This original theory of Debye was extended to include electrophoretic, relaxation and pressure gradient forces [23, 24].

A much stronger effect can be observed in colloidal dispersions. The sound waves transmitted by the suspension of charged particles generate an electric field because the relative motion of the two phases is different. The displacement of a charged parti- cle from its environment by the ultrasound waves generates an alternating potential, termed colloidal vibration potential (CVP). The IVP and CVP are both called ultra- sound vibration potential (UVP).

The converse effect, namely the generation of sound waves by an alternating elec- tric field [25] in a colloidal dispersion, can be measured and is termed the electroki- netic sonic amplitude (ESA). The theory for the ESA effect has been developed by O'Brian and co-workers [26–29]. Dynamic mobility can be determined by measuring either UVP or ESA, although in general the ESA is the preferred method. Several com- mercial instruments are available for measuring dynamic mobility: (i) the ESA-8000 system from Matec Applied Sciences that can measure both CVP and ESA signals; (ii) the Pen Kem System 7000 Acoustophoretic titrator that measures the CVP, con- ductivity, pH, temperature, pressure amplitude and sound velocity.

In the ESA system (from Matec) and the AcoustoSizer (from Colloidal Dynamics) the dispersion is subjected to a high frequency alternating field and the ESA signal is measured. The ESA-8000 operates at constant frequency of $\sim 1\,MHz$ and the dynamic mobility and zeta potential (but not particle size) are measured. The AcoustoSizer op- erates at various frequencies of the applied electric field and can measure the particle mobility, zeta potential and particle size.

The frequency synthesizer feeds a continuous sinusoidal voltage into a grated am- plifier that creates a pulse of sinusoidal voltage across the electrodes in the dispersion. The pulse generates sound waves which appear to emanate from the electrodes. The oscillation, the back-and-forth movement of the particle caused by an electric field is the product of the particle charge times the applied field strength. When the direction of the field is alternating, particles in the suspension between the electrodes are driven away towards the electrodes. The magnitude and phase angle of the ESA signal created is measured with a piezoelectric transducer mounted on a solid nonconductive (glass) rod attached to the electrode as illustrated in Fig. 4.6. The purpose of this nonconduc- tive acoustic delay line is to separate the transducer from the high-frequency electric field in the cell. Three pulses of the voltage signal are recorded as schematically shown in Fig. 4.7. The first pulse of the signal, shown on the left, is generated when the voltage pulse is applied to the sample and is unrelated to the ESA effect. This first pulse of the signal is received before the sound has sufficient time to pass down the glass rod and is electronic cross-talk deleted from data processing. The second and third pulses are

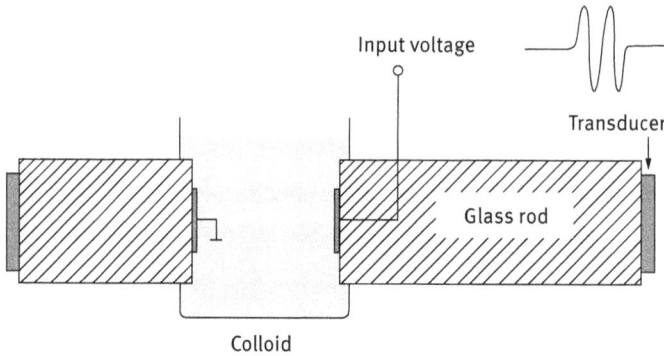

Fig. 4.6: Schematic representation of the AcoustoSizer cell.

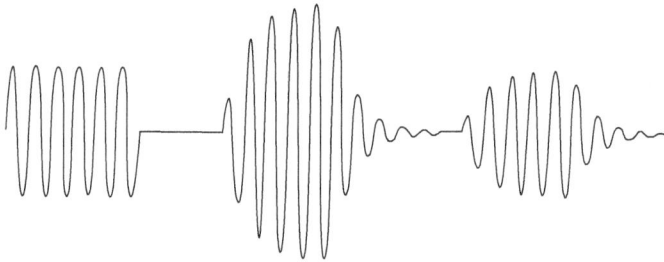

Fig. 4.7: Signals from the right-hand transducer.

ESA signals. The second pulse is detected by the nearest electrode. This pulse is used for data processing to determine the particle size and zeta potential. The third pulse originates from the other electrode and is deleted.

In addition to the electrodes, the sample cell of the ESA instrument also houses sensors for pH, conductivity and temperature measurements. It is also equipped with a stirrer and the system is linked to a digital titrator for dynamic mobility and zeta potential measurements as a function of pH.

To convert the ESA signal to dynamic mobility one needs to know the density of the disperse phase and dispersion medium, the volume fraction of the particles and the velocity of sound in the solvent. As shown before, to convert mobility to zeta potential one needs to know the viscosity of the dispersion medium and its relative permittivity. Because of inertia effects in dynamic mobility measurements, the weight average particle size has to be known.

For dilute suspensions with a volume fraction $\phi = 0.02$, the dynamic mobility u_d can be calculated from the electrokinetic sonic amplitude $A_{ESA}(\omega)$ using the following expression [26, 27]:

$$A_{ESA}(\omega) = Q(\omega)\phi(\Delta\rho/\rho)(u_d), \qquad (4.54)$$

where ω is the angular frequency of the applied field, $\Delta\rho$ is the density difference between the particle (with density ρ) and the medium. $Q(\omega)$ is an instrument-related coefficient independent of the system being measured.

For a dilute dispersion of spherical particles with $\phi < 0.1$, a thin double layer ($\kappa R > 50$) and narrow particle size distribution (with standard deviation $< 20\%$ of the mean size), u_d can be related to the zeta potential ζ by equation (4.55) [26],

$$u_d = \frac{2\varepsilon\zeta}{3\eta} G\left(\frac{\omega R^2}{v}\right)[1 + f(\lambda, \omega)], \qquad (4.55)$$

where ε is the permittivity of the liquid (that is equal to $\varepsilon_r\varepsilon_0$, defined before), R is the particle radius, η is the viscosity of the medium, λ is the double layer conductance and v is the kinematic viscosity ($= \eta/\rho$). G is a factor that represents particle inertia, which reduces the magnitude of u_d and increases the phase lag in a monotonic fashion as the frequency increases. This inertia factor can be used to calculate the particle size from electroacoustic data. The factor $[1 + f(\lambda, \omega)]$ is proportional to the tangential component of the electric field and dependent on the particle permittivity and a surface conductance parameter λ. For most suspensions with large κR, the effect of surface conductance is insignificant and the particle permittivity/liquid permittivity $\varepsilon_p/\varepsilon$ is small. In most cases where the ionic strength is at least 10^{-3} mol dm^{-3} and a zeta potential < 75 mV, the factor $[1 + f(\lambda, \omega)]$ assumes the value 0.5. In this case the dynamic mobility is given by the simple expression,

$$u_d = \frac{\varepsilon\zeta}{\eta} G(\alpha). \qquad (4.56)$$

Equation (4.56) is identical to the Smoluchowski equation, except for the inertia factor $G(\alpha)$.

The equation for converting the ESA amplitude, A_{ESA}, to dynamic mobility is given by

$$u_d = \frac{A_{ESA}}{\phi v_s \Delta\rho} G(\alpha)^{-1}. \qquad (4.57)$$

The zeta potential ζ is given by

$$\zeta = \frac{u_d \eta}{\varepsilon} G(\alpha)^{-1} = \frac{A_{ESA}}{\phi v_s \Delta\rho} G(\alpha)^{-1}. \qquad (4.58)$$

For a polydisperse system $\langle u_d \rangle$ is given by

$$\langle u_d(\omega)\rangle = \int_0^\infty u(\omega, R)\, p(R)\, dR, \qquad (4.59)$$

where $u(\omega, R)$ is the average dynamic mobility of particles with radius R at a frequency ω, and pR dR is the mass fraction of particles with radii in the range $R \pm dR/2$.

ESA measurements can also be applied for determining particle size from particle mobilities in a suspension. The electric force acting upon a particle is opposed by the hydrodynamic friction and inertia of the particles. At low frequencies of alternating electric field, the inertial force is insignificant and the particle moves in the alternating electric field with the same velocity as it would have moved in a constant field. Particle mobility at low frequencies can be measured to calculate the zeta potential. At high frequencies the inertia of the particle increases causing the velocity of the particle to decrease and the movement of the particle to lag behind the field. This is illustrated in Fig. 4.8 which shows the variation of applied field and particle velocity with time. Since inertia depends on particle mass, both of these effects depend on the particle mass and consequently on its size. Hence both zeta potential and particle size can be determined from the ESA signal, if the frequency of the alternating field is sufficiently high. This is the method that is provided by the AcoustoSizer from Colloidal Dynamics.

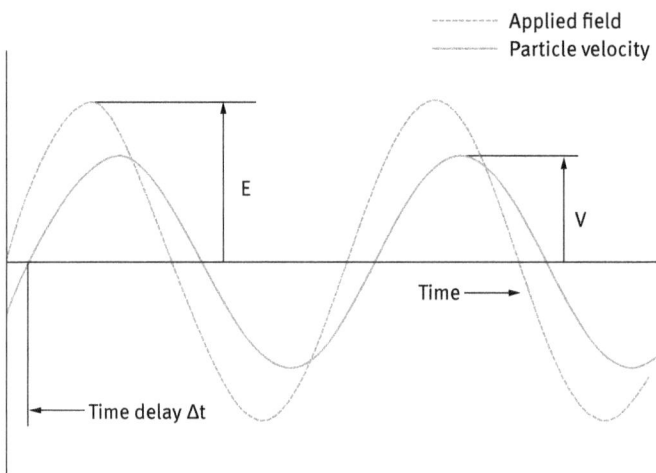

Fig. 4.8: Variation of applied field and particle velocity with time at high frequency.

Several variables affect ESA measurements and these are listed below:

(i) Particle concentration range: Very dilute suspensions generate a weak signal and are not suitable for ESA measurements. The magnitude of the ESA signal is proportional to average particle mobility, the volume fraction of the particles ϕ and the density difference between the particles and the medium $\Delta\rho$. To obtain a signal that is at least one order of magnitude higher than the background electrical noise (~ 0.002 mPa M/V) the concentration and/or the density difference have to be sufficiently large. If the density difference between the particles and medium is small, e.g. polystyrene latex with $\Delta\rho \sim 0.05$, then a sufficiently high concentration ($\phi > 0.02$) is needed to obtain a reasonably strong ESA signal. The accuracy of the ESA measurement is also not good at high ϕ values. This is due to the nonlinearity of the ESA amplitude-ϕ relationship at high ϕ values. Such deviation becomes appreciable at $\phi > 0.1$. However, reasonable values of zeta potential can be obtained from ESA measurements up to $\phi = 0.2$. Above this concentration, the measurements are not sufficiently accurate and the results obtained can only be used for qualitative assessment.

(ii) Electrolyte effects: Ions in the dispersion generate electroacoustic (IVP) potential and the ESP signal is therefore a composite of the signals created by the particles and ions. However, the ionic contribution is relatively small, unless the particle concentration is low, their zeta potential is low and the ionic concentration is high. The ESA system is therefore not suitable for dynamic mobility and zeta potential measurements in systems with electrolyte concentration higher than $0.3\,\text{mol}\,\text{dm}^{-3}$ KCl.

(iii) Temperature: Since the viscosity of the dispersion decreases by $\sim 2\%$ per °C and its conductivity increases by about the same amount, it is important that temperature be accurately controlled using a Peltier device. Temperature control should also be maintained during sample preparation, for example when the suspension is sonicated. To avoid overheating, the sample should be cooled in an ice bath at regular intervals during sonication.

(iv) Calibration and accuracy: The electroacoustic probe should be calibrated using a standard reference dispersion such as polystyrene latex or colloidal silica (Ludox). The common sources of error are unsuitable particle concentration (too low or too high), irregular particle shape, polydispersity, electrolyte signals, temperature variations, sedimentation, coagulation and entrained air bubbles. The latter in particular can cause erroneous ESA signal fluctuations resulting from weakening of the sound by the air bubbles. In many cases the zeta potential results obtained using the ESA method do not agree with those obtained using other methods such as microelectrophoresis or laser velocimetry. However, the difference seldom exceeds 20 % and this makes the ESA method more convenient for measurement of many industrial methods. The main advantages are the speed of measurement and the dispersion does not need to be diluted which could change the state of the suspension.

References

[1] Philipse, A., "Particulate Colloids: Aspects of Preparation and Characterisation", in "Fundamentals of Interface and Colloid Science", Vol. IV, J. Lyklema (ed.), Elsevier, Amsterdam (2005).
[2] Tadros, Th. F., "Rheology of Dispersions", Wiley-VCH, Germany (2010).
[3] Stober, W., Fink, A. and Bohn, E., J. Colloid Interface Sci., **26**, 62 (1968).
[4] Gibbs, J. W., Collected Work, Vol. I, Longman, New York, p. 219
[5] Smoluchowski, M.v., Z. Phys. Chem., **92**, 129 (1927).
[6] Lifshitz, I. M. and Slesov, V. V., Sov. Phys. JETP, **35**, 331 (1959).
[7] Wagner, C., Z. Electrochem., **35**, 581 (1961).
[8] Capek, I., in "Encyclopedia of Colloid and Interface Science", Th. F. Tadros (ed.), Springer (2013), p. 748.
[9] Tadros, Th. F., Izquierdo, P., Esquena, J. and Solans, C., Advances Colloid and Interface Science, **108–109**, 303 (2004).

[10] Eastoe, J., Hopkins Hatzopolous, M. and Tabor, R., in "Encyclopedia of Colloid and Interface Science", Th. F. Tadros (ed.), Springer (2013), p. 688.

[11] Singh, L. P., Bhattacharyya, S. K., Kumar, R., Mishra, G., Sharma, U., Singh, G. and Ahalawat, S., Advances Colloid Interface Science, **214**, 17 (2014).

[12] Blakely, D. C., "Emulsion Polymerization", Elsevier, Applied Science, London (1975).

[13] Litchi, G., Gilbert, R. G. and Napper, D. H., J. Polym. Sci., **21**, 269 (1983).

[14] Feeney, P. J., Napper, D. H. and Gilbert, R. G., Macromolecules, **17**, 2520 (1984); **20**, 2922 (1987).

[15] Nestor, J., Esquena, J., Solans, C., Luckham, P. F., Levecke, B. and Tadros, Th. F., J. Colloid Interface Sci., **311**, 430 (2007).

[16] Leong, Y. S. and Candau, F., J. Phys. Chem., **86**, 2269 (1982).

[17] Candau, F., Leong, Y. S., Candau, G. and Candau, S., in Progr. SIF Course XV, V. Degiorgi and M. Corti (eds.), (1985), p. 830.

[18] Larpent, C. and Tadros, Th. F., Colloid Polym. Sci., **269**, 1171 (1991).

[19] Girard, N., Tadros, Th. F. and Bailey, A. I., Colloid Polym. Sci., **276**, 999 (1998).

[20] Tadros, Th. F., "Dispersion of Powders in Liquids and Stabilisation of Suspensions", Wiley-VCH, Germany (2012) Chapter 11.

[21] Debye, P., J. Chem. Phys., **1**, 13 (1933).

[22] Bugosh, J., Yeager, E. and Hovarka, F., J. Chem. Phys., **15**, 542 (1947).

[23] Yeager, E., Bugosh, J., Hovarka, F. and McCarthy, J., J. Chem. Phys., **17**, 411 (1949).

[24] Dukhin, A. S. and Goetz, P. J., Colloids and Surfaces, **144**, 49 (1998).

[25] Oja, T., Petersen, G. L. and Cannon, D. C., US Patent 4,497,208 (1985).

[26] O'Brian, R. W., J. Fluid Mech., **190**, 71 (1988).

[27] O'Brian, R. W., J. Fluid Mech., **212**, 81 (1990).

[28] O'Brian, R. W., Garaside, P. and Hunter, R. J., Langmuir, **10**, 931 (1994).

[29] O'Brian, R. W., Cannon, D. W. and Rowlands, W. N., J. Colloid Interface Sci., **173**, 406 (1995).

[30] Rowlands, W. N. and O'Brian, R. W., J. Colloid Interface Sci., **175**, 190 (1995).

5 Preparation of nanosuspensions using the top-down process

As mentioned before, in the top-down process one starts with the bulk material (which may consist of aggregates and agglomerates) which is dispersed into single particles (using a wetting/dispersing agent) followed by subdivision of the large particles into smaller units that fall within the required nanosize [1, 2]. This process requires the application of intense mechanical energy that can be applied using bead milling, high pressure homogenization and/or application of ultrasonics. Finally, the resulting nanodispersion must remain colloidally stable under all conditions (such as temperature changes, vibration, etc.) with absence of any flocculation and/or crystal growth.

A schematic representation of the dispersion process is shown in Fig. 5.1.

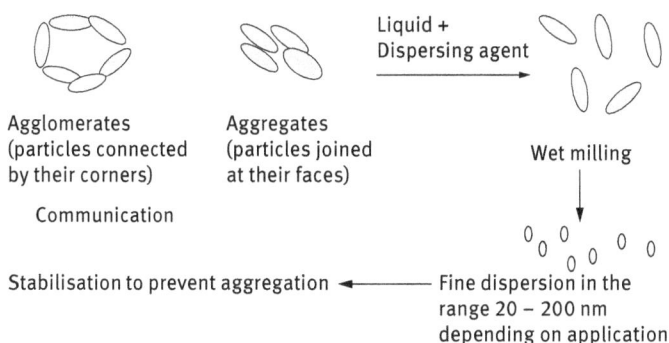

Fig. 5.1: Schematic representation of the dispersion process.

5.1 Wetting of the bulk powder

Most chemicals are supplied as powders consisting of aggregates, in which the particles are joined together with their "faces" (compact structures), or agglomerates, in which the particles are connected at their corners (loose aggregates) as illustrated in Fig. 5.1. It is essential to wet both the external and internal surface (in the pores within the aggregate or agglomerate structures) and this requires the use of an effective wetting agent (surfactant) [1, 2]. Wetting of a solid by a liquid (such as water) requires the replacement of the solid/vapour interfacial tension, γ_{SV}, by the solid/liquid interfacial tension, γ_{SL}.

A useful parameter to describe wetting is the contact angle θ of a liquid drop on a solid substrate [3, 4]. If the liquid makes no contact with the solid, i.e. $\theta = 180°$, the solid is referred to as non-wettable by the liquid in question. This may be the case for a perfectly hydrophobic surface with a polar liquid such as water. However, when

$180° > \theta > 90°$, one may refer to a case of poor wetting. When $0° < \theta < 90°$, partial (incomplete) wetting is the case, whereas when $\theta = 0°$ complete wetting occurs and the liquid spreads on the solid substrate forming a uniform liquid film.

The utility of contact angle measurements depends on equilibrium thermodynamic arguments (static measurements) using the well-known Young's equation [3]. The value depends on: (i) The history of the system; (ii) whether the liquid is tending to advance across or recede from the solid surface (Advancing angle θ_A, Receding angle θ_R; usually $\theta_A > \theta_R$). Under equilibrium, the liquid drop takes the shape that minimizes the free energy of the system. Three interfacial tensions can be identified: γ_{SV}, solid/vapour area A_{SV}; γ_{SL}, solid/liquid area A_{SL}; γ_{LV}, liquid/vapour area A_{LV}. A schematic representation of the balance of tensions at the solid/liquid/vapour interface is shown in Fig. 5.2. The contact angle is that formed between the planes tangent to the surfaces of the solid and liquid at the wetting perimeter. Here, solid and liquid are simultaneously in contact with each other and the surrounding phase (air or vapour of the liquid). The wetting perimeter is referred to as the three phase line or wetting line. In this region there is an equilibrium between vapour, liquid and solid.

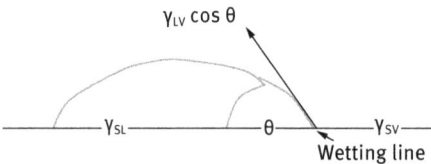

Fig. 5.2: Schematic representation of the contact angle and wetting line.

$\gamma_{SV} A_{SV} + \gamma_{SL} A_{SL} + \gamma_{LV} A_{LV}$ should be a minimum at equilibrium and this leads to the well-known Young's equation [3],

$$\gamma_{SV} = \gamma_{SL} + \gamma_{LV} \cos \theta, \tag{5.1}$$

$$\cos \theta = \frac{\gamma_{SV} - \gamma_{SL}}{\gamma_{LV}}. \tag{5.2}$$

The contact angle θ depends on the balance between the solid/vapour (γ_{SV}) and solid/liquid (γ_{SL}) interfacial tensions. The angle which a drop assumes on a solid surface is the result of the balance between the adhesion force between solid and liquid and the cohesive force in the liquid,

$$\gamma_{LV} \cos \theta = \gamma_{SV} - \gamma_{SL}. \tag{5.3}$$

Wetting of a powder is achieved by the use of surface active agents (wetting agents) of the ionic or nonionic type which are capable of diffusing quickly (i.e. lower the dynamic surface tension) to the solid/liquid interface and displacing the air entrapped by rapid penetration through the channels between the particles and inside any "capillaries". For wetting of hydrophobic powders into water, anionic surfactants, e.g. alkyl sulphates or sulphonates or nonionic surfactants of the alcohol ethoxylates are usually used [1, 2].

A useful concept for choosing wetting agents from the ethoxylated surfactants is the hydrophilic-lipophilic balance (HLB) concept,

$$HLB = \frac{\% \text{ of hydrophilic groups}}{5}. \tag{5.4}$$

Most wetting agents of this class have an HLB number in the range 7–9.

The process of wetting a solid of unit surface area by a liquid involves three types of wetting [1, 2]: adhesion wetting, W_a; immersion wetting, W_i; spreading wetting, W_s. In every step one can apply Young's equation,

$$W_a = \gamma_{SL} - (\gamma_{SV} + \gamma_{LV}) = -\gamma_{LV}(\cos\theta + 1), \tag{5.5}$$

$$W_i = 4\gamma_{SL} - 4\gamma_{SV} = -4\gamma_{LV}\cos\theta, \tag{5.6}$$

$$W_s = (\gamma_{SL} + \gamma_{LV}) - \gamma_{SV} = -\gamma_{LV}(\cos\theta - 1). \tag{5.7}$$

The work of dispersion of a solid with unit surface area W_d is the sum of W_a, W_i, and W_s,

$$W_d = W_a + W_i + W_s = 6\gamma_{SV} - \gamma_{SL} = -6\gamma_{LV}\cos\theta. \tag{5.8}$$

Wetting and dispersion depend on: γ_{LV}, liquid surface tension; θ, contact angle between liquid and solid. W_a, W_i, and W_s are spontaneous when $\theta < 90°$. W_d is spontaneous when $\theta = 0$. Since surfactants are added in sufficient amounts ($\gamma_{dynamic}$ is lowered sufficiently) spontaneous dispersion is the rule rather than the exception.

The work of dispersion of a powder with surface area A, W_d, is given by [1, 2],

$$W_d = A(\gamma_{SL} - \gamma_{SV}). \tag{5.9}$$

Using Young's equation,

$$\gamma_{SV} = \gamma_{SL} + \gamma_{LV}\cos\theta, \tag{5.10}$$

where γ_{LV} is the liquid/vapour interfacial tension and θ is the contact angle of the liquid drop at the wetting line.

$$W_d = -A\gamma_{LV}\cos\theta. \tag{5.11}$$

Equation (5.11) shows that W_d depends on γ_{LV} and θ both of which are lowered by addition of surfactants (wetting agents). If $\theta < 90°$, W_d is negative and dispersion is spontaneous.

Wetting of the internal surface requires penetration of the liquid into channels between and inside the agglomerates. The process is similar to forcing a liquid through fine capillaries. To force a liquid through a capillary with radius r, a pressure p is required that is given by

$$p = -\frac{2\gamma_{LV}\cos\theta}{r} = \left[\frac{-2(\gamma_{SV} - \gamma_{SL})}{r\gamma_{LV}}\right]. \tag{5.12}$$

γ_{SL} has to be made as small as possible; rapid surfactant adsorption to the solid surface, low θ. When $\theta = 0$, p $\propto \gamma_{LV}$. Thus for penetration into pores one requires a high γ_{LV}. Thus, wetting of the external surface requires low contact angle θ and low surface tension γ_{LV}. Wetting of the internal surface (i.e. penetration through pores) requires low θ but high γ_{LV}. These two conditions are incompatible and a compromise has to be made: $\gamma_{SV} - \gamma_{SL}$ must be kept at a maximum. γ_{LV} should be kept as low as possible but not too low.

The above conclusions illustrate the problem of choosing the best dispersing agent for a particular powder. This requires measurement of the above parameters as well as testing the efficiency of the dispersion process.

The contact angle of liquids on solid powders can be measured by application of the Rideal–Washburn equation. For horizontal capillaries (gravity neglected), the depth of penetration l in time t is given by the Rideal–Washburn equation [5, 6],

$$ 1 = \left[\frac{rt\gamma_{LV}\cos\theta}{2\eta} \right]^{1/2} . \tag{5.13}$$

To enhance the rate of penetration, γ_{LV} has to be made as high as possible, θ as low as possible and η as low as possible. For dispersion of powders into liquids one should use surfactants that lower θ while not reducing γ_{LV} too much. The viscosity of the liquid should also be kept at a minimum. Thickening agents (such as polymers) should not be added during the dispersion process. It is also necessary to avoid foam formation during the dispersion process.

For a packed bed of particles, r may be replaced by K, which contains the effective radius of the bed and a tortuosity factor, which takes into account the complex path formed by the channels between the particles, i.e.

$$ 1^2 = \frac{Kt\gamma_{LV}\cos\theta}{2\eta} . \tag{5.14}$$

Thus a plot of 1^2 versus t gives a straight line and from the slope of the line one can obtain θ. The Rideal–Washburn equation can be applied to obtain the contact angle of liquids (and surfactant solutions) in powder beds. K should first be obtained using a liquid that produces a zero contact angle. A packed bed of powder is prepared, say in a tube fitted with a sintered glass at the end (to retain the powder particles). It is essential to pack the powder uniformly in the tube (a plunger may be used in this case). The tube containing the bed is immersed in a liquid that gives spontaneous wetting (e.g. a lower alkane), i.e. the liquid gives a zero contact angle and $\cos\theta = 1$. By measuring the rate of penetration of the liquid (this can be carried out gravimetrically using for example a microbalance or a Kruss instrument) one can obtain K. The tube is then removed from the lower alkane liquid and left to stand for evaporation of the liquid. It is then immersed in the liquid in question and the rate of penetration is measured again as a function of time. Using equation (5.14), one can calculate $\cos\theta$ and hence θ.

For efficient wetting of hydrophobic solids in water, a surfactant is needed that lowers the surface tension of water very rapidly (within few ms) and quickly adsorbs at the solid/liquid interface [1, 2]. To achieve rapid adsorption the wetting agent should be either a branched chain with central hydrophilic group or a short hydrophobic chain with hydrophilic end group. The most commonly used wetting agents are the following:

Aerosol OT (diethylhexyl sulphosuccinate)

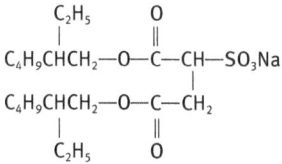

$$
\begin{array}{ll}
\mathrm{C_2H_5} & \mathrm{O} \\
| & \| \\
\mathrm{C_4H_9CHCH_2{-}O{-}C{-}CH{-}SO_3Na} \\
\mathrm{C_4H_9CHCH_2{-}O{-}C{-}CH_2} \\
| & \| \\
\mathrm{C_2H_5} & \mathrm{O}
\end{array}
$$

The above molecule has a low critical micelle concentration (cmc) of $0.7\,\mathrm{g\,dm^{-3}}$ and at and above the cmc the water surface tension is reduced to $\sim 25\,\mathrm{mN\,m^{-1}}$ in less than 15 s.

Several nonionic surfactants, such as the alcohol ethoxylates, can also be used as wetting agents. These molecules consist of a short hydrophobic chain (mostly C_{10}) which is also branched. A medium chain polyethylene oxide (PEO) mostly consisting of 6 EO units or lower is used. These molecules also reduce the dynamic surface tension within a short time (< 20 s) and they have reasonably low cmc.

In all cases one should use the minimum amount of wetting agent to avoid interference with the dispersant that needs to be added to maintain the colloid stability during dispersion and on storage.

5.2 Breaking of aggregates and agglomerates into individual units

This usually requires the application of mechanical energy. High speed mixers (which produce turbulent flow) of the rotor-stator type [7] are efficient in breaking up the aggregates and agglomerates, e.g. Silverson mixers, Ultra-Turrax. The mixing conditions have to be optimized: heat generation at high stirring speeds must be avoided. This is particularly the case when the viscosity of the resulting dispersion increases during dispersion (note that the energy dissipation as heat is given by the product of the square of the shear rate and the viscosity of the suspension). One should avoid foam formation during dispersion; proper choice of the dispersing agent is essential and antifoams (silicones) may be applied during the dispersion process.

Rotor-stator mixers can be characterized as energy-intensive mixing devices. The main feature of these mixers is their ability to focus high energy/shear in a small volume of fluid. They consist of a high speed rotor enclosed in a stator, with the gap between them ranging from 100 to 3000 µm. Typically, the rotor speed is between 10 and

$50\,\text{m s}^{-1}$, which, in combination with a small gap, generates very high shear rates. By operating at high speed, the rotor-stator mixers can significantly reduce the processing time. In terms of energy consumption per unit mass of product, the rotor-stator mixers require high power input over a relatively short time. However, as the energy is uniformly delivered and dissipated in a relatively small volume, each element of the fluid is exposed to a similar intensity of processing. Frequently, the quality of the final product is strongly affected by its structure/morphology and it is essential that the key ingredients are uniformly distributed throughout the whole mixer volume.

The most common application of rotor-stator mixers is in emulsification and dispersion of powders in liquids and they are used in manufacture of particle-based products with sizes between 1 and $20\,\mu\text{m}$, e.g. in pharmaceuticals, paints, agrochemicals and cosmetics.

There are a wide range of designs of rotor-stator mixers, of which the Ultra-Turrax (IKA Works, Germany) and Silverson (UK) are the most commonly used. They are broadly classified according to their mode of operation such as batch or in-line (continuous) mixers. In-line radial-discharge mixers are characterized by high throughput and good pumping capacity at low energy consumption. The disperse phase can be injected directly into the high shear/turbulent zone, where mixing is much faster than by injection into the pipe or into the holding tank. They are used for manufacturing very fine solid particles of relatively narrow dispersed size distribution. They are typically supplied with a range of interchangeable screens, making them reliable and versatile in different applications. Toothed devices are available as in-line as well as batch mixers. Due to their open structure they have a relatively good pumping capacity and they frequently do not need an additional impeller to induce bulk flow even in relatively large vessels.

In rotor-stator mixers, both shear rate in laminar flow and energy dissipate flow depend on the position inside the mixer. In laminar flow in stirred vessels, the average shear rate is proportional to the rotor speed N with the proportionality constant K dependent on the type of the impeller [7],

$$\dot{\gamma} = KN. \tag{5.15}$$

In stirred vessels the proportionality constant cannot be calculated and has to be determined experimentally. In rotor-stator mixers, the average shear rate in the gap between the rotor and stator can be calculated if the rotor speed and geometry of the mixer are known,

$$\dot{\gamma} = \frac{\pi DN}{\delta} = K_1 N, \tag{5.16}$$

where D is the outer rotor diameter and δ is the rotor-stator gap width.

The average energy dissipation rate ε in turbulent flow in rotor-stator mixers can be calculated from [7],

$$\varepsilon = \frac{P}{\rho_c V}, \tag{5.17}$$

where P is the power draw, V is the swept rotor volume and ρ_c is the continuous phase density.

The power draw in batch rotor-stator mixers is calculated in the same way as in stirred vessels,

$$P = P_o \rho_c N^3 D^5 , \qquad (5.18)$$

where P_o is the power number constant for in-line rotor-stator mixers zero flow.

The power draw in in-line rotor-stator mixers in turbulent flow is given by,

$$P = P_{oz} \rho_c N^3 D^5 + k_1 M N^2 D^2 + P_L , \qquad (5.19)$$

where M is the mass flow rate and P_L is the power losses term. The first term in equation (5.19) is analogous to power consumption in a batch rotor-stator mixer and the second term takes into account the effect of pumping action on total power consumption. The third term accounts for mechanical losses and is typically a few percent, and therefore can be ignored.

While in turbulent flow, P_{oz} in equation (5.19) is approximately independent of the Reynolds number Re, in laminar flow there is a strong dependence of power number on Re and in this case the power draw can be calculated from

$$P = k_o N^2 D^3 \eta_c + k_1 M N^2 D^2 + P_L , \qquad (5.20)$$

where η_c is the viscosity of the continuous phase and k_o is a constant that depends on the Reynolds number Re,

$$k_o = P_{oz} Re . \qquad (5.21)$$

From equations (5.17)–(5.21), the average energy dissipation rate in the rotor-stator mixer can be calculated.

5.3 Wet milling or comminution

The primary dispersion (sometimes referred to as the mill base) may be subjected to a bead milling process to produce nanoparticles. Subdivision of the primary particles into much smaller units in the nanosize range (10–100 nm) requires application of intense energy. In some cases high pressure homogenizers (such as the Microfluidizer, USA) may be sufficient to produce nanoparticles. This is particularly the case with many drugs. In some cases, the high pressure homogenizer is combined with application of ultrasound to produce the nanoparticles [8]. It has been shown that high pressure homogenization is a simple technique, well established on large scale for the production of fine suspensions and already available in the pharmaceutical industry. High pressure homogenization is also an efficient technique that has been utilized to prepare stable nanosuspensions of several drugs such as carbazepin, bupravaquone, aphidicolin, cyclosporine, paclitaxel, prednisolone, etc. During homogenization, cavitation forces as well as collision and shear forces determine breakdown of the drug particles down to the nanometer range. Process conditions lead to an average particle size that remains constant as a result of continuous fragmentation and reaggregation processes. These high energetic forces can also induce a change of crystal structure and/or partial or total amorphization of the sample, which further enhances the solubility. For long-term storage stability of the nanosuspension formulation, the crystal structure modification must be maintained over the storage time.

Microfluidization is a milling technique which results in minimal product contamination. Besides minimal contamination, this technique can be easily scaled up. In this method a sample dispersion containing large particles is made to pass through specially designed interaction chambers at high pressure. In the interaction chambers the liquid feed is divided into two parts. The specialized geometry of the chambers along with the high pressure causes the liquid streams to reach extremely high velocities and these streams then impinge against each other and against the walls of the chamber resulting in particle size reduction. The shear forces developed at high velocities due to attrition of particles against one another and against the chamber walls, as well as the cavitation fields generated inside the chamber are the main mechanisms of particle size reduction with this technique [8].

The process of microfluidization for the preparation of nanosuspensions varies in a complex way with the various critical processes and formulation parameters. Milling time, microfluidization pressure, stabilizer type, processing temperature and stabilizer concentration were identified as critical parameters affecting the formation of stable nanoparticles. Both ionic as well as steric stabilization were effective in stabilizing the nanosuspensions. Microfluidization and precipitation under sonication can also be used for nanosuspension preparation.

The extreme transient conditions generated in the vicinity and within the collapsing cavitational bubbles have been used for size reduction of the material to the nanoscale. Nanoparticle synthesis techniques include sonochemical processing and cavitation processing. In sonochemistry, an acoustic cavitation process can generate a transient localized hot zone with extremely high temperature gradient and pressure. Such sudden changes in temperature and pressure assist the destruction of the sonochemical precursor and the formation of nanoparticles [8].

A dimensionless number known as the cavitation number (C_v) is used to relate the flow conditions with the cavitation intensity [8],

$$C_v = \frac{(P_2 - P_v)}{(0.5 \rho V_0^2)} \tag{5.22}$$

where P_2 is the recovered downstream pressure; P_v is the vapour pressure of the liquid, ρ is the density of dispersed media and V_0 is the liquid velocity at the orifice. The cavitation number at which the inception of cavitation occurs is known as the cavitation inception number C_{vi}. Ideally speaking, cavitation inception should occur at 1.0. It was also reported that generally the inception of cavitation occurs from 1.0 to 2.5. This has been attributed to the presence of dissolved gases in the flowing liquid. C_v is a function of the flow geometry and usually increases with an increase in the size of the opening in a constriction such as an orifice in a flow.

Cavitation can be used, for example, for the formation of iron oxide nanoparticles. Iron precursor, either as a neat liquid or in a decalin solution, can be sonicated and this produces 10–20 nm size amorphous iron particles. Similar experiments have been reported for the synthesis of nanoparticles of many other inorganic materials using acoustic cavitation. To understand the mechanism of the formation of nanoparticles during the cavitation phenomenon, the hot-spot theory has been successfully applied. It explains the adiabatic collapse of a bubble, producing the hot spots. This theory claims that very high temperatures (5000–25 000 K) are obtained upon the collapse of the bubble. Since this collapse occurs in few microseconds, very high cooling rates have been obtained. These high cooling rates hinder the organization and crystallization of the products. While the explanation for the creation of amorphous products is well understood, the reason for the formation of nanostructured products under cavitation is not yet clear. The products are sometimes nanoamorphous particles, and in other cases, nanocrystalline. This depends on the temperature in the fluid ring region where the reaction takes place. The temperature in this liquid ring is lower than that inside the collapsing bubble, but higher than the temperature of the bulk liquid. In summary, in the sonochemical reactions leading to inorganic products, nanomaterials have been obtained. They vary in size, shape, structure, and in their solid phase (amorphous or crystalline), but they were always of nanometer size. Since cavitation is a nuclei dominated (statistical in nature) phenomenon, such variations are expected. In hydrodynamic cavitation, nanoparticles are generated through the creation and release of gas bubbles inside the sol-gel solution. By rapidly pressurizing it in a supercritical drying chamber and exposing it to the cavitational disturbance and high temperature heating, the sol-gel solution is rapidly mixed. The erupting hydrodynamically generated cavitating bubbles are responsible for the nucleation, the growth of the nanoparticles, and also for their quenching to the bulk operating temperature. Particle size can be controlled by adjusting the pressure and the solution retention time in the cavitation chamber. Cavitation methods can be used to reduce the size of rubber latex particles (styrene butadiene rubber, SBR) present in the form of aqueous suspension with micrometer particle initial size to the nanoscale [8].

An alternative method of size reduction to produce nanoparticles, which is commonly used in many industrial applications, is through wet milling. This is also referred to as comminution (the generic term for size reduction) and is a complex process with little fundamental information on its mechanism. For the breakdown of single crystals or particles into smaller units, mechanical energy is required. This energy in a bead mill is supplied by impaction of the glass or ceramic beads with the particles. As a result permanent deformation of the particles and crack initiation result. This will eventually lead to the fracture of particles into smaller units. Since the milling conditions are random, some particles receive impacts far in excess of those required for fracture whereas others receive impacts that are insufficient to initiate the fracture process. This makes the milling operation grossly inefficient and only a small fraction of the applied energy is used in comminution. The rest of the energy is dissipated as heat, vibration, sound, interparticulate friction, etc.

The role of surfactants and dispersants on grinding efficiency is far from being understood. In most cases the choice of surfactants and dispersant is made by trial and error until a system is found that gives the maximum grinding efficiency. Rehbinder and his collaborators [9] investigated the role of surfactants in the grinding process. As a result of surfactant adsorption at the solid/liquid interface, the surface energy at the boundary is reduced and this facilitates the process of deformation or destruction. The adsorption of surfactants at the solid/liquid interface in cracks facilitates their propagation. This mechanism is referred to as the Rehbinder effect.

Several factors affect the efficiency of dispersion and milling [10]: (i) The volume concentration of dispersed particles (i.e. the volume fraction). (ii) The nature of the wetting/dispersing agent. (iii) The concentration of wetter/dispersant (which determines the adsorption characteristics).

For optimization of the dispersion/milling process the above parameters need to be systematically investigated. From the wetting performance of a surfactant, that can be evaluated using contact angle measurements, one can establish the nature and concentration of the wetting agent. The nature and concentration of the dispersing agent required is determined by adsorption isotherm and rheological measurements.

Once the concentration of wetting/dispersing agent is established dispersions are prepared at various volume fractions keeping the ratio of wetting/dispersing agent to the solid content constant. Each system is then subjected to the dispersion/milling process keeping all parameters constant: (i) Speed of the stirrer (normally one starts at lower speed and gradually increases the speed in increments at fixed time). (ii) Volume and size of beads relative to the volume of the dispersion (an optimum value is required). (iii) Speed of the mill.

The change of average particle size with time of grinding is established using, for example, the Mastersizer (Malvern, UK). Figure 5.3 shows a schematic representation of the reduction of particle size with grinding time in minutes using a typical bead mill (see below) at various volume fractions.

The presentation in Fig. 5.3 is only schematic and is not based on experimental data. It shows the expected trend. When the volume fraction ϕ is below the optimum (in this case the relative viscosity of the dispersion is low) one requires a long time to achieve size reduction. In addition, the final particle size may be large and outside the nanorange. When ϕ is above the optimum value the dispersion time is prolonged (due to the relatively high relative viscosity of the system) and the grinding time is also longer. In addition, the final particle size is larger than that obtained at the optimum ϕ. At the optimum volume fraction both the dispersion and grinding time is shorter and also the final particle size is smaller [10].

For preparation of nanosuspensions, bead mills are most commonly used. The beads are mostly made of glass or ceramics (which are preferred due to minimum

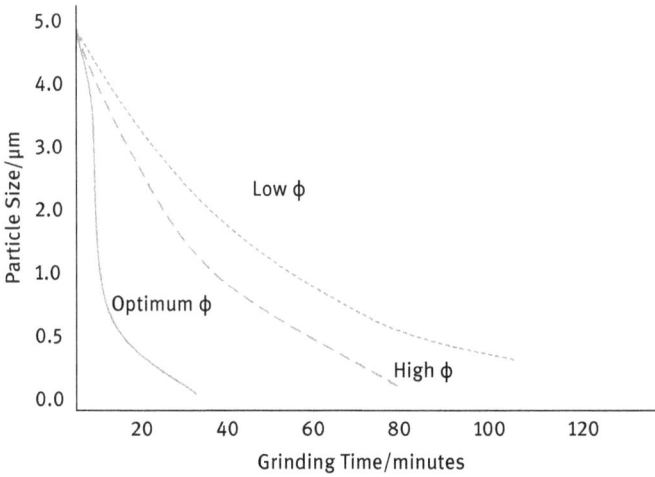

Fig. 5.3: Variation of particle size with grinding time in a typical bead mill.

contamination). The operating principle is to pump the premixed, preferably predis-persed (using a high speed mixer), millbase through a cylinder containing a specified volume of say ceramic beads (normally 0.5–1 mm diameter to achieve nanosize parti-cles). The dispersion is agitated by a single or multidisc rotor. The disc may be flat or perforated. The millbase passing through the shear zone is then separated from the beads by a suitable screen located at the opposite end of the feedport [10].

Generally speaking, bead mills may be classified to two types: (i) Vertical mills with open or closed top. (ii) Horizontal mills with closed chambers. The horizontal mills are more efficient and the most commonly used one are: Netzsch (Germany) and Dyno Mill (Switzerland). These bead mills are available in various sizes from 0.5 to 500 liters. The factors affecting the general dispersion efficiency are known reason-ably well (from the manufacturer). The selection of the correct diameter for the beads is important for maximum utilization. In general, the smaller the size of the beads and the higher their density, the more efficient the milling process [10].

To understand the principle of operation of the bead mill, one must consider the centrifugal force transmitted to the grinding beads at the tip of the rotating disc which increases considerably by its weight. This applies greater shear to the mill base. This explains why the more dense beads are more efficient in grinding. The speed transmit-ted to the individual chambers of the beads at the tip of the disc assumes that speed and the force can be calculated [10].

The centrifugal force F is simply given by

$$F = \frac{v^2}{rg},\tag{5.23}$$

where v is the velocity, r is the radius of the disc and g is the acceleration due to gravity.

5.4 Stabilization of the suspension during dispersion and milling and the resulting nanosuspension

In order to maintain the particles as individual units during dispersion and milling, it is essential to use a dispersing agent that provides an effective repulsive barrier preventing aggregation of the particles by van der Waals forces. This dispersing agent must be strongly adsorbed on the particle surface and should not be displaced by the wetting agent. As discussed in detail in Chapter 2, the repulsive barrier can be electrostatic in nature, whereby electrical double layers are formed at the solid/liquid interface [11, 12]. These double layers must be extended (by maintaining low electrolyte concentration) and strong repulsion occurs on double layer overlap. Alternatively, the repulsion can be produced by the use of nonionic surfactant or polymer layers which remain strongly hydrated (or solvated) by the molecules of the continuous medium [13]. On approach of the particles to a surface-to-surface separation distance that is lower than twice the adsorbed layer thickness strong repulsion occurs as a result of two main effects: (i) Unfavourable mixing of the layers when these are in good solvent conditions. (ii) Loss of configurational entropy on significant overlap of the adsorbed layers. This process is referred to as steric repulsion [13]. A third repulsive mechanism is that whereby both electrostatic and steric repulsion are combined, for example when using polyelectrolyte dispersants.

The particles of the resulting nanosuspension may undergo aggregation (flocculation) on standing as a result of the universal van der Waals attraction. This was discussed in detail in Chapter 2 and only a summary is given in this chapter. This attractive energy becomes very large at short distances of separation between the particles. The attractive energy, G_A, is given by the following expression:

$$G_A = -\frac{A_{11(2)}R}{12h},$$

(5.24)

where $A_{11(2)}$ is the effective Hamaker constant of two identical particles with Hamaker constant A_{11} in a medium with Hamaker constant A_{22}. The Hamaker constant of any material is given by the following expression:

$$A = \pi q^2 \beta.$$

(5.25)

q is the number of atoms or molecules per unit volume, and β is the London dispersion constant. Equation (5.24) shows that A_{11} has the dimension of energy.

As mentioned in Chapter 2, to overcome the everlasting van der Waals attraction energy, it is essential to have repulsive energy between the particles. The first mechanism is electrostatic repulsive energy produced by the presence of electrical double layers around the particles produced by charge separation at the solid/liquid interface. The dispersant should be strongly adsorbed to the particles, produce high charge (high surface or zeta potential) and form an extended double layer (that can be achieved at low electrolyte concentration and low valency) [11, 12].

When charged colloidal particles in a dispersion approach each other such that the double layers begin to overlap (particle separation becomes less than twice the double layer extension), repulsion occurs. The individual double layers can no longer develop unrestrictedly, since the limited space does not allow complete potential decay [11, 12]. The potential $\psi_{H/2}$ half way between the plates is no longer zero (as would be the case for isolated particles at $x \to \infty$). For two spherical particles of radius R and surface potential ψ_o and condition $\kappa R < 3$ (where κ is the reciprocal Debye length), the expression for the electrical double layer repulsive interaction is given by [11, 12],

$$G_{elec} = \frac{4\pi\varepsilon_r\varepsilon_o R^2 \psi_o^2 \exp -(\kappa h)}{2R + h}, \tag{5.26}$$

where h is the closest distance of separation between the surfaces.

Expression (5.26) shows an exponential decay of G_{elec} with h. The higher the value of κ (i.e. the higher the electrolyte concentration), the steeper the decay. This means that at any given distance h, the double layer repulsion decreases with increasing electrolyte concentration.

Combining G_{elec} and G_A results in the well-known theory of stability of colloids (DLVO theory) [11, 12],

$$G_T = G_{elec} + G_A. \tag{5.27}$$

A plot of G_T versus h is shown in Fig. 5.4, which represents the case at low electrolyte concentrations, i.e. strong electrostatic repulsion between the particles. G_{elec} decays exponentially with h, i.e. $G_{elec} \to 0$ as h becomes large. G_A is $\propto 1/h$, i.e. G_A does not decay to 0 at large h.

At long distances of separation, $G_A > G_{elec}$ resulting in a shallow minimum (secondary minimum), which for nanosuspensions is very low ($< kT$). At very short distances, $G_A \gg G_{elec}$, resulting in a deep primary minimum. At intermediate distances, $G_{elec} > G_A$ resulting in energy maximum, G_{max}, whose height depends on ψ_o (or ψ_d) and the electrolyte concentration and valency. At low electrolyte concentrations ($< 10^{-2}$ mol dm^{-3} for a 1 : 1 electrolyte), G_{max} is high ($> 25kT$) and this prevents particle aggregation into the primary minimum. The higher the electrolyte concentration (and the higher the valency of the ions), the lower the energy maximum.

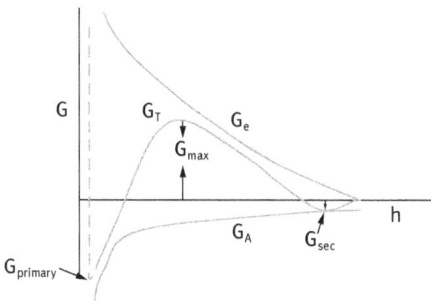

Fig. 5.4: Schematic representation of the variation of G_T with h according to DLVO theory.

The second stabilization mechanism is referred to as steric repulsive energy, produced by the presence of adsorbed (or grafted) layers of surfactant or polymer molecules [13]. In this case the nonionic surfactant or polymer (referred to as polymeric surfactant) should be strongly adsorbed to the particle surface and the stabilizing chain should be strongly solvated (hydrated in the case of aqueous suspensions) by the molecules of the medium [13]. The most effective polymeric surfactants are those of the A–B, A–B–A block or BA_n graft copolymer. The "anchor" chain B is chosen to be highly insoluble in the medium and has strong affinity to the surface. The A stabilizing chain is chosen to be highly soluble in the medium and strongly solvated by the molecules of the medium. For nanosuspensions of hydrophobic solids in aqueous media, the B chain can be polystyrene, poly(methylmethacrylate) or poly(propylene oxide). The A chain could be poly(ethylene oxide) which is strongly hydrated by the medium.

When two particles each with a radius R and containing an adsorbed polymer layer with a hydrodynamic thickness δ_h, approach each other to a surface-surface separation distance h that is smaller than $2\delta_h$, the polymer layers interact with each other resulting in two main situations [13]: (i) The polymer chains may overlap with each other. (ii) The polymer layer may undergo some compression. In both cases, there will be an increase in the local segment density of the polymer chains in the interaction region. The real situation is perhaps in between the above two cases, i.e. the polymer chains may undergo some interpenetration and some compression.

Provided the dangling chains (the A chains in A–B, A–B–A block or BA_n graft copolymers) are in a good solvent, this local increase in segment density in the interaction zone will result in strong repulsion as a result of two main effects [13]: (i) An increase in the osmotic pressure in the overlap region as a result of the unfavourable mixing of the polymer chains, when these are in good solvent conditions. This is referred to as osmotic repulsion or mixing interaction and it is described by a free energy of interaction G_{mix}. (ii) Reduction of the configurational entropy of the chains in the interaction zone; this entropy reduction results from the decrease in the volume available for the chains when these are either overlapped or compressed. This is referred to as volume restriction interaction, entropic or elastic interaction and it is described by a free energy of interaction G_{el}.

The combination of G_{mix} and G_{el} is usually referred to as the steric interaction free energy, G_s, i.e.

$$G_s = G_{mix} + G_{el} . \tag{5.28}$$

The sign of G_{mix} depends on the solvency of the medium for the chains. If in a good solvent, i.e. the Flory–Huggins interaction parameter χ is less than 0.5, then G_{mix} is positive and the mixing interaction leads to repulsion (see below). In contrast, if $\chi >$ 0.5 (i.e. the chains are in a poor solvent condition), G_{mix} is negative and the mixing interaction becomes attractive. G_{el} is always positive and hence in some cases one can produce stable nanosuspensions in a relatively poor solvent (enhanced steric stabilization).

The expression for G_{mix} is

$$G_{mix} = \left(\frac{2V_2^2}{V_1}\right) v_2 \left(\frac{1}{2} - \chi\right)\left(3R + 2\delta + \frac{h}{2}\right). \tag{5.29}$$

G_{el} is given by the following expression:

$$\frac{G_{el}}{kT} = -2v_2 \ln\left[\frac{\Omega(h)}{\Omega(\infty)}\right] = 2v_2 R_{el}(h), \tag{5.30}$$

where v is the number of chains per unit area, $\Omega(h)$ is the number of configurations at a separation distance h, $\Omega(\infty)$ is the number of configurations at infinite distance between the surfaces, $R_{el}(h)$ is a geometric function whose form depends on the segment density distribution. It should be stressed that G_{el} is always positive and could play a major role in steric stabilization. It becomes very strong when the separation distance between the particles becomes comparable to the adsorbed layer thickness δ.

Combining G_{mix} and G_{el} with G_A gives the total energy of interaction G_T (assuming there is no contribution from any residual electrostatic interaction), i.e.

$$G_T = G_{mix} + G_{el} + G_A. \tag{5.31}$$

A schematic representation of the variation of G_{mix}, G_{el}, G_A, and G_T with surface-surface separation distance h is shown in Fig. 5.5. G_{mix} increases very sharply with decreasing h when $h < 2\delta$. G_{el} increases very sharply with decreasing h when $h < \delta$. G_T versus h shows a minimum, G_{min}, at separation distances comparable to 2δ. When $h < 2\delta$, G_T shows a rapid increase with decreasing h. The depth of the minimum depends on the Hamaker constant A, the particle radius R and adsorbed layer thickness δ. G_{min} decreases with decreasing A and R. At a given A and R, G_{min} decreases with increasing δ (i.e. with an increase in the molecular weight, M_w, of the stabilizer).

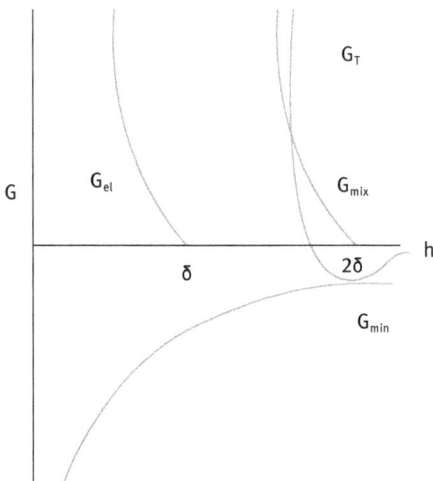

Fig. 5.5: Energy-distance curves for sterically stabilized systems.

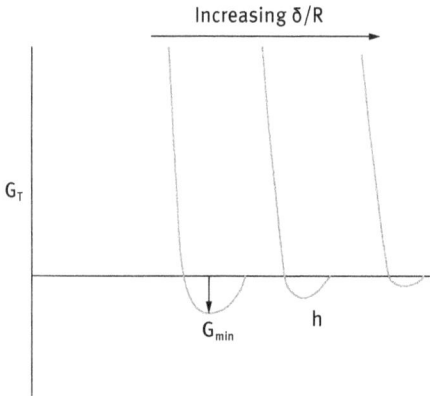

Fig. 5.6: Variation of G_{min} with δ/R.

This is illustrated in Fig. 5.6 which shows the energy-distance curves as a function of δ/R. The larger the value of δ/R, the smaller the value of G_{min}. In this case the system may approach thermodynamic stability as is the case with nanosuspensions.

5.5 Prevention of Ostwald ripening (crystal growth)

As discussed in Chapter 3, the driving force for Ostwald ripening is the difference in solubility between the small and large particles (the smaller particles have higher solubility than the larger ones). The difference in chemical potential between different sized particles was given by Lord Kelvin [14]:

$$S(r) = S(\infty) \exp\left(\frac{2\gamma V_m}{rRT}\right),$$

(5.32)

where $S(r)$ is the solubility of a particle with radius r and $S(\infty)$ is the solubility of a particle with infinite radius (the bulk solubility), γ is the S/L interfacial tension, R is the gas constant and T is the absolute temperature. Equation (5.32) shows a significant increase in solubility of particles with a reduction of particle radius, particularly when the latter becomes significantly smaller than $1\,\mu m$.

For two particles with radii r_1 and r_2 ($r_1 < r_2$),

$$\frac{RT}{V_m} \ln\left[\frac{S(r_1)}{S(r_2)}\right] = 2\gamma\left[\frac{1}{r_1} - \frac{1}{r_2}\right].$$

(5.33)

Equation (5.33) shows that the larger the difference between r_1 and r_2, the higher the rate of Ostwald ripening.

Ostwald ripening can be quantitatively assessed from plots of the cube of the radius versus time t [15, 16],

$$r^3 = \frac{8}{9}\left[\frac{S(\infty)\gamma V_m D}{\rho RT}\right] t.$$

(5.34)

D is the diffusion coefficient of the disperse phase in the continuous phase.

Several factors affect the rate of Ostwald ripening and these are determined by surface phenomena, although the presence of surfactant micelles in the continuous phase can also play a major role. Trace amounts of impurities that are highly insoluble in the medium and have strong affinity to the surface can significantly reduce Ostwald ripening by blocking the active sites on the surface on which the molecules of the active ingredient can deposit. Many polymeric surfactants, particularly those of the block and graft copolymer types can also reduce the Ostwald ripening rate by strong adsorption on the surface of the particles, thus making it inaccessible for molecular deposition. Surfactant micelles that can solubilize the molecules of the active ingredient may enhance the rate of crystal grow by increasing the flux of transport by diffusion.

References

[1]   Tadros, Th. F., "Dispersions of Powders in Liquids and Stabilisation of Suspensions", Wiley-VCH, Germany (2012).
[2]   Tadros, Th. F., "Formulation of Disperse Systems", Wiley-VCH, Germany (2014).
[3]   Young, T., Phil. Trans. Royal Soc. (London), **95**, 65 (1805).
[4]   Blake, T. B., "Wetting", in "Surfactants", Th. F. Tadros (ed.), Academic Press, London, (1984).
[5]   Rideal, E. K., Phil. Mag., **44**, 1152 (1922).
[6]   Washburn, E. D., Phys. Rev., **17**, 273 (1921).
[7]   Pacek, A. W., Hall, S., Cooke, M. and Kowalski, A. J., "Emulsification in Rotor-Stator Mixers", in "Emulsion Formation and Stability", Th. F. Tadros (ed.), Wiley-VCH, Germany (2013).
[8]   Capek, I., in "Encyclopedia of Colloid and Interface Science", Th. F. Tadros (ed.), Springer (2013), p. 748.
[9]   Rehbinder, P. A., Colloid J. USSR, **20**, 493 (1958).
[10] Tadros, Th. F., "Colloids in Paints", Wiley-VCH, Germany (2010).
[11] Deryaguin, B. V. and Landau, L., Acta Physicochem. USSR, **14**, 633 (1941).
[12] Verwey, E. J. W. and Overbeek, J. Th. G., "Theory of Stability of Lyophobic Colloids", Elsevier, Amsterdam (1948).
[13] Napper, D. H., "Polymeric Stabilisation of Colloidal Dispersions", Academic Press, London (1983).
[14] Thomson, W. (Lord Kelvin), Phil. Mag., **42**, 448 (1871).
[15] Lifshitz, I. M. and Slesov, V. V., Sov. Phys. JETP, **35**, 331 (1959).
[16] Wagner, C., Z. Electrochem., **35**, 581 (1961).

6 Industrial application of nanosuspensions

6.1 Introduction

Nanosuspensions have wide applications in various industrial fields of which one can mention applications in drug delivery systems of poorly insoluble compounds, where reduction of particle size to nanoscale dimensions enhances drug bioavailability.

Another important application of nanosuspensions is in the field of cosmetics and personal care, in particular in sunscreens for UV protection. These systems use semiconductor inorganic particles of titanium dioxide and zinc oxide that are able to absorb UV light and maximum attenuation is obtained with particle sizes in the range of 30–50 nm.

Several other applications of nanosuspensions can be listed such as preparation of nanopolymer particles, clays and composites, metal nanoparticles, nanotubes and dispersions of carbon black for printing applications. All these systems must be stabilized against aggregation, Ostwald ripening (crystal growth) and sintering.

In this chapter I will only give examples of applications in three fields, namely pharmaceuticals, cosmetics and paints and coatings.

6.2 Application of nanosuspensions for drug delivery

At present, the small molecular entities produced by current pharmaceutical discoveries show an increasing trend to be highly water insoluble [1, 2]. This low water solubility is a challenge to achieve adequate bioavailability [3] for oral administration. It also limits types of formulation suitable for parenteral administration [4]. In recent years nanocrystalline suspensions have been applied for drug delivery of highly water insoluble ingredients (APIs) [5]. By reducing the particle size of the API, the rate of the dissolution dC/dt which is directly proportional to the surface specific area A, is increased as described by the Noyes and Whitney equation [6],

$$\frac{dC}{dt} = KA(C_s - C),\qquad (6.1)$$

where K is a constant and C_s is the saturation solubility.

The solubility of the API can be significantly enhanced. This is due to the increase of solubility of the active ingredient on reduction of particle radius as given by the Kelvin equation [7],

$$S(r) = S(\infty) \exp\left(\frac{2\gamma V_m}{rRT}\right),\qquad (6.2)$$

where S(r) is the solubility of a particle with radius r and S(∞) is the solubility of a particle with infinite radius (the bulk solubility), γ is the S/L interfacial tension, R is the gas constant and T is the absolute temperature.

Kelvin Equation

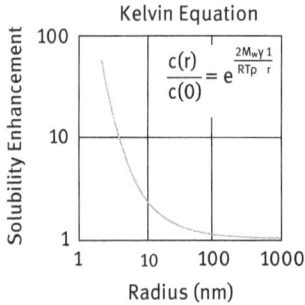

$$\frac{c(r)}{c(0)} = e^{\frac{2M_w\gamma}{RT\rho}\frac{1}{r}}$$

Solubility Enhancement

Radius (nm)

Fig. 6.1: Solubility enhancement with decreasing particle radius.

Equation (6.2) shows a significant increase in solubility of the particle with reduction of particle radius, particularly when the latter becomes significantly smaller than 1 μm. This was illustrated in Fig. 3.1 of Chapter 3 which is reproduced here for clarity.

It can be seen from Fig. 6.1 that the solubility of nanodispersion particles increases very rapidly with decreasing radius, particularly when r < 100 nm. This means that a particle with a radius of say 4 nm will have its solubility enhanced about 10 times compared say with a particle with 10 nm radius whose solubility is only enhanced 2 times.

Significant increases in solubility are typically observed when r is less than 200 nm. In addition, the injectable dose can be increased for the parenteral administration since nanoparticle formulations are made essentially of pure drug and typically use a small amount of excipients. In contrast, standard formulations using solvents such as polysorbate limit the dose due to poor tolerability of the excipient.

For the preparation of nanocrystalline suspensions, particle size reduction by means of a top-down process is the most commonly used method due to the possibility to control particle size by proper choice of wetting/dispersing agent, as well as by control of milling conditions. The wetting agent is essential to prevent aggregates and agglomerates of particles in the formulation. A wet milling process is applied to reduce API particles' size. The nanocrystalline particles produced in the formulation must be stabilized against flocculation and crystal growth [8].

6.2.1 Preparation of drug nanosuspensions using the top-down process

Recently, Nakach et al. [9] investigated the methods that can be applied for selecting the appropriate wetting/dispersing agent in a top-down process. Wetting can be assessed by using the sinking time test method, as well as by measuring contact angle using direct observation of a sessile drop of liquid on a powder compact or by measuring the rate of penetration of surfactant solution through a powder plug [10]. The ability of the dispersant to reduce or eliminate flocculation of the nanodispersion can be assessed by measuring the average particle size as a function of time after milling.

Flocculation results in an increase in the average particle size on storage since the size of floc produced is larger than the size of the single particle. Even in the absence of flocculation the average particle size may increase with time as a result of Ostwald ripening [10]. The driving force of the latter process is the higher solubility of the smaller particles when compared with larger ones [7]. This results in a shift of the particle size distribution to larger values when the nanosuspension is stored, particularly at higher temperature. When using ionic dispersant, the efficiency of electrostatic repulsion can be assessed from a knowledge of the ionic concentration and ion valency, as well as measurement of the zeta potential of the particles [10]. It is well known that electrostatic repulsion increases with decreasing electrolyte concentration, decreasing ion valency and increasing zeta potential [10]. Nonionic dispersants reduce flocculation through steric repulsion [11]. These agents, mostly polymers, form adsorbed layers with thickness δ which is strongly hydrated in water. When two particles, each having an adsorbed layer of thickness δ, approach each other at a surface-to-surface distance h that is smaller than 2δ, strong repulsion occurs as a result of two phenomena: (i) Unfavourable mixing of the stabilizing chains when these are in good solvent. (ii) Reduction of configurational entropy on considerable overlap of the stabilizing chains [11].

To apply the above principles, a model hydrophobic highly insoluble API provided by Sanofi (Paris) was micronized by jet milling before use. The physicochemical properties of the API are given in Tab. 6.1.

Tab. 6.1: Physicochemical properties of the API.

Average particle diameter	5 μm
Specific surface area ($m^2\ g^{-1}$)**	1.5
Molecular weight (g/mol)	497.4
Water solubility (μg/ml)	0.2
pK_a	No pK_a
log P*	6.9
Density (g/ml)	1.42
Melting point (°C)	156.7

* P is the partition coefficient between Octanol and water
** measurement done using the Blaine method [12].

Several dispersing/wetting agents were used for the investigation ranging from cellulose derivatives, polyvinyl pyrrolidone, phospholipids, poloxamers (A–B–A block copolymers of polyethylene oxide A and polypropylene oxide B), polyethylene glycol and derivatives. For the screening of dispersant/wetting agents low shear milling was applied using 20 % (w/w) of API, 3 % (w/w) of dispersant/wetting agents, and 77 % (w/w) of water for injection (WFI). An aliquot of 10 ml suspension and 20 ml of zirconium oxide beads (700 μm diameter supplied by Netzsch, Germany) were introduced into a 30 ml vial. The vial was agitated in an orbital roller mill for 5 days at 0.03 m/s and at room temperature.

For the assessment of process ability using high shear milling, a suspension containing 20 % (w/w) of API, 3 % dispersant/wetting agent and 77 % (w/w) of WFI was prepared. An aliquot of 50 ml suspension and 50 ml of Polymill® Cross-linked Polystyrene beads milling media (500 μm diameter) supplied by Alkermes, Inc. (Waltham, MA, USA) were introduced into a NanoMill® 01 milling system (Annular mill purchased from Alkermes, Inc. (Waltham, MA, USA), having a stator of 80 mm diameter and rotor of 73 mm). The mill was operated during 1 h at 20°C and 3 m/s.

For the optimization of the dispersant/wetting agent content a suspension was prepared using 20 % (w/w) of API, the dispersant/wetting agents concentration was varied from 0.3 to 3 % (w/w) and WFI was varied accordingly from 79.3 to 77 % (w/w). An aliquot of 50 ml suspension and 50 ml of Polymill® Cross-linked Polystyrene beads milling media (500 μm diameter supplied by Alkermes, Inc. (Waltham, MA, USA) were introduced into a NanoMill® 01 milling system (Annular mill purchased from Alkermes, Inc. (Waltham, MA, USA), having a stator of 80 mm diameter and rotor of 73 mm). The mill was operated at 20 °C and 3 m/s. the milling operation was performed during 105–240 min. The resulting nanosuspension was characterized by using several techniques briefly described below.

6.2.1.1 Particle size measurement

Particle size measurement was performed using two methods:

(i) Dynamic light scattering, referred to as photon correlation spectroscopy (PCS), using Coulter N4+ equipment (supplied by Beckman Coulter, France). The method is based on measuring the intensity fluctuation of scattered light as the particles undergo Brownian diffusion. From the intensity fluctuation the diffusion coefficient D can be calculated. From this, the particle radius, r, is estimated using the Stokes–Einstein equation [13]. The measurements were carried out using a scattering angle of 90°. The refractive index was fixed at 1.332 and the temperature at 20 °C. The suspension was diluted from 20 % (w/w) to 0.1 % (w/w) with distilled water. 10 μl of diluted suspension was added to 1 ml distilled water.

(ii) Laser diffraction using a Malvern Mastersizer 2000. This method is based on measuring the angle of light diffracted by particles, which depends on the particle radius using Fraunhofer diffraction theory. This method can measure particle sizes down to 1 μm. For smaller particles, forward light scattering is measured by applying the Mie theory of light scattering. By combining results obtained with light diffraction and forward light scattering, particle size distributions in the range 0.02 to 10 μm can be obtained.

6.2.1.2 Scanning electron microscopy (SEM) evaluation

The suspensions were diluted 10 000 times using WFI. Then 1 ml of the obtained suspension was filtered through Millipore filter Isopore 0.1 μm. The filter was then rinsed 3 times with 1 ml of WFI for each rinse. The filter was then bonded to an aluminium pad using conductive adhesive on both sides and metalized with gold using the metallizer Xenosput XE200 EDWARDS. The gold deposit was approximately 1.5 to 2 nm thick. Nanoparticles were observed at 15 kV using JOEL JSM-6300F field emission SEM. The observation was done at several magnifications (× 1000, × 5000, × 10 000, × 20 000) for an overview and detailed views.

6.2.1.3 Short-term stability assessment

The short-term stability was assessed by measuring the particle size right after milling, and then after 7 and 15 days storage at ambient temperature. For the selected formulations, the stability was assessed for a period of 8 weeks at ambient temperature.

6.2.1.4 Zeta potential measurement

A ZetaSizer Nano ZS from Malvern, UK, which applies the M3-PALS technique, a combination of laser Doppler velocimetry (LDV) and phase analysis light scattering (PALS), was used for the zeta potential measurements. The equipment uses an He-Ne laser (red light of 633 nm wavelength) which first splits into two, providing an incident and a reference beam.

From the electrophoretic mobility, μ, zeta potential, ζ, is calculated using the Smoluchowski equation [14], that is valid when $\kappa r \gg 1$ (where κ^{-1} is the Debye length and r is the particle radius). For the case of small particles and low electrolyte concentration, the Huckel equation [14] is applicable for the calculation of zeta potential.

6.2.1.5 Rheological measurement

Study state, shear stress vs. shear rate curves, was carried out using a HAAKE VT550 (Germany) Rheometer. A concentric cylinder device was used for this measurement. The measurement was carried out at 20 °C. The shear rate was gradually increased from 0 to 1500 s^{-1} (up curve) over a period of 2 min and decreased from 1500 to 0 s^{-1} (down curve) over another period of 2 min. The test samples were 25 ml of unmilled suspension, which contained 20 % (w/w) API, 3 % (w/w) stabilizer, and 77 % (w/w) of WFI. Those samples were homogenized using an Ultra-Turrax for 10 min at 6000 rpm. When the system is Newtonian, the shear stress increases linearly with the applied shear rate and the slope of the line gives the viscosity of the suspension. In this case the up and down curves coincide with each other. When the system is non-Newtonian, the viscosity of the suspension decreases with the applied shear rate. When the system is thixotropic, the down curve is below the up curve showing hysteresis. The latter could be assessed by measuring the area under the loop. In summary, Newtonian, non-Newtonian, as well as thixotropy fluids, can be distinguished from the shear stress and shear rate curves.

6.2.1.6 Surface tension measurement

The surface tension γ of the selected dispersant/wetting agent was measured using a KRUSS K12 tensiometer (Germany). In these measurements, the Wilhelmy plate method was applied under quasi-equilibrium conditions. Therefore, the force required to detach the plate from the interface was accurately determined. From the γ versus log C, where C is the total surfactant concentration curves, the critical micelle concentration (cmc) was determined.

6.2.1.7 Evaluation of wetting/dispersant agent

Wetting was assessed by measuring the rate of penetration of surfactant solution through a powder plug. The result shows a linear relationship between the rate of penetration and time. From the slope of the line a wettability factor can be calculated using equation (6.3):

$$H^2 = \frac{\gamma}{2\eta} CR \cos \theta , \qquad (6.3)$$

where H is the height of liquid penetrated within the powder plug, θ is the contact angle, γ the surface tension of the liquid, η is the liquid viscosity, R is the mean radius of the capillary within the powder plug, C is the tortuosity factor, and t is the time. Since all powder plugs are prepared at the same compression pressure, the parameter C can be assumed to be a constant.

To calculate H^2, the mass of the liquid penetrated within the powder plug was measured using a microbalance. The relationship of the mass (m) and the height of the liquid penetrated within the powder plug can be expressed by the equation:

$$m = HS\varepsilon\rho , \qquad (6.4)$$

where m is the mass of the liquid penetrated within the powder plug, H is the height of the liquid penetrated within the powder plug, ρ is the volumetric mass of the liquid, S is the surface of the powder plug, and ε is the fraction of the dead volume of the powder.

Combining equations (6.3) and (6.4), the following equation is obtained:

$$m^2 = \frac{\gamma\rho^2}{\eta} \frac{s^2}{2} CR\varepsilon^2 \cos \theta . \qquad (6.5)$$

From a plot of m^2 versus time (linear curve), the slope $(d(m^2)/dt)$ can be determined and the wettability factor can be calculated from a knowledge of the surface tension (γ) and the viscosity (η) of the liquid.

The wettability factor can be expressed by the equation

$$\frac{d(m^2)}{dt}\left(\frac{\eta}{\gamma}\right) = K = \frac{s^2\rho^2 CR\cos\theta}{2}.$$

(6.6)

6.2.1.8 Adsorption isotherm measurement

The adsorption isotherm of the model dispersant, namely PVP, was measured at room temperature. Known amounts of API were equilibrated at room temperature with various concentrations of PVP dispersant. Then, the bottles containing the various dispersions were rotated from several hours to up to 15 hours until equilibrium was achieved. Then the particles were removed from the dispersant solution by centrifugation. The dispersant concentration in the supernatant was analytically determined using UV spectrometry by Cary 50 at the wavelength of 200 nm. To obtain the amount of adsorption per unit area of the powder (Γ), the specific surface area of the powder (A) in m^2/g was determined using the gas flow method (Blaine).

6.2.1.9 Methodology for selection of wetting/dispersant agent

Two criteria were used to select the optimum stabilizer. The first criterion is that the API particle diameter has to be in the range of 100–500 nm after milling. The second criterion is that the formulation should be free of flocculation after at least two weeks of storage at room temperature.

The selection was performed by the following step by step approach:
1. The suspension prepared after using the roller mill is assessed using visual observation, particle size measurement and stability after two weeks at room temperature. This step eliminated wetting dispersion agents that gave particle size above 500 nm and did not prevent flocculation within a two week period.
2. Measuring the viscosity as a function of shear rate as well as thixotropy. The samples that gave viscosity greater than 15 mPa s at shear rate of $1000\,s^{-1}$ were rejected. This criterion is essential to ensure faster milling kinetics as well as manufacturability at industrial scale.
3. Milling ability using the high shear mill, namely NanoMill® 01 milling system. This step is essential to ensure preparation of the nanosuspension at industrial scale using high speed milling. All samples that gave particle size greater than 500 nm or showed instability due to flocculation or Ostwald ripening were rejected.

The combination of SDS/PVP appeared to be superior to the other tested agents. Therefore, they were selected to be further evaluated as follows:
1. Assessment of wettability. To ensure that the combined system of SDS/PVP gives better wettability than that of the wetting agent SDS alone. The synergistic ef-

fect obtained using the mixture was confirmed by measuring the surface tension as well as the critical micelle concentration (cmc), and that of the wetting agent alone. Furthermore, the optimum concentration of the SDS/PVP system was used to confirm the milling ability using the high shear mill.

2. Measurement of the adsorption isotherm to ensure the strong adsorption of the dispersant (PVP) on the particles' surface.
3. Measurement of zeta potential to ensure the electrostatic stabilization. An absolute value greater than 20 mV is usually required for electrostatic repulsion to offer the overall stability because the electrostatic repulsion is proportional to the square of zeta potential.
4. Measurement of the long-term physical stability of the selected formulation. This was assessed by measuring the particle size distribution as a function of time over a period of eight weeks at room temperature.

6.2.1.10 Assessment of milling ability using a low shear mill (the roller miller)
After roller milling, all samples were inspected for API suspendability, HPC, Cyclodextrin, PEG, Montanov 68 and sodium polyacrylate showed obvious flocculation and the appearance of a "dry" sample. Therefore, they are not included for further evaluation. The remaining samples were assessed by measuring the particle size at time 0, after 7 and 14 days. The results are shown in Fig. 6.4. Suspensions with a particle size greater than 500 nm and/or showing flocculation after 7 days are discontinued for further evaluation. These discarded samples are: HPMC, Poloxamer188, Poloxamer407, PVP-SDS (50-50 % w/w), PVP-SDS (30-70 % w/w) and SDS.

6.2.1.11 Assessment using the rheological behaviour of the suspension
Figures 6.2 and 6.3 show typical flow curves for unmilled suspensions prepared using Solutol HS15 (hydroxystearate) and Phosal 50 PG (phospholipid). The suspension prepared using Solutol shows Newtonian behaviour with a low viscosity of 4.8 mPa s. In contrast, the suspension using Phosal 50 PG gives non-Newtonian behaviour with clear thixotropy, indicating flocculation of the suspension.

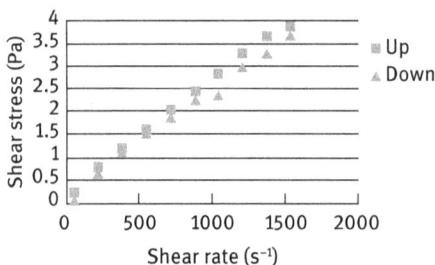

Fig. 6.2: Shear stress-shear rate curves for unmilled suspensions using Solutol HS15.

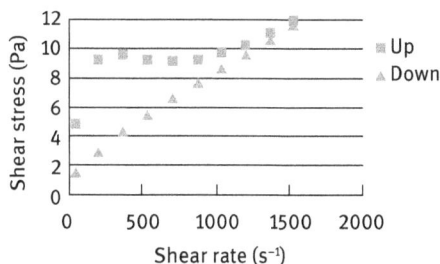

Fig. 6.3: Shear stress-shear rate curves for unmilled suspensions using Phosal 50 PG.

Suspensions with a high viscosity greater than 15 mPa s at shear rate of $1000\,s^{-1}$ were excluded from further evaluation.

6.2.1.12 Assessment of milling ability using a high shear mill (NanoMill® 01 milling system)

After high shear milling, the suspensions were assessed by measuring the particle size at time 0, after 7 and 14 days. The results are shown in Fig. 6.4. Two systems, PVP/SDS at the ratio of 60/40 or vitamin E TPGS offered the best stabilization of the nanocrystalline formulations. Confirmation of these results was obtained by SEM measurement as illustrated in Fig. 6.5 for suspensions prepared using PVP-SDS and Montanox (ethoxylated sorbitan ester). These SEM pictures show large differences between unstable formulation based on Montanox (needle-shaped particles) and stable formulation based on PVP-SDS (small but irregular shaped particles). When using Montanox the suspension shows Ostwald ripening and formation of needle-shaped crystals. This may be due to the specific adsorption of the Montanox molecules on certain crystal faces allowing growth to occur on the other faces and hence the formation of needles.

6.2.2 Optimization of wetting/dispersant agent using PVP-SDS as model

The selection of the final formulation should be based on the following criteria: wettability evaluation, adsorption isotherm measurement, and stress tests evaluation (heating, freezing-thawing stability, centrifugation, ionic strength, dilution in bio-relevant media). PVP-SDS was selected as a model to exemplify the methodology of wetting/dispersant agent selection.

6.2.2.1 Wettability measurement

Figure 6.6 shows the γ-log C curve for a typical SDS-PVP mixture (80-20 % w/w). This graph shows a typical behaviour with γ decreasing with increasing log C until the critical micelle concentration (cmc) is reached after which γ shows only a small decrease with increasing log C.

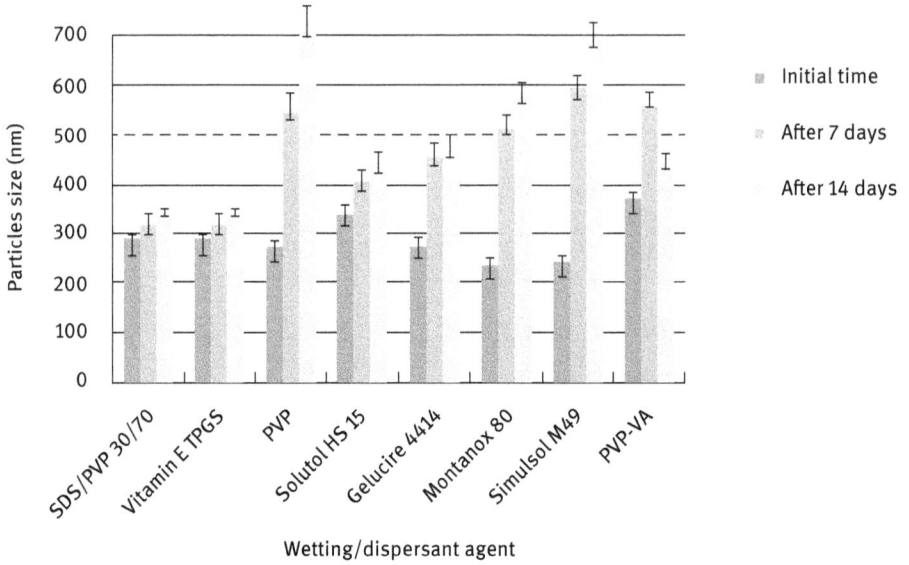

Fig. 6.4: Particle size results after high shear milling using different dispersants.

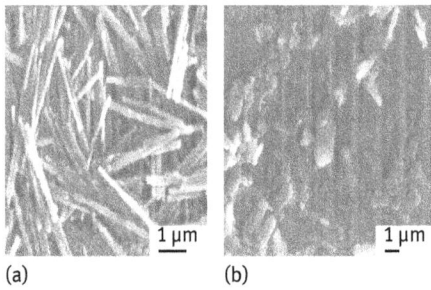

(a)   (b)

Fig. 6.5: SEM pictures of milled particles using PVP/SDS (a) and Montanex (b).

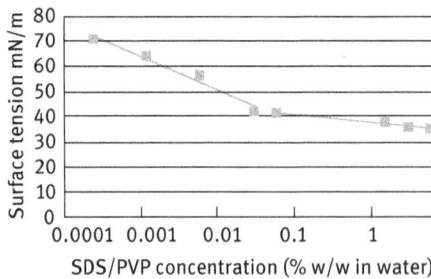

Fig. 6.6: γ-log C curves for SDS/PVP mixtures (80/20 % wt/wt).

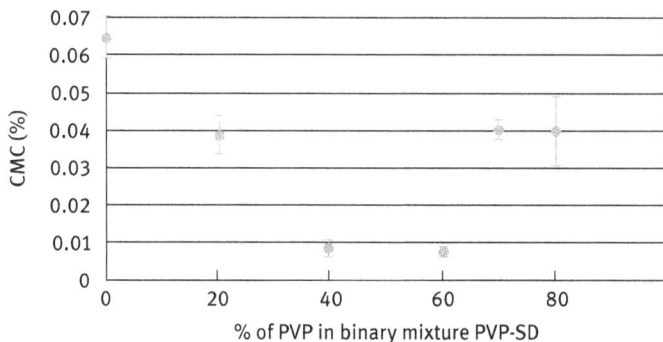

Fig. 6.7: Variation of cmc with % PVP in the binary mixture of PVP-SDS.

A plot of cmc versus % of PVP (Fig. 6.7) in the binary mixture shows a minimum at 50 to 70 % of PVP above which the cmc increases. This result implies a maximum of surface activity between 50 to 70 % in the binary mixture. To obtain maximum wetting of the API particles, a PVP-SDS mixture containing 60 % PVP in the minimum region of the cmc was chosen. Under this condition, maximum reduction in surface energy can be expected for the powder-liquid interface, which will offer enhanced crack propagation (Rehbinder effect), and enhanced breakage of the particles during the wet milling process.

Figure 6.8 shows a plot of wettability factor K vs. PVP-SDS concentration. For comparison, the results obtained using SDS alone are shown in the same graph. It can be seen from Fig. 6.8 that K increases with increasing surfactant concentration, reaching a plateau at a certain surfactant concentration. For the PVP-SDS system, this plateau is reached at 1.2 % consisting of 0.72 % PVP and 0.48 % (w/w). Using the same con-

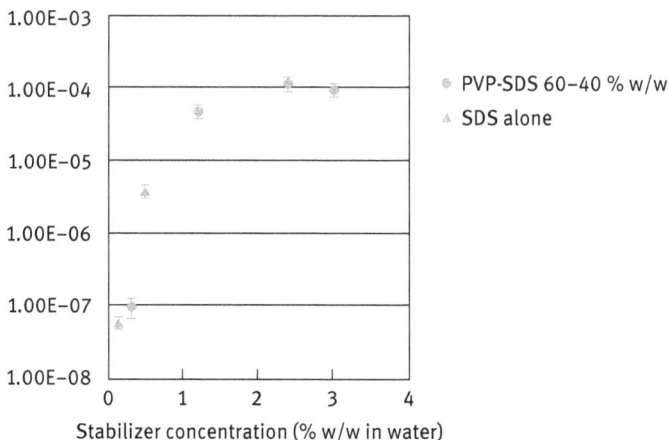

Fig. 6.8: Wettability factor versus surfactant concentration.

centration of SDS alone (0.48 % w/w), the K value is much lower than that obtained with the combined system. This clearly demonstrates the synergistic effect obtained when a polymer surfactant mixture is used. The latter is a much more effective wetting system when compared with the individual components.

6.2.2.2 Milling ability as function of % stabilizer (PVP-SDS (60-40 % w/w))

Milling ability was investigated using a kinetic experiment where the reduction in particle size or the equivalent increase in surface specific area was measured as a function of milling time. A typical result is shown in Fig. 6.9 at 1.2 % w/w PVP-SDS stabilizer. The results obtained show an exponential increase in the surface area (or decrease in particle size) reaching a plateau at a certain milling time (Fig. 6.9). The results follow first-order kinetics that can be represented by the equation:

$$\frac{6}{d_{50}} = \left(\frac{6}{d_{50}}\right)_{\infty} (1 - e^{\frac{-t}{\tau}}), \tag{6.7}$$

where $6/d_{50}$ is the implicit specific surface area, d_{50} is the particle's diameter at time t, $(d_{50})_{\infty}$ is the plateau value, and τ is the duration to reach 63 % of the maximum surface are. Values for $(6/d_{50})_{\infty}$ and τ were obtained at various stabilizer concentrations and the results are shown in Fig. 6.10. The results show an initial increase in $(6/d_{50})_{\infty}$ and τ with increasing stabilizer concentration reaching a plateau value at 1.2 %. These results are consistent with those obtained using wettability evaluation. It is clear that a minimum of 1.2 % stabilizer concentration is required to obtain the smallest particle size. Below this stabilizer concentration, there is not enough power to completely saturate the particles with surfactant molecules and this may result in rejoining of the small particles after their formation during the milling process.

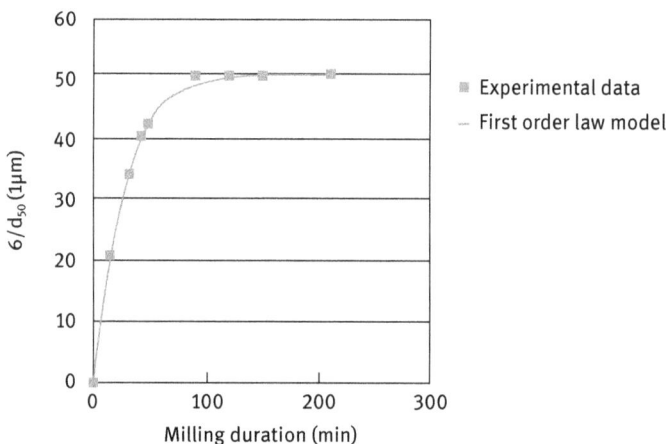

Fig. 6.9: Surface area versus milling time at 1.2 % PVP-SDS stabilizer.

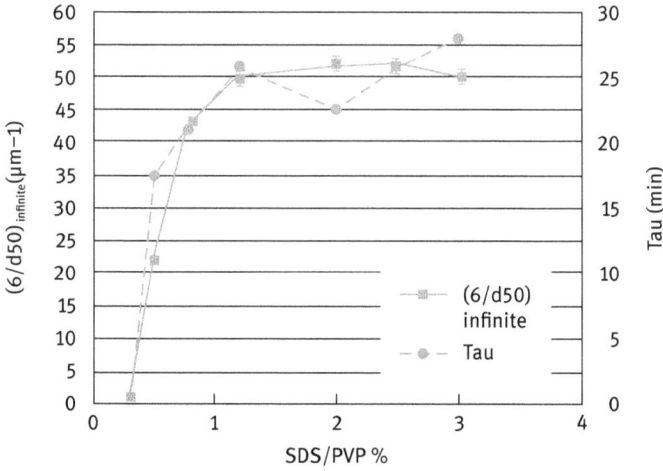

Fig. 6.10: Variation of surface area and duration τ to reach 63 % of maximum surface area with SDS/PVP %.

6.2.2.3 Adsorption isotherm measurement of PVP

Figure 6.11 shows the adsorption isotherm of PVP alone on the API powder surface. The results show the high affinity type isotherm, as indicated by the complete adsorption of the first added PVP molecules. The results obtained at high PVP concentration show a great deal of scatter, which is likely due to the possible error of the UV method for determining the remaining PVP concentration. At high PVP concentration, one measures the difference between two large quantities. Any uncertainty in the estimated concentration using the UV method can produce a large error in the amount adsorbed. It is therefore difficult to ascertain an exact plateau value of the isotherm which appears to be between 0.6 and 0.9 mg/m². Assuming a plateau value of 0.7 mg/m², the concentration of PVP required to completely saturate the particles can be roughly es-

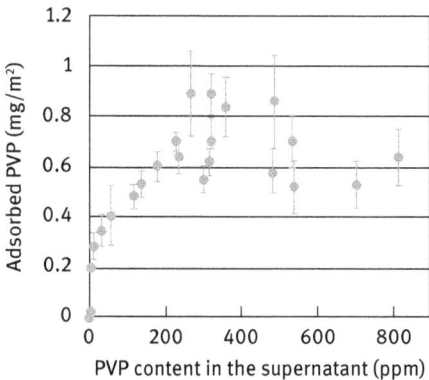

Fig. 6.11: Adsorption isotherm of PVP on API.

timated. From Fig. 6.9 the smallest particle diameter obtained is about 120 nm. This gives a surface specific area of 42.8 m²/g. For a 20 % suspension the total surface area was calculated as 704 m² using the equation,

$$\text{surface area} = \frac{6 \times 20}{\rho d_{50}}. \tag{6.8}$$

The total surface area coverage requires 493 mg or 0.493 % of PVP which corresponds to 0.82 % of PVP-SDS 60-40 % w/w. The results were in good agreement with the values obtained in milling ability and wettability tests.

6.2.2.4 Zeta potential results

Figure 6.12 shows the variation of zeta potential for the PVP-SDS system as a function of SDS concentration. In the absence of SDS, PVP alone gave a low negative zeta potential of −20 mV, which is insufficient to give electrostatic stabilization. In this case, the main stability arises from steric repulsion due to the adsorbed loops and tails of PVP molecules. Upon addition of SDS (30-70 SDS-PVP), the zeta potential increases sharply to −50 mV, which contributes to stability through electrostatic repulsion. With further increase of SDS concentration to 40-60 SDS-PVP, the zeta potential increases further to −54 mV and remains almost constant with a further increase in SDS concentration. Thus, when using a mixture of SDS and PVP the stabilizing mechanism is a combination of electrostatic repulsion, which shows an energy maximum at intermediate separation distance, and steric repulsion that occurs at shorter distances of separation comparable to twice the adsorbed layer thickness. This combined stabilization mechanism is referred to as electrosteric.

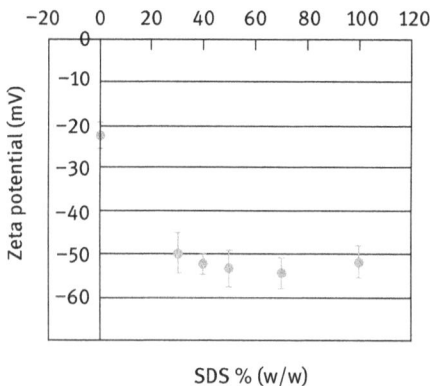

Fig. 6.12: Zeta potential as a function of SDS concentration.

6.2.2.5 Long-term stability results

Using the optimum PVP-SDS ratio of 60-40 at concentration of 1.2%, long-term stability results were obtained by following the particle size as a function of time at room temperature for the 20% w/w API nanosuspension. Figure 6.13 shows the variation of d_{10}, d_{50}, and d_{90} with storage duration over a period of 57 days. It can be seen from this figure that no change in particle size is observed during this period. This further confirmed the high colloidal stability of the nanosuspension that was prepared using the method described above.

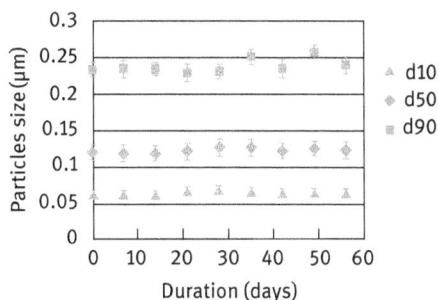

Fig. 6.13: Variation of particle size with storage time.

6.2.3 Protocol for preparation of nanosuspensions of water insoluble drugs

Using a colloidal and interfacial fundamental approach, an optimum wetting/dispersant agent can be selected for preparation of nanosuspensions with a d_{50} lower than 150 nm. These nanosuspensions can be prepared using a simple milling procedure, namely a roller mill combined with particle size measurement. This procedure is exemplified using a model hydrophobic drug (API) and nanosuspensions could be prepared using a dispersing/wetting agent of PVP-SDS mixture. The results clearly showed an optimum ratio of 60-40 PVP-SDS and a minimum total concentration of 1.2%. This composition gave the maximum wettability, the best milling results and the maximum stability. This approach can help the formulator to select the best wetting/dispersant system for any API. A step forward would be to introduce additional stress tests to assess the formulation robustness such as thermal stability, freeze-thaw stability and effect of other ingredients in the formulation such as electrolytes and nonelectrolytes.

6.3 Application of nanosuspensions in cosmetics

One of the main applications of nanosuspensions in cosmetics is in the area of sunscreens. Sunscreen dispersions of semiconductor TiO_2 particles require particles in the range of 30–50 nm which need to remain stable against aggregation in the formulation and on application. This is essential for the required UV protection. Inorganics have several benefits over organics in that they are capable of absorbing over a broad spectrum of wavelengths and they are mild and nonirritant. Both of these advantages are becoming increasingly important as the demand for *daily* UV protection against both UVB and UVA radiation increases.

The ability of fine particle inorganics to absorb radiation depends upon their refractive index. For inorganic semiconductors such as titanium dioxide and zinc oxide this is a complex number indicating their ability to absorb light. The band gap in these materials is such that UV light up to around 405 nm can be absorbed. They can also scatter light due to their particulate nature and their high refractive indices make them particularly effective scatterers. Both scattering and absorption depend critically on particle size [15]. Particles of around 250 nm for example are very effective at scattering visible light and TiO_2 of this particle size is the most widely used white pigment. At smaller particle sizes absorption and scattering maxima shift to the UV region and at 30–50 nm UV attenuation is maximized.

The use of TiO_2 as a UV attenuator in cosmetics was, until recently, largely limited to baby sun protection products due to its poor aesthetic properties (viz; scattering of visible wavelengths results in whitening). Recent advances in particle size control and coatings have enabled formulators to use fine particle titanium dioxide and zinc oxide in daily skincare formulations without compromising the cosmetic elegance [16].

The benefits of a pre-dispersion of inorganic sunscreens are widely acknowledged. However, an understanding of the nature of colloidal stabilization is required in order to optimize this pre-dispersion (for both UV attenuation and stability) and to exceed the performance of powder-based formulations. Dispersion rheology and its dependence on interparticle interactions is a key factor in this optimization. Optimization of sunscreen actives however does not end there; an appreciation of the end application is crucial to maintaining performance. Formulators need to incorporate the particulate actives into an emulsion, mousse or gel with due regard to aesthetics (skin feel and transparency), stability and rheology.

The present section is aimed at applying colloid and interface science principles for optimization of inorganic sunscreen dispersions. These are usually formulated using dispersants that provide effective steric stabilization to avoid flocculation particularly on application. Maintenance of particle size is essential for effective sunscreens. In addition, these colloidally stable nanoparticles can provide transparency and hence good aesthetic characteristics. The theory of steric stabilization with particular reference to the importance of solvation of the polymer chain by the medium molecules has been discussed in Chapter 2. Results are presented for the adsorption isotherms of typical dispersants that are used in nonaqueous media. The dispersing power of these polymeric surfactants is assessed using rheological measurements. UV absorbance of these dispersions is measured to evaluate the effectiveness of the sunscreen dispersions and finally the ability of colloidally stable dispersions to deliver SPF when incorporated into a skincare formulation is summarized.

Dispersions of surface modified TiO$_2$ in alkyl benzoate and hexamethyltetra-cosane (squalane) were prepared at various solids loadings using a polymeric/oligo-meric polyhydroxystearic acid (PHS) surfactant of molecular weight 2500 (PHS2500) and 1000 (PHS1000). For comparison, results were also obtained using a low molecu-lar weight (monomeric) dispersant, namely isostearic acid, ISA. The titania particles had been coated with alumina and/or silica. The electron micrograph in Fig. 6.14 shows the typical size and shape of these rutile particles. The surface area and particle size of the three powders used are summarized in Tab. 6.2.

Fig. 6.14: Transmission electron micrograph of titanium dioxide particles.

Tab. 6.2: Surface modified TiO$_2$ powders.

Powder	Coating	Surface Area* / m^2/g	Particle size** / nm
A	Alumina/silica	95	40–60
B	Alumina/stearic acid	70	30–40
C	Silica/stearic acid	65	30–40

* BET N2   ** equivalent sphere diameter, X-ray disc centrifuge

Dispersions of the surface modified TiO$_2$ powder, dried at 110 °C, were prepared by milling (using a horizontal bead mill) in polymer solutions of different concentrations for 15 minutes and were then allowed to equilibrate for more than 16 hours at room temperature before making measurements. Adsorption isotherms were obtained by preparing dispersions of 30 % w/w TiO$_2$ at different polymer concentration (C$_0$, mg/l). The particles and adsorbed dispersant were removed by centrifugation at 20 000 rpm (~ 48 000 g) for 4 hours, leaving a clear supernatant. The concentration of the polymer in the supernatant was determined by acid value titration. Isotherms were calculated by mass balance to determine the amount of polymer adsorbed at the particle surface (Γ, mg/m^2) of a known mass of particulate material (m, g) relative to that equilibrated in solution (C_e, mg/l),

$$\Gamma = \frac{(C_0 - C_e)}{mA_s}.$$ (6.9)

The surface area of the particles (A_s, m^2/g) was determined by the BET nitrogen adsorption method. Dispersions of various solids loadings were obtained by milling at progressively increasing TiO_2 concentration at an optimum dispersant/solids ratio. The dispersion stability was evaluated by viscosity measurement and by attenuation of UV-vis radiation. The viscosity of the dispersions was measured by subjecting the dispersions to an increasing shear stress, from 0.03 Pa to 200 Pa over 3 minutes at 25 °C using a Bohlin CVO rheometer. It was found that the dispersions exhibited shear thinning behaviour and the zero shear viscosity, identified from the plateau region at low shear stress (where viscosity was apparently independent of the applied shear stress), was used to provide an indication of the equilibrium energy of interaction that had developed between the particles.

UV-vis attenuation was determined by measuring transmittance of radiation between 250 nm and 550 nm. Samples were prepared by dilution with a 1% w/v solution of dispersant in cyclohexane to approximately 20 mg/l and placed in a 1 cm path length cuvette in a UV-vis spectrophotometer. The sample solution extinction ε ($1 g^{-1} cm^{-1}$) was calculated from Beer's Law,

$$\varepsilon = \frac{A}{cl},$$ (6.10)

where A is absorbance, c is concentration of attenuating species (g/l), l is path length (cm).

The dispersions of powders B and C were finally incorporated into typical water-in-oil sunscreen formulations at 5% solids with an additional 2% of organic active (butyl methoxy dibenzoyl methane) and assessed for efficacy, SPF (sun protection factor) as well as stability (visual observation, viscosity). SPF measurements were made on an Optometrics SPF-290 analyzer fitted with an integrating sphere, using the method of Diffey and Robson [10].

6.3.1 Adsorption isotherms

Figure 6.15 shows the adsorption isotherms of ISA, PHS1000 and PHS2500 on TiO_2 in alkyl benzoate (Fig. 6.15 (a)) and in squalane (Fig. 6.15 (b)). The adsorption of the low molecular weight ISA from alkyl benzoate is of low affinity (Langmuir type) indicating reversible adsorption (possibly physisorption). In contrast, the adsorption isotherms for PHS100 and PHS2500 are of the high affinity type indicating irreversible adsorption and possible chemisorption due to acid-base interaction. From squalane, all adsorption isotherms show the high affinity type and they show higher adsorption values when compared with the results using alkyl benzoate. This reflects the difference in solvency of the dispersant by the medium as will be discussed below.

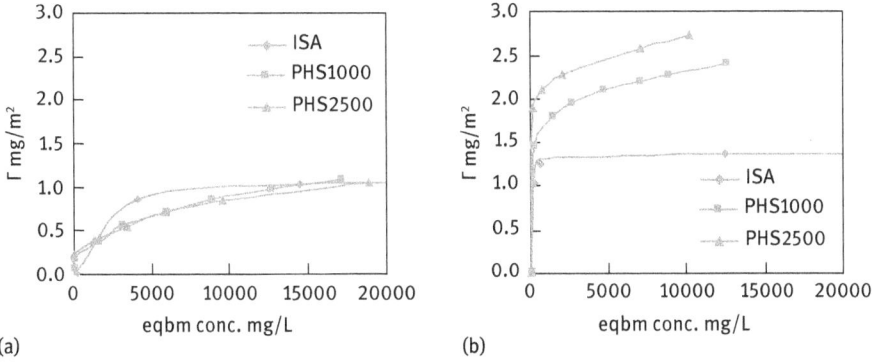

Fig. 6.15: Adsorption isotherms in (a) alkyl benzoate and (b) squalane.

6.3.2 Dispersant demand

Figure 6.16 shows the variation of zero shear viscosity with dispersant loading % on solid for a 40 % dispersion. It can be seen that the zero shear viscosity decreases very rapidly with increasing dispersant loading and eventually the viscosity reaches a minimum at an optimum loading that depends on the solvent used as well as the nature of the dispersant. With the molecular dispersant ISA, the minimum viscosity that could be reached at high dispersant loading was very high (several orders of magnitude more than the optimized dispersions) indicating poor dispersion of the powder in both solvents. Even reducing the solids content of TiO_2 to 30 % did not result in a low viscosity dispersion. With PHS1000 and PHS2500, a low minimum viscosity could be reached at 8–10 % dispersant loading in alkyl benzoate and 18–20 % dispersant loading in squalane. In the latter case the dispersant loading required for reaching a viscosity minimum is higher for the higher molecular weight PHS.

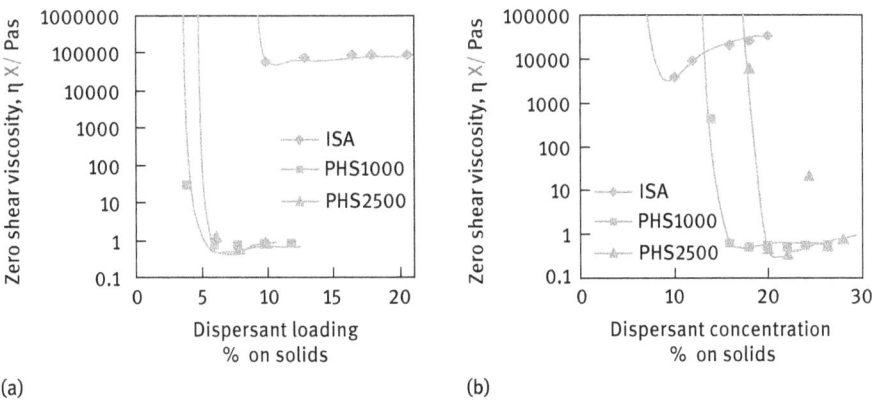

Fig. 6.16: Dispersant demand curves in (a) alkyl benzoate and (b) squalane.

6.3.3 Quality of dispersion UV-vis attenuation

At very low dispersant concentration a high solids dispersion can be achieved by simple mixing but the particles are aggregated as demonstrated by the UV-vis curves (Fig. 6.17).These large aggregates are not effective as UV attenuators. As the PHS dispersant level is increased, UV attenuation is improved and above 8 % wt dispersant on particulate mass, optimized attenuation properties (high UV, low visible attenuation) are achieved (for the PHS1000 in alkyl benzoate). However milling is also required to break down the aggregates into their constituent nanoparticles and a simple mixture which is unmilled has poor UV attenuation even at 14 % dispersant loading.

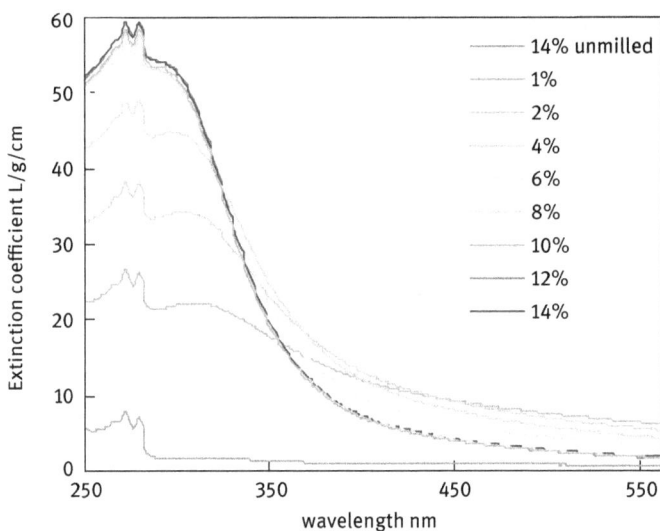

Fig. 6.17: UV-vis attenuation for milled dispersions with 1–14 % PHS1000 dispersant and unmilled at 14 % dispersant on solids.

The UV-vis curves obtained when monomeric isostearic acid was incorporated as a dispersant (Fig. 6.18) indicate that these molecules do not provide a sufficient barrier to aggregation, resulting in relatively poor attenuation properties (low UV, high visible attenuation).

6.3.4 Solids loading

The steric layer thickness δ could be varied by altering the dispersion medium and hence the solvency of the polymer chain. This had a significant effect upon dispersion rheology. Solids loading curves (Fig. 6.19) demonstrate the differences in effective vol-

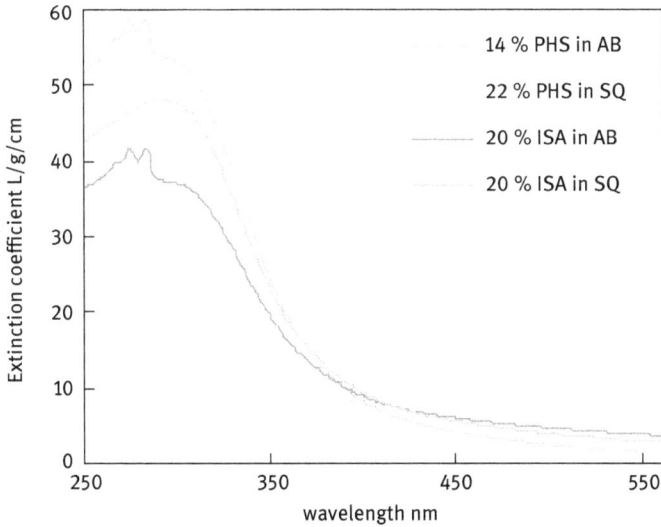

Fig. 6.18: UV-vis attenuation for dispersions in squalane (SQ) and in alkyl benzoate (AB) using 20 % isostearic acid (ISA) as dispersant compared to optimized PHS1000 dispersions in the same oils.

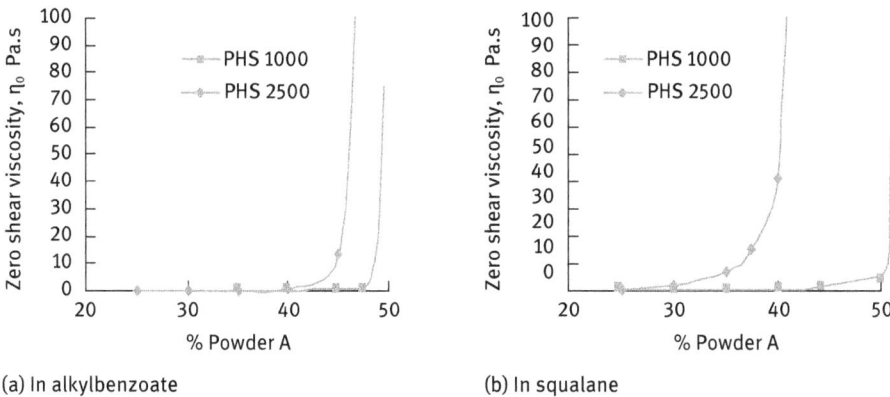

(a) In alkylbenzoate

(b) In squalane

Fig. 6.19: Zero shear viscosity dependence on solids loading: (a) alkyl benzoate; (b) squalane.

ume fraction ϕ_{eff} due to the adsorbed layer,

$$\phi_{eff} = \phi\left(1 + \frac{\delta}{R}\right),\tag{6.11}$$

where ϕ is the core volume fraction and R is the particle radius.

In the poorer solvent case (squalane) the effective volume fraction and adsorbed layer thickness showed a strong dependence upon molecular weight with solids loading becoming severely limited above 35 % for the higher molecular weight whereas ~ 50 % could be reached for the lower molecular weight polymer. In alkyl benzoate no strong dependence was seen with both systems achieving more than 45 % solids. Solids weight fraction above 50 % resulted in very high viscosity dispersions in both solvents.

6.3.5 SPF Performance in emulsion preparations

The same procedure described above enabled optimized dispersion of equivalent particles with alumina and silica inorganic coatings (powders B and C). Both particles additionally had the same level of organic (stearate) modification. These optimized dispersions were incorporated into water-in-oil formulations and their stability/efficacy monitored by visual observation and SPF measurements (Tab. 6.3).

Tab. 6.3: Sunscreen emulsion formulations from dispersions of powders B and C.

Emulsion	Visual observation	SPF	Emulsifier level
Powder B emulsion 1	Good homogenous emulsion	29	2.0 %
Powder C emulsion 1	Separation, inhomogeneous	11	2.0 %
Powder C emulsion 2	Good homogeneous emulsion	24	3.5 %

The formulation was destabilized by the addition of the powder C dispersion and poor efficacy was achieved despite an optimized dispersion before formulation. When emulsifier concentration was increased from 2 to 3.5 % (emulsion 2) the formulation became stable and efficacy was restored.

The anchor of the chain to the surface (described qualitatively through the adsorption energy per segment χ_s) is very specific and this could be illustrated by silica coated particles which showed lower adsorption of the PHS (Fig. 6.20).

In addition, when a quantity of emulsifier was added to an optimized dispersion of powder C (silica surface) the acid value of the equilibrium solution was seen to rise indicating some displacement of the PHS2500 by the emulsifier.

6.3.6 Criteria for preparation of a stable sunscreen dispersion

The dispersant demand curves (Fig. 6.16) and solids loading curves (Fig. 6.19) show that one can reach a stable dispersion using PHS1000 or PHS2500 both in alkyl benzoate and in squalane. These can be understood in terms of the stabilization produced when using these polymeric dispersants. Addition of sufficient dispersant enables coverage of the surface and results in a steric barrier (Fig. 6.21) preventing aggregation due to van der Waals attraction. Both molecular weight oligomers were able to achieve stable dispersions. The much smaller molecular weight "monomer", isostearic acid

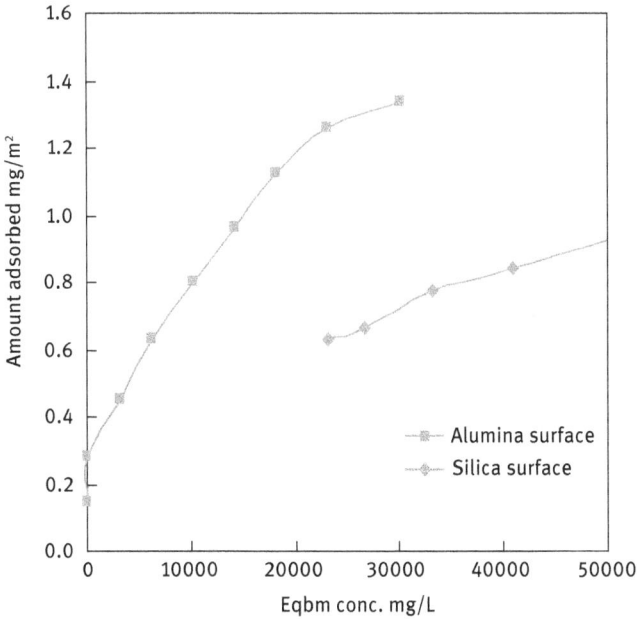

Fig. 6.20: Adsorption isotherms for PHS2500 on powder B (alumina surface) and powder C (silica surface).

however is insufficient to provide this steric barrier and dispersions were aggregated, leading to high viscosities, even at 30 % solids. UV-vis curves confirm that these dispersions are not fully dispersed since their full UV potential is not realized (Fig. 6.18). Even at 20 % isostearic acid the dispersions are seen to give a lower E_{max} and increased scattering at visible wavelengths indicating a partially aggregated system.

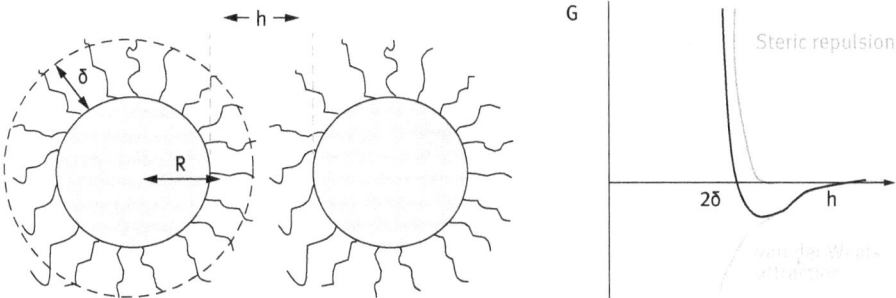

Fig. 6.21: Schematic representation of adsorbed polymer layers and resultant interaction energy G on close approach at distance h < 2R.

The differences between alkyl benzoate and squalane observed in the optimum dispersant concentration required for maximum stability can be understood by examining the adsorption isotherms in Fig. 6.15. The nature of the steric barrier depends on the solvency of the medium for the chain, and is characterized by the Flory–Huggins interaction parameter χ. Information on the value of χ for the two solvents can be obtained from solubility parameter calculations.

One of the most useful concepts for assessing solvation of any polymer by the medium is to use the Hildebrand's solubility parameter δ^2 which is related to the heat of vaporization ΔH by the following equation,

$$\delta^2 = \frac{\Delta H - RT}{V_M} , \tag{6.12}$$

where R is the gas constant, T is the absolute temperature and V_M is the molar volume of the solvent.

Hansen [17] first divided Hildebrand's solubility parameter into three terms,

$$\delta^2 = \delta_d^2 + \delta_p^2 + \delta_h^2 , \tag{6.13}$$

where δ_d, δ_p and δ_h correspond to London dispersion effects, polar effects and hydrogen bonding effects, respectively.

Hansen and Beerbower [18] developed this approach further and proposed a stepwise approach such that theoretical solubility parameters can be calculated for any solvent or polymer based upon its component groups. In this way one can arrive at theoretical solubility parameters for dispersants and oils. In principle, solvents with a similar solubility parameter to the polymer should also be a good solvent for it (low χ).

The results of these calculations are given in Tab. 6.4 for PHS, alkyl benzoate and squalane.

Tab. 6.4: Hansen and Beerbower solubility parameters for the polymer and both solvents.

	δ_T	δ_d	δ_p	δ_h	$\Delta\delta_T$
PHS	19.00	18.13	0.86	5.60	
Alkyl benzoate	17.01	19.13	1.73	4.12	1.99
Squalane	12.9	15.88	0	0	6.1

It can be seen that both PHS and alkyl benzoate have polar and hydrogen bonding contributions to the solubility parameter δ_T. In contrast, squalane, which is nonpolar, has only a dispersion component to δ_T. The difference in the total solubility parameter $\Delta\delta_T$ value is much smaller for alkyl benzoate when compared with squalane. Thus one can expect that alkyl benzoate is a better solvent for PHS when compared with squalane. This explains the higher adsorption amounts of the dispersants in squalane when compared with alkyl benzoate (Fig. 6.15). The PHS finds adsorption at the particle surface energetically more favourable than remaining in solution. The adsorption values at the plateau for PHS in squalane ($> 2\,\text{mg}\,\text{m}^{-2}$ for PHS1000 and $> 2.5\,\text{mg}\,\text{m}^{-2}$ for PHS2500) are more than twice the values obtained in alkyl benzoate ($1\,\text{mg}\,\text{m}^{-2}$ for both PHS1000 and PHS2500). It should be mentioned, however, that both alkyl benzoate and squalane will have χ values less than 0.5, i.e., good solvent conditions and a positive steric potential. This is consistent with the high dispersion stability produced in both solvents. However, the relative difference in solvency for PHS between alkyl benzoate and squalane is expected to have a significant effect on the conformation of the adsorbed layer. In squalane, a poorer solvent for PHS, the polymer chain is denser when compared with the polymer layer in alkyl benzoate. In the latter case a diffuse layer that is typical for polymers in good solvents is produced. This is illustrated in Fig. 6.22 (a) which shows a higher hydrodynamic layer thickness for the higher molecular weight PHS2500. A schematic representation of the adsorbed layers in squalane is shown in Fig. 6.22 (b) which also shows a higher thickness for the higher molecular weight PHS2500.

PHS 1000   PHS 2500   PHS 1000   PHS 2500

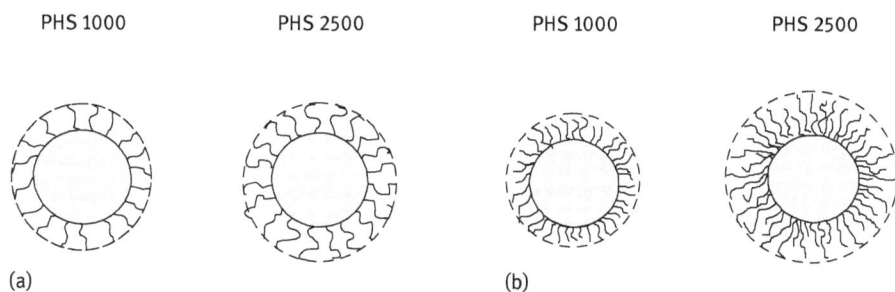

(a)   (b)

Fig. 6.22: (a) Well solvated polymer results in diffuse adsorbed layers (alkyl benzoate). (b) Polymers are not well solvated and form dense adsorbed layers (squalane).

In squalane the dispersant adopts a close packed conformation with little solvation and high amounts are required to reach full surface coverage ($\Gamma > 2\,\text{mg}\,\text{m}^{-2}$). It seems also that in squalane there is much more dependence of the amount of adsorption on the molecular weight of PHS than in the case of alkyl benzoate. It is likely that with the high molecular weight PHS2500 in squalane the adsorbed layer thickness can reach higher values when compared with the results in alkyl benzoate. This larger layer thickness increases the effective volume fraction and this restricts the total solids that can be dispersed. This is clearly shown from the results of Fig. 6.19 which show a rapid increase in zero shear viscosity at a solids loading $> 35\,\%$. With the lower molecular weight PHS1000, with smaller adsorbed layer thickness, the effective volume fraction is lower and high solids loading ($\sim 50\,\%$) can be reached. The solids loading that can be reached in alkyl benzoate when using PHS2500 is higher ($\sim 40\,\%$) than that obtained in squalane. This implies that the adsorbed layer thickness of PHS2500 is smaller in alkyl benzoate when compared with the value in squalane as schematically shown in Fig. 6.22. The solids loading with PHS1000 in alkyl benzoate is similar to that in squalane, indicating a similar adsorbed layer thickness in both cases.

The solids loading curves demonstrate that with an extended layer such as that obtained with the higher molecular weight (PHS2500) the maximum solids loading becomes severely limited as the effective volume fraction (eq. (6.11)) is increased.

In squalane the monomeric dispersant, isostearic acid, shows a high affinity adsorption isotherm with a plateau adsorption of $1\,\text{mg}\,\text{m}^{-2}$ but this provides an insufficient steric barrier (δ/R too small) to ensure colloidal stability.

6.3.7 Competitive interactions in formulations

On addition of the sunscreen dispersion to an emulsion to produce the final formulation, one has to consider the competitive adsorption of the dispersant/emulsifier system. In this case the strength of adsorption of the dispersant to the surface modified TiO_2 particles must be considered. As shown in Fig. 6.20 the silica coated particles (C) show lower PHS2500 adsorption compared to the alumina coated particles (B). However, the dispersant demand for the two powders to obtain a colloidally stable dispersion was similar in both cases (12–14 % PHS2500). This appears at first sight to indicate similar stabilities. However, when added to a water-in-oil emulsion prepared using an A–B–A block copolymer of PHS-PEO-PHS as emulsifier, the system based on the silica coated particles (C) became unstable showing separation and coalescence of the water droplets. SPF performance also dropped drastically from 29 to 11. In contrast, the system based on alumina coated particles (B) remained stable showing no separation as illustrated in Tab. 6.3. These results are consistent with the stronger adsorption (higher χ_s) of PHS2500 on the alumina coated particles. With the silica coated particles, it is likely that the PHS-PEO-PHS block copolymer becomes adsorbed on the particles thus depleting the emulsion interface from the polymeric emulsifier and this is the cause of coalescence. It is well known that molecules based on PEO can adsorb on silica surfaces [11]. By addition of more emulsifier (increasing its concentration from 2 to 3.5 %) the formulation remained stable as is illustrated in Tab. 6.3.

This final set of results demonstrates how a change in surface coating can alter the adsorption strength which can have consequences for the final formulation. The same optimization process used for powder A enabled stable dispersions to be formed from powders B and C. Dispersant demand curves showed optimized dispersion rheology at similar added dispersant levels of 12–14 % PHS2500. To the dispersion scientist these appeared to be stable TiO_2 dispersions. However, when the optimized dispersions were formulated into the external phase of a water-in-oil emulsion differences were observed and alterations in formulation were required to ensure emulsion stability and performance.

The above results show that the application of colloid and interface science principles give a sound basis on which to carry out true optimization of consumer acceptable sunscreen formulations based upon particulate TiO_2. It was found that both dispersion stability and dispersion rheology depended upon adsorbed amount Γ and steric layer thickness δ (which in turn depends on oligomer molecular weight M_n and solvency χ) but that in order to optimize formulation, the adsorption strength χ_s must also be considered. The nature of interaction between particles, dispersant, emulsifiers and thickeners must be considered with regard to competitive adsorption and/or interfacial stability if a formulation is to deliver its required protection when spread on the skin.

6.4 Application of nanosuspensions in paints and coatings

Paints or surface coatings are complex multiphase colloidal systems that are applied as a continuous layer to a surface [19, 20]. A paint usually contains pigmented materials to distinguish it from clear films that are described as lacquers or varnishes. The main purpose of a paint or surface coating is to provide aesthetic appeal as well as to protect the surface. For example, a motor car paint can enhance the appearance of the car body by providing colour and gloss and it also protects the car body from corrosion.

When considering a paint formulation one must know the specific interaction between the paint components and substrates. This subject is of particular importance when one considers the deposition and adhesion of the components to the substrate. The latter can be wood, plastic, metal, glass, etc. The interaction forces between the paint components and the substrate must be considered when formulating any paint. In addition the method of application can vary from one substrate to another.

For many applications it has been recognized that to achieve the required property, such as durability, strong adhesion to the substrate, opacity, colour, gloss, mechanical properties, chemical resistance, corrosion protection, etc., requires the application of more than one coat. The first two or three coats (referred to as the primer and undercoat) are applied to seal the substrate and provide strong adhesion to the substrate. The topcoat provides the aesthetic appeal such as gloss, colour, smoothness, etc. This clearly explains the complexity of paint systems which require fundamental understanding of the processes involved such as particle-surface adhesion, colloidal interaction between the various components, mechanical strength of each coating, etc.

The main objective of the present section is to consider the colloidal phenomena involved in a paint system, its flow characteristics or rheology, its interaction with the substrate and the main criteria that are needed to produce a good paint for a particular application.

To obtain the fundamental understanding of the above basic concepts one must consider first the paint components. Most paint formulations consist of disperse systems (solid in liquid dispersions). The disperse phase consists of primary pigment particles (organic or inorganic) which provide the opacity, colour and other optical effects. These are usually in the submicron range. Other coarse particles (mostly inorganic) are used in the primer and undercoat to seal the substrate and enhance adhesion of the top coat. The continuous phase consist of a solution of polymer or resin which provides the basis of a continuous film that seals the surface and protects it from the outside environment. Most modern paints contain latexes which are used as film formers. These latexes (with a glass transition temperature mostly below ambient temperature) coalesce on the surface and form a strong and durable film. Other components may be present in the paint formulation such as corrosion inhibitors, driers, fungicides, etc.

The primary pigment particles (normally in the nanosize range) are responsible for the opacity, colour and anti-corrosive properties. The principal pigment in use is titanium dioxide and due to its high refractive index is the one used to produce white paint. To produce maximum scattering, the particle size distribution of titanium dioxide has to be controlled within a narrow limit. Rutile with a refractive index of 2.76 is preferred over anatase that has a lower refractive index of 2.55. Thus, the primary pigment particles (normally in the submicron range) are responsible for the opacity, colour and anti-corrosive properties. Rutile gives the possibility of higher opacity than anatase and it is more resistant to chalking on exterior exposure. To obtain maximum opacity the particle size of rutile should be within 220–140 nm. The surface of rutile is photoactive and it is surface coated with silica and alumina in various proportions to reduce its photoactivity.

Coloured pigments may consist of inorganic or organic particles. For a black pigment one can use carbon black, copper carbonate, manganese dioxide (inorganic) or aniline black (organic). For yellow one can use lead, zinc, chromates, cadmium sulphide, iron oxides (inorganic) or nickel azo yellow (organic). For blue/violet one can use ultramarine, Prussian blue, cobalt blue (inorganic) or phthalocyanin, indanthrone blue, carbazol violet (organic). For red one can use red iron oxide, cadmium selenide, red lead, chrome red (inorganic) or toluidine red, quinacridones (organic).

The colour of a pigment is determined by the selective absorption and reflection of the various wavelengths of visible light (400–700 nm) which impinges on it. For example a blue pigment appears so because it reflects the blue wavelengths in the incident white light and absorbs the other wavelengths. Black pigments absorb all the wavelengths of incident light almost totally, whereas a white pigment reflects all the visible wavelengths.

The primary shape of pigmented particles is determined by their chemical nature, their crystalline structure (or lack of it) and the way the pigment is created in nature or made synthetically. Pigments as primary particles may be spherical, nodular, needle or rod-like, or plate-like (lamellar).

Pigments are usually supplied in the form of aggregates (whereby the particles are attached at their faces) or agglomerates (where the particles are attached at their corners). When dispersed in the continuous phase, these aggregates and agglomerates must be dispersed into single units. This requires the use of an effective wetter/dispersant as well as application of mechanical energy. This process of dispersion was discussed in detail in Chapter 5.

In paint formulations, secondary pigments are also used. These are referred to as extenders, fillers and supplementary pigments. They are relatively cheaper than the primary pigments and they are incorporated in conjunction with the primary pigments for a variety of reasons such as cost effectiveness, enhancement of adhesion, reduction of water permeability, enhancement of corrosion resistance, etc. For example, in primer or undercoat (matt latex paint), coarse particle extenders such as calcium carbonate are added in conjunction with TiO_2 to achieve whiteness and opacity in a matt or semi-matt product. The particle size of extenders ranges from submicron (nanosize range) to few tens of microns. Their refractive index is very close to that of the binder and hence they do not contribute to the opacity from light scattering. Most extenders used in the paint industry are naturally occurring materials such as barytes (barium sulphate), chalk (calcium carbonate), gypsum (calcium sulphate) and silicates (silica, clay, talc or mica). However, more recently synthetic polymeric extenders have been designed to replace some of the TiO_2. A good example is spindrift which consists of polymer beads that consist of spherical particles (up to 30 μm in diameter) containing submicron air bubbles and a small proportion of TiO_2. The small air bubbles (< 0.1 μm) reduce the effective refractive index of the polymer matrix, thus enhancing the light scattering of TiO_2.

The refractive index (RI) of any material (primary or secondary pigment) is a key to its performance. As is well known, the larger the difference in refractive index between the pigment and the medium in which it is dispersed, the greater the opacity effect. A summary of the refractive indices of various extender and opacifying pigments is given in Tab. 6.5.

Tab. 6.5: Refractive indices (RI) of extenders and opacifying pigments.

Extender Pigments	RI	Opacifying white pigments	RI
Calcium carbonate	1.58	Zinc sulphide	1.84
China clay	1.56	Zinc oxide	2.01
Talc	1.55	Zinc sulphide	2.37
Barytes	1.64	TiO_2 anatase	2.55
		TiO_2 rutile	2.76

The refractive index of the medium in which the pigment is dispersed ranges from 1.33 (for water) to 1.4–1.6 (for most film formers). Thus rutile will give the highest opacity, whereas talc and calcium carbonate will be transparent in fully bound surface coatings. Another important fact that affects light scattering is the particle size and hence to obtain the maximum opacity from rutile an optimum particle size of 250 nm is required. This explains the importance of good dispersion of the powder in the liquid that can be achieved by a good wetting/dispersing agent as well as application of sufficient milling efficiency to obtain nanoparticles.

For coloured pigments, the refractive index of the pigment in the nonabsorbing, or highly reflecting part of the spectrum affects the performance as an opacifying material. For example, Pigment Yellow 1 and Arylamide Yellow G give lower opacity than Pigment Yellow 34 Lead Chromate. Most suppliers of coloured pigments attempt to increase the opacifying effect by controlling the particle size.

The nature of the pigment's surface plays a very important role in its dispersion in the medium as well as its affinity to the binder. For example, the polarity of the pigment determines its affinity for alkyds, polyesters, acrylic polymers and latexes that are commonly used as film formers. In addition, the nature of the pigment's surface determines its wetting characteristics in the medium in which it is dispersed (which can be aqueous or nonaqueous) as well as the dispersion of the aggregates and agglomerates into single particles. It also affects the overall stability of the liquid paint. Most pigments are surface treated by the manufacturer to achieve optimum performance. As mentioned above, the surface of rutile particles is treated with silica and alumina in various proportions to reduce its photoactivity. If the pigment has to be used in a nonaqueous paint, its surface is also treated with fatty acids and amines to make it hydrophobic for incorporation in an organic medium. This surface treatment enhances the dispersibility of the paint, its opacity and tinting strength, its durability (glass retention, resistance to chalking and colour retention). It can also protect the binder in the paint formulation.

The dispersion of the pigment powder in the continuous medium requires several processes, namely wetting of the external and internal surface of the aggregates and agglomerates, separation of the particles from these aggregates and agglomerates by application of mechanical energy, displacement of occluded air and coating of the particles with the dispersion resin. It is also necessary to stabilize the particles against flocculation either by electrostatic double layer repulsion and/or steric repulsion. The process of wetting and dispersion of pigments was described in detail in Chapter 5, whereas the eminence of colloid stability (lack of aggregation) was discussed in Chapter 2.

The dispersion medium can be aqueous or nonaqueous depending on application. It consists of a dispersion of the binder in the liquid (which is sometimes referred to as the diluent). The term solvent is frequently used to include liquids that do not dissolve the polymeric binder. Solvents are used in paints to enable the paint to be made and they enable application of the paint to the surface. In most cases the solvent is removed after application by simple evaporation and if the solvent is completely removed from the paint film it should not affect the paint film's performance. However, in the early life of the film, solvent retention can affect hardness, flexibility and other film properties. In water-based paints, the water may act as a true solvent for some of the components but it should be a nonsolvent for the film former. This is particularly the case with emulsion paints.

With the exception of water, all solvents, diluents and thinners used in surface coatings are organic liquids with low molecular weight. Two types can be distinguished, hydrocarbons (both aliphatic and aromatic) and oxygenated compounds such as ethers, ketones, esters, ether alcohols, etc. Solvents, thinners and diluents control the flow of the wet paint on the substrate to achieve a satisfactory smooth, even, thin film, which dries in a predetermined time. In most cases mixtures of solvents are used to obtain the optimum condition for paint application. The main factors that must be considered when choosing solvent mixtures are their solvency, viscosity, boiling point, evaporation rate, flash point, chemical nature, odour and toxicity.

The solvent power or solvency of a given liquid or mixture of liquids determines the miscibility of the polymer binder or resin. It has also a big effect on attraction between particles in a paint formulation as was discussed in detail in Chapter 4. A very useful parameter that describes solvency is the Hildebrand solubility parameter δ which is related to the energy of association of molecules in the liquid phase, in terms of "cohesive energy density". The latter is simply the ratio of the energy required to vaporize $1\,cm^3$ of liquid ΔE_v to its molar volume V_m. The solubility parameter δ is simply the square root of that ratio,

$$\delta = \left(\frac{\Delta E_v}{V_m} \right)^{1/2}.$$
(6.14)

Liquids with similar values of δ are miscible, whereas those with significantly different values are immiscible. The solubility parameters of liquids can be determined experimentally by measuring the energy of vaporization. For polymers, one can determine the solubility parameter using an empirical approach by contacting the polymer with liquids with various δ values and observing whether or not dissolution occurs. The solubility parameter of the polymer is taken as the average of two δ values for two solvents that appear to dissolve the polymer. A better method is to calculate the solubility parameter from the "molar attraction constant" G of the constituent parts of the molecule [18],

$$\delta = \left(\frac{\rho \sum G}{M} \right),$$
(6.15)

where ρ is the density of the polymer and M is its molecular weight.

As mentioned before, Hansen [17] extended Hildebrand's concept by considering three components for the solubility parameter: a dispersion component δ_d, a polar component δ_p, and a hydrogen bonding component δ_h as given by equation (6.14). Values of δ and its components are tabulated in the book by Barton [18].

As mentioned above, the dispersion medium consists of a solvent or diluent and the film former. The latter is also sometimes referred to as a "binder", since it functions by binding the particulate components together and this provides the continuous film- forming portion of the coating. The film former can be a low molecular weight polymer (oleoresinous binder, alkyd, polyurethane, amino resins, epoxide resin, unsaturated polyester), a high molecular weight polymer (nitrocellulose, solution vinyls, solution acrylics), an aqueous latex dispersion (polyvinyl acetate, acrylic or styrene/butadiene) or a nonaqueous polymer dispersion (NAD). In this section, I will only briefly describe film formers based on polymer solutions. The subject of polymer latexes and nonaqueous dispersions was dealt with in Chapter 4. The polymer solution may exist in the form of a fine particle dispersion in a nonsolvent. In some cases the system may be mixed solution/dispersion implying that the solution contains both single polymer chains and aggregates of these chains (sometimes referred to as micelles) which are in the nanosize range. A striking difference between a polymer that is completely soluble in the medium and that which contains aggregates of that polymer is the viscosity reached in both cases. A polymer that is completely soluble in the medium will show a higher viscosity at a given concentration compared to another polymer (at the same concentration) that produces aggregates. Another important difference is the rapid increase in the solution viscosity with increasing molecular weight for a completely soluble polymer. If the polymer makes aggregates in solution, an increase in molecular weight of the polymer does not show a dramatic increase in viscosity.

The earliest film forming polymers used in paints were based on natural oils, gums and resins. Modified natural products are based on cellulose derivatives such as nitrocellulose which is obtained by nitration of cellulose under carefully specified conditions. Organic esters of cellulose such as acetate and butyrate can also be produced. Another class of naturally occurring film formers are those based on vegetable oils and their derived fatty acids (renewable resource materials). Oils used in coatings include linseed oil, soya bean oil, coconut oil and tall oil. When chemically combined into resins, the oil contributes flexibility and with many oils oxidative crosslinking potential. The oil can also be chemically modified, for example the hydrogenation of castor oil can be combined with alkyd resins to produce some specific properties of the coating.

Another early binder used in paints are the oleoresinous vehicles that are produced by heating together oils and either natural or certain preformed resins, so that the resin dissolves or disperses in the oil portion of the vehicle. However these oleoresinous vehicles have been replaced by alkyd resins which are probably one of the first applications of synthetic polymers in the coating industry. These alkyd resins are polyesters obtained by reaction of vegetable oil triglycerides, polyols (e.g. glycerol) and dibasic acids or their anhydrides. These alkyd resins enhanced the mechanical strength, drying speed and durability over and above those obtained using the oleoresinous vehicles. The alkyds were also modified by replacing part of the dibasic acid with a diisocyanate (such toluene diisocyanate, TDI) to produce greater toughness and quicker drying characteristics.

Another type of binder is based on polyester resins (both saturated and unsaturated). These are typically composed mainly of co-reacted di- or polyhydric alcohols and di- or tribasic acid or acid anhydride. They have also been modified using silicone to enhance their durability.

More recently, acrylic polymers have been used in paints due to their excellent properties of clarity, strength and chemical and weather resistance. Acrylic polymers refer to systems containing acrylate and methylacrylate esters in their structure along with other vinyl unsaturated compounds. Both thermoplastic and thermosetting systems can be made, the latter are formulated to include monomers possessing additional functional groups that can further react to give crosslinks following the formation of the initial polymer structure. These acrylic polymers are synthesized by radical polymerization. The main polymer-forming reaction is a chain propagation step which follows an initial initiation process. A variety of chain transfer reactions are possible before chain growth ceases by a termination process.

Radicals produced by transfer, if sufficiently active, can initiate new polymer chains where a monomer is present which is readily polymerized. Radicals produced by chain transfer agents (low molecular weight mercaptans, e.g. primary octyl mercaptan) are designed to initiate new polymer chains. These agents are introduced to control the molecular weight of the polymer.

The monomers used for preparation of acrylic polymers vary in nature and can generally be classified as "hard" (such as methylmethacrylate, styrene and vinyl acetate) or "soft" (such as ethyl acrylate, butyl acrylate, 2-ethyl hexyl acrylate). Reactive monomers may also have hydroxyl groups (such as hydroxy ethyl acrylate). Acidic monomers such as methacrylic acid are also reactive and may be included in small amounts so that the acid groups may enhance pigment dispersion. Practical coating systems are usually copolymers of "hard" and "soft". The polymer hardness is characterized by its glass transition temperature, T_g. The T_g (K) of the copolymer can be estimated from the T_g of the individual T_g (K) of the homopolymers with weight fractions W_1 and W_2,

$$\frac{1}{T_g} = \frac{W_1}{T_{g1}} + \frac{W_2}{T_{g2}} . \tag{6.16}$$

The vast majority of acrylic polymers consist of random copolymers. By controlling the proportion of "hard" and "soft" monomers and the molecular weight of the final copolymer one arrives at the right property that is required for a given coating. As mentioned above, two types of acrylic resins can be produced, namely thermoplastic and thermosetting. The former find application in automotive topcoats although they suffer from some disadvantages like cracking in cold conditions and this may require a process of plasticization. These problems are overcome by using thermosetting acrylics which improve the chemical and alkali resistance. Also it allows one to use higher solid contents in cheaper solvents. Thermosetting resins can be self-crosslinking or may require a co-reacting polymer or hardener.

In a paint film the pigment particles need to undergo a process of deposition to the surfaces (that is governed by long range forces such as van der Waals attraction and electrical double layer repulsion or attraction). This process of deposition is also affected by polymers (nonionic, anionic or cationic) which can enhance or prevent adhesion. Once the particles reach the surface they have to adhere strongly to the substrate. This process of adhesion is governed by short-range forces (chemical or nonchemical). The same applies to latex particles which also undergo a process of deposition, adhesion and coalescence.

Control of the flow characteristics of paints is essential for their successful application. All paints are complex systems consisting of various components such as pigments, film formers, latexes and rheology modifiers. These components interact with each other and the final formulation becomes non-Newtonian showing complex rheological behaviour. The paint is usually applied in three stages, namely transfer of the paint from the bulk container, transfer of the paint from the applicator (brush or roller) to the surface to form a thin even film and flow-out of film surface, coalescence of polymer particles (latexes) and loss of the medium by evaporation. During each of these processes the flow characteristics of the paint and its time relaxation produce interesting rheological responses. To understand the rheological behaviour of a paint system, one must start with the basic knowledge of rheology [21].

References

[1] Lee, E. M., "Nanocrystals: Resolving pharmaceutical formulation issues associated with poorly water-soluble compounds", in: J. J. Marty (ed.), "Particles", Marcel Dekker, Orlando (2002).

[2] Sharma, D., Soni, M., Kumar, S. and Gupta, G., Res. J. Pharm. Technol., **2**, 220 (2009).

[3] Kipp, J., Int. J. Pharm., **284**, 109 (2004).

[4] Wong, J., Brugger, A., Khare, A., Chaubal, M., Papadopoulos, P., Rabinow, B., Kipp, J. and Ning, J., J., Advanced Drug Delivery Reviews, **60**, 939 (2008).

[5] Shegokar, R. and Müller, R. H., Internat. J. Pharm., **399**, 129 (2010).

[6] Noyes, A. A. and Whitney, W. R., "The rate of solution of solid substances in their own solutions", Journal of the American Chemical Society, **19**, 930 (1897).

[7] Thompson W. (Lord Kelvin), Phil. Mag., **42**, 448 (1871).

[8] Tadros, Th. F., "Formulation of Disperse Systems, Wiley-VCH, Germany (2014).

[9] Nakach, M., Authelin, J.-R., Tharwat Tadros, Th. F., Galet, L. and Chamayou, A., Internat. J. Pharm., **476**, 277 (2014).

[10] Tadros, Th. F., "Dispersion of Powders in Liquids and Stabilisation of Suspensions", Wiley-VCH, Germany (2012).

[11] Napper, D. H., "Polymeric Stabilisation of Colloidal Dispersions", Academic Press, London (1983).

[12] Kaye, B. H., Powder Technol., 1 (1967).

[13] Pecora, R., "Dynamic Light Scattering: Applications of Photon Correlation Spectroscopy", Springer, Germany (1985).

[14] Hunter, R. J., "Zeta Potential in Colloid Science: Principles and Application", Academic Press, London (1988).

[15] Robb, J. L., Simpson, L. A. and Tunstall, D. F., Scattering and absorption of UV radiation by sunscreens containing fine particle and pigmentary titanium dioxide, Drug. Cosmet. Ind., **March**, 32–39 (1994).

[16] Hewitt, J. P., Soap Perfum. Cosmet. **75(3)**, 47–50 (2002).

[17] Hansen, C. M., J. Paint Technology, **39**, 104–117, 505–514 (1967).
[18] Barton, A. F. M., "Handbook of Solubility Parameters and Other Cohesion Parameters", Boca Raton, Florida, CRC Press, Inc. (1983).
[19] Lambourne, R. (ed.), "Paint and Surface Coating", Ellis Horwood, Chichester (1987).
[20] Tadros, Th. F., "Colloids in Paints", Wiley-VCH, Germany (2010).
[21] Tadros, Th. F., "Rheology of Dispersions", Wiley-VCH, Germany (2010).

7 Nanoparticles as drug carriers

7.1 Introduction

The concept of delivering a drug to its pharmaceutical site of action in a controlled manner has attracted the interest of the pharmaceutical industry in recent years. A great deal of research is being carried out because the site delivery of a drug can be controlled at a rate and concentration that optimizes the therapeutic activity, while minimizing the adverse toxic effects [1, 2].

The use of biodegradable colloidal nanoparticles offers a number of advantages over more conventional dosage forms [2, 3]. Due to their small size (20–200 nm) they are suitable for intravenous administration, they can be applied as long-circulating drug depots and for targeting specific organs or sites. Several other advantages of nanoparticles can be listed: protection of drugs against metabolism or recognition by the immune system, reduction of toxic effects especially for chemotherapeutic drugs and improved patient compliance by avoiding repetitive administration [4].

Various biodegradable colloidal drug carriers have been developed, of which liposomes (and vesicles) and polymeric nanoparticles are the most widely used systems. Liposomes are spherical phospholipid liquid crystalline phases (smectic mesophases) that are simply produced by dispersion of phospholipid (such as lecithin) in water by simple shaking. This results in the formation of multilayer structures consisting of several bilayers of lipids (several μm). When sonicated, these multilayer structures produce unilamellar structures (with size range of 25–50 nm) that are referred to as vesicles. A schematic picture of liposomes and vesicles is given in Fig. 7.1. Glycerol containing phospholipids are used for the preparation of liposomes and vesicles: phosphatidylcholine; phosphatidylserine; phosphatidylethanolamine; phosphatidylinositol; phosphatidylglycerol; phosphatidic acid; cholesterol. In most preparations, a mixture of lipids is used to obtain the most optimum structure.

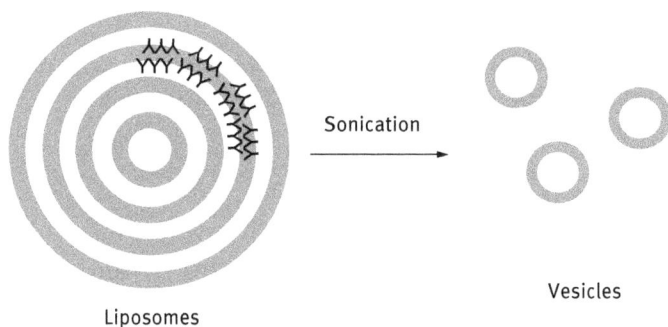

Fig. 7.1: Schematic representation of liposomes and vesicles.

It should be mentioned, however, that the nomenclature for phospholipid vesicles is far from being clear. It is now generally accepted that "All types of lipid bilayers surrounding an aqueous phase are in the general category of liposomes" [5, 6]. The term "liposome" is usually reserved for vesicles composed, even partly, of phospholipids. The more generic term "vesicle" is to be used to describe any structure consisting of one or more bilayers of various other surfactants. In general the names "liposome" and "phospholipid vesicle" are used interchangeably. Liposomes are classified in terms of the number of bilayers, as multilamellar vesicles (MLVs > 400 nm), large unilamellar vesicles (LUVs > 100 nm) and small unilamellar vesicles (SUVs < 100 nm). Other types reported are the giant vesicles (GV), which are unilamellar vesicles of diameter between 1–5 μm and large oligolamellar vesicles (LOV) where a few vesicles are entrapped in the LUV or GV.

The most widely used polymeric nanoparticles are those of the A–B and A–B–A block copolymer type. Block copolymers of the B–A and B–A–B types are known to form micelles that can be used as drug carriers [4]. These block copolymers consist of a hydrophobic B block that is insoluble in water and one or two A blocks which are very soluble in water and strongly hydrated by its molecules. In aqueous media the block copolymer will form a core of hydrophobic chains and a shell of the hydrophilic chains. These self-assembled structures are referred to as micelles and they are schematically illustrated in Fig. 7.2. The core-shell structure is ideal for drug delivery where the water insoluble drug is incorporated in the core and the hydrophilic shell provides effective steric stabilization thus minimizing adsorption of the blood plasma components and preventing adhesion to phagocytic cells.

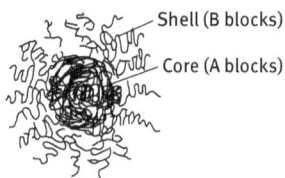

Shell (B blocks)
Core (A blocks)

Fig. 7.2: Core-shell structure of block copolymers.

In this chapter I will describe the above two biodegradable nanoparticles separately with emphasis on their formation, stability and application for drug delivery.

7.2 Liposomes as drug carriers

As mentioned in the introduction, liposomes and vesicles can be prepared using biodegradable lipids. The structure of some lipids is shown in Fig. 7.3. The most widely used lipid for drug delivery is phosphatidylcholine that can be obtained from eggs or soybean. These liposome bilayers can be considered as mimicking models of biological membranes. They can solubilize both lipophilic drug molecules in the lipid bilayer phase, as well as hydrophilic molecules in the aqueous layers between the lipid bilayers and in the inner aqueous phase. Due to this ability, liposomes have been used to deliver enzymes [5], genetic material [5] and various anticancer drugs [5]. Liposomes have also proved particularly useful as general vaccine additives [5], for example liposome-based vaccines against hepatitis A. Another very useful application of liposomes is for new-born babies suffering from lung surfactant deficiency [5]. In addition, liposomes are frequently used in cosmetic formulations for enhancement of the penetration of anti-wrinkle agents [5].

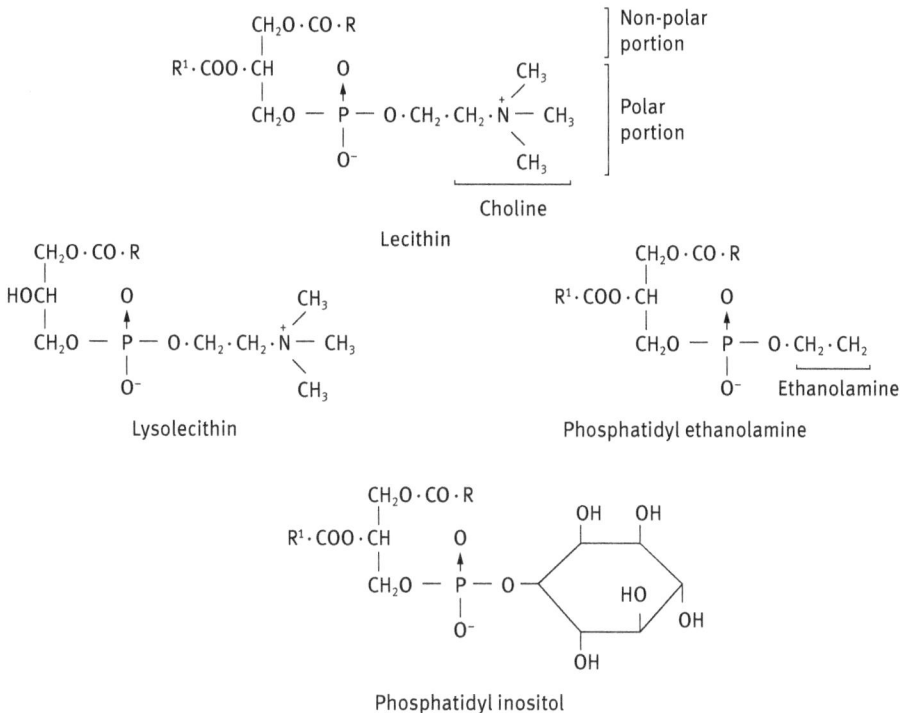

Fig. 7.3: Structure of lipids.

The driving force for formation of vesicles has been described in detail by Israelachvili et al. [7–9]. From equilibrium thermodynamics, small aggregates, or even monomers, are entropically favoured over larger ones. This entropic force explains the aggregation of single-chain amphiphiles into small spherical micelles instead to bilayers or cylinders, as the aggregation number of the latter aggregates is much higher. Israelachvili et al. [7–9] attempted to describe the thermodynamic drive for vesicle formation by biological lipids. From equilibrium thermodynamics of self-assembly, the chemical potential of all molecules in a system of aggregated structures such as micelles or bilayers will be the same,

$$\mu_N^o + \frac{kT}{N} \ln\left(\frac{X_N}{N}\right) = \text{const.}; \quad N = 1, 2, 3, \ldots, \tag{7.1}$$

where μ_N^o is the free energy per molecule in the aggregate, X_N is the mole fraction of molecules incorporated into the aggregate, with an aggregation number N, k is the Boltzmann constant and T is the absolute temperature.

For monomers in solution with $N = 1$,

$$\mu_N^o + \frac{kT}{N} \ln\left(\frac{X_N}{N}\right) = \mu_1^o + kT \ln X_1. \tag{7.2}$$

Equation (7.1) can be written as,

$$X_N = N\left(\frac{X_M}{M}\right)^{N/M} \exp\left(\frac{N(\mu_M^o - \mu_N^o)}{kT}\right), \tag{7.3}$$

where M is any arbitrary state of reference of aggregation number N.

The following assumptions are made to obtain the free energy per molecule: (i) the hydrocarbon interior of the aggregate is considered to be in a fluid-like state; (ii) geometric consideration and packing constraints in term of aggregate formation are excluded; (iii) strong long-range forces (van der Waals and electrostatic) are neglected. By considering the "opposing forces" approach of Tanford [10], the contributions to the chemical potential, μ_N^o, can be estimated. A balance exists between the attractive forces mainly of hydrophobic (and interfacial tension) nature and the repulsive forces due to steric repulsion (between the hydrated head group and alkyl chains), electrostatic and other forces [11]. The free energy per molecule is thus,

$$\mu_N^o = \gamma a + \frac{C}{a}. \tag{7.4}$$

The attractive contribution (the hydrophobic free energy contribution) to μ_N^o is γa where γ is the interfacial free energy per unit area and a is the molecular area measured at the hydrocarbon/water interface. C/a is the repulsive contribution where C is a constant term used to incorporate the charge per head group, e, and includes terms such as the dielectric constant at the head group region, ε, and curvature corrections.

This fine balance yields the optimum surface area, a_o, for the polar head groups of the amphiphile molecules at the water interface, at which the total interaction free energy per molecule is a minimum,

$$\mu_N^o(\text{min}) = \gamma a + \frac{C}{a} = 0, \tag{7.5}$$

$$\frac{\partial \mu_N^o}{\partial a} = \gamma - \frac{C}{a^2} = 0, \tag{7.6}$$

$$a = a_o = \left(\frac{C}{\gamma}\right)^{1/2}. \tag{7.7}$$

Using the above equations, the general form relating the free energy per molecule μ_N^o with a_o can be expressed as,

$$\mu_N^o = \gamma\left(a + \frac{a_o^2}{a}\right) = 2a_o\gamma + \frac{\gamma}{a}(a - a_o)^2 \tag{7.8}$$

Equation (7.8) shows that: (i) μ_N^o has a parabolic (elastic) variation about the minimum energy; (ii) amphiphilic molecules, including phospholipids, can pack in a variety of structures in which their surface areas will remain equal or close to a_o. Both single-chain and double-chain amphiphiles have very much the same optimum surface area per head group ($a_o \sim 0.5$–$0.7\ nm^2$), i.e. a_o is not dependent on the nature of the hydrophobe. Thus, by considering the balance between entropic and energetic contributions to the double-chain phospholipid molecule one arrives at the conclusion that the aggregation number must be as low as possible and a_o for each polar group is of the order of 0.5–$0.7\ nm^2$ (almost the same as that for a single-chain amphiphile). For phospholipid molecules containing two hydrocarbon chains of 16–18 carbon atoms per chain, the volume of the hydrocarbon part of the molecule is double the volume of a single-chain molecule, while the optimum surface area for its head group is of the same order as that of a single-chain surfactant ($a_o \sim 0.5$–$0.7\ nm^2$). Thus the only way for this double-chain surfactant is to form aggregates of the bilayer sheet or the close bilayer vesicle type. This will be further explained using the critical packing parameter concept (CPP) described by Israelachvili et al. [7–9]. The CPP is a geometric expression given by the ratio of the cross-sectional area of the hydrocarbon tail(s) a to that of the head group a_o. a is equal to the volume of the hydrocarbon chain(s) v divided by the critical chain length l_c of the hydrocarbon tail. Thus the CPP is given by [12]

$$CPP = \frac{v}{a_o l_c} . \tag{7.9}$$

Regardless of shape, any aggregated structure should satisfy the following criterion: no point within the structure can be farther from the hydrocarbon-water surface than l_c which is roughly equal, but less than the fully extended length l of the alkyl chain.

For a spherical micelle, the radius $r = l_c$ and from simple geometry $CPP = v/a_o l_c \leq \frac{1}{3}$. Once $v/a_o l_c > \frac{1}{3}$, spherical micelles cannot be formed and when $\frac{1}{2} \geq CPP > \frac{1}{3}$ cylindrical micelles are produced. When the CPP $> \frac{1}{2}$ but < 1, vesicles are produced. These vesicles will grow until CPP ~ 1 when planer bilayers will start forming. A schematic representation of the CPP concept is given in Tab. 7.1.

According to Israelachvili et al. [7–9], the bilayer sheet lipid structure is energetically unfavourable to the spherical vesicle, because of the lower aggregation number of the spherical structure. Without introduction of packing constraints (described above), the vesicles should shrink to such a small size that they would actually form micelles. For double-chain amphiphiles three considerations must be borne in mind: (i) an optimum a_o (almost the same as that for single-chain surfactants) must be achieved by considering the various opposing forces; (ii) structures with minimum

Tab. 7.1: CPP concept and various shapes of aggregates.

Lipid	Critical packing parameter $v/a_0 l_c$	Critical packing shape	Structures formed
Single-chained lipids (surfactants) with large head-group areas: – *SDS in low salt*	< 1/3	Cone	Spherical micelles
Single-chained lipids with small head-group areas: – *SDS and CTAB in high salt* – *nonionic lipids*	1/3–1/2	Truncated cone	Cylindrical micelles
Double-chained lipids with large head-group areas, fluid chains: – *Phosphatidyl choline (lecithin)* – *phosphatidyl serine* – *phosphatidyl glycerol* – *phosphatidyl inositol* – *phosphatidic acid* – *sphingomyelin, DGDG[a]* – *dihexadecyl phosphate* – *dialkyl dimethyl ammonium* – *salts*	1/2–1	Truncated one	Flexible bilayers, vesicles
Double-chained lipids with small head-group areas, anionic lipids in high salt, saturated frozen chains: – *phosphatidyl ethanaiamine* – *phosphatidyl serine + Ca[2+]*	~1	Cylinder	Planar bilayers
Double-chained lipids with small head-group areas, nonionic lipids, poly(cis) unsaturated chains, high T: – *unsat. phosphatidyl ethanolamine* – *cardiolipin + Ca[2+]* – *phosphatidic acid + Ca[2+]* – *cholesterol, MGDG[b]*	> 1	Inverted truncated cone or wedge	Inverted micelles

a DGDG: digalactosyl diglyceride, diglucosyldiglyceride
b MGDG: monogalactosyl diglyceride, monoglucosyl diglyceride

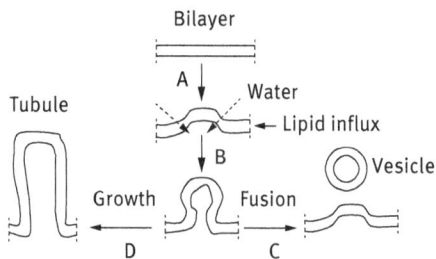

Fig. 7.4: Bilayer vesicle and tubule formation [12].

aggregation number N must be formed; (iii) aggregates into bilayers must be the favourite structure. A schematic picture of the formation of bilayer vesicle and tubule structures was introduced by Israelachvili and Mitchell [12] and is shown in Fig. 7.4.

Israelachvili et al. [7–10] believe that steps A and B are energetically favourable. They considered step C to be governed by packing constraints and thermodynamics in terms of the least aggregation number. They concluded that the spherical vesicle is an equilibrium state of the aggregate in water and it is certainly more favoured over extended bilayers.

The main drawback to the application of liposomes as drug delivery systems is their metastability. On storage, the liposomes tend to aggregate and fuse to form larger polydisperse systems and finally the system reverses into a phospholipid lamel-lar phase in water. This process takes place relatively slowly because of the slow exchange between the lipids in the vesicle and the monomers in the surrounding medium. Therefore, it is essential to investigate both the chemical and physical sta-bility of the liposomes. Examination of the process of aggregation can be obtained by measuring their size as a function of time. Maintenance of the vesicle structure can be assessed using freeze fracture and electron microscopy. The influence of biological fluids on the liposome integrity and permeability must also be investigated. Due to these instability problems, the most common method for their storage for commercial purposes is by freeze drying them.

Several methods have been applied to increase the rigidity and physicochemical stability of the liposome bilayer of which the following methods are the most com-monly used: hydrogenation of the double bonds within the liposomes, polymerization of the bilayer using synthesized polymerizable amphiphiles and inclusion of choles-terol to rigidify the bilayer [5].

Other methods to increase the stability of the liposomes include modification of the liposome surface, for example by physical adsorption of polymeric surfactants onto the liposome surface (e.g. proteins and block copolymers). Another approach is to covalently bond the macromolecules to the lipids, subsequently forming vesicles. A third method is to incorporate the hydrophobic segments of the polymeric surfac-tant within the lipid bilayer. This latter approach has been successfully applied by Kostarelos et al. [6] who used A–B–A block copolymers of polyethylene oxide (A) and polypropylene oxide (PPO), namely poloxamers (Pluronics). Two different techniques of adding the copolymer were attempted [6]. In the first method (A), the block copolymer was added after formation of the vesicles. In the second method, the phospholipid and copolymer are first mixed together and this is followed by hydration and formation of SUV vesicles. These two methods are briefly described below.

The formation of small unilamellar vesicles (SUVs) was carried out by sonication of 2 % w/w of the hydrated lipid (for about 4 hours). This produced SUV vesicles with a mean vesicle diameter of 45 nm (polydispersity index of 1.7–2.4). This is followed by the addition of the block copolymer solution and dilution of × 100 times to obtain a lipid concentration of 0.02 % (method (A)). In the second method (I) SUV vesicles were prepared in the presence of the copolymer at the required molar ratio.

In method (A), the hydrodynamic diameter increases with increasing block copolymer concentration, particularly for those with high PEO content, reaching a plateau at a certain concentration of the block copolymer. The largest increase in hydrodynamic diameter (from ~ 43 nm to ~ 48 nm) was obtained using Pluronic F127 (that contains a molar mass of 8330 PPO and molar mass of 3570 PEO). In method (I) the mean vesicle diameter showed a sharp increase with increase in % w/w copolymer reaching a maximum at a certain block copolymer concentration, after which a further increase in polymer concentration showed a sharp reduction in average diameter. For example with Pluronic F127, the average diameter increased from ~ 43 nm to ~ 78 nm at 0.02 % w/w block copolymer and then it decreased sharply with a further increase in polymer concentration, reaching ~ 45 nm at 0.06 % w/w block copolymer. This reduction in average diameter at high polymer concentration is due to the presence of excess micelles of the block copolymer.

A schematic representation of the structure of the vesicles obtained on addition of the block copolymer using methods (A) and (I) is shown in Fig. 7.5.

With method (A), the triblock copolymer is adsorbed on the vesicle surface by both PPO and PEO blocks. These "flat" polymer layers are prone to desorption due to the weak binding onto the phospholipid surface. In contrast, with the vesicles prepared using method (I) the polymer molecules are more strongly attached to the lipid bilayer

TRIBLOCK COPOLYMER ADSORBED   TRIBLOCK COPOLYMER INCORPORATED

(A) Versicle system   (I) Versicle system

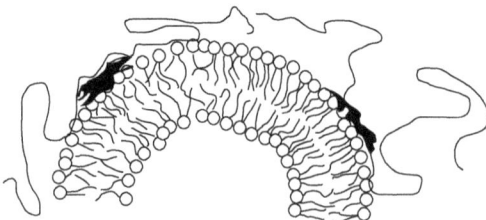 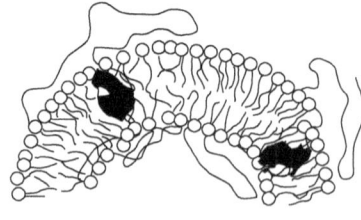

Fig. 7.5: Schematic representation of vesicle structure in the presence of triblock copolymer prepared using method (A) and (I) [5].

with PPO segments "buried" in the bilayer environment surrounded by the lipid fatty acids. The PEO chains remain at the vesicle surfaces free to dangle in solution and attain the preferred conformation. The resulting sterically stabilized vesicles [(I) system] have several advantages over the (A) system with the copolymer simply coating their outer surface. The anchoring of the triblock copolymer using method (I) results in irreversible adsorption and lack of desorption. This is confirmed by dilution of both systems. With (A), dilution of the vesicles results in reduction of the diameter to its original bare liposome system, indicating polymer desorption. In contrast, dilution of the vesicles prepared by method (I) showed no significant reduction in diameter size indicating strong anchoring of the polymer to the vesicle. A further advantage of constructing the vesicles with bilayer-associated copolymer molecules is the possibility of increased rigidity of the lipid-polymer bilayer [5, 6].

7.3 Polymeric nanoparticles

Polymeric nanoparticles, with the drug entrapped within the polymer matrix have some advantages in terms of their stability both in storage and in vivo application [4]. The choice of a polymer is restricted by its biodegradability. Both model non-biodegradable and biodegradable systems and these nanoparticles have been used for studies on their use of these as drug carriers.

Regardless of the type of nanoparticle, these colloidal systems are recognized as foreign bodies after administration to the systemic circulation. They can be quickly removed by the phagocytic cells (macrophages) of the reticuloendothelial system (RES), in particular by the Kupffer cells of the liver. The main approach is to design nanoparticles that avoid RES recognition. This can be achieved by controlling the size and surface properties of the nanoparticle. If the nanoparticles remain in circulation for a prolonged period of time, and avoid liver deposition, there is the possibility for redirecting the particles to other organs/tissues. Long-circulating nanoparticles can potentially be actively directed to a particular site by the use of targeting moieties such as antibodies or sugar residues that can be specifically recognized by cell-surface receptors [4].

As mentioned above, following intravenous (i.v.) injection, the colloidal nanoparticles are recognized as foreign bodies and they may be removed from the blood circulation by the phagocyte cells of the RES. Within 5 minutes after i.v. injection ~ 60–90 % of the nanoparticles can be phagocytosed by the macrophages of the liver and spleen. Site specific delivery to other organs must avoid this process taking place. The design of any nanoparticle system with long-circulation requires understanding of the mechanism of phagocytosis. The clearance of nanoparticles is mediated by adsorption of blood components to the surface of the particles, a process referred to as opsonization that is described below.

The adsorption of proteins (a component of blood) at the surface of the nanoparticles can result in the surface becoming hydrophobic and this may lead to enhanced phagocytosis. The hydrophobic segments of a protein molecule may adsorb on a hydrophobic surface. While on the surface, the protein may be denatured due to the loss of configurational liability. There may be also a gain in configurational entropy on going from a globular to a more extended state. However, the process of protein adsorption is quite complex due to the presence of more than one type in the blood plasma. The process of protein adsorption is summarized below.

Opsinons refer to proteins that enhance phagocytosis, whereas dyopsinons are molecules that suppress phagocytosis. This depends on the hydrophobic/hydrophilic nature of the protein. Opsinons are immunoglobulin molecules that adsorb on the particle surface, thus making them more "palatable" to macrophages. Dyopsinons are immunoglobulin molecules that render the surface of the particles more hydrophilic, thus suppressing phagocytosis. The interaction of the blood components with the nanoparticles is a complex process, although its control is the key to avoid phagocytosis [4].

When considering nanoparticles as drug delivery systems one must consider three main characteristics: (i) Particle size, which determines the deposition of colloidal nanoparticles containing the drug following intravenous administration. (ii) Surface charge, which determines the interaction between the nanoparticles and the macrophages. (iii) Surface hydrophobicity, which determines the interaction of the serum components with the nanoparticle surface. This determines the degree of opsinization. A description of each of the above characteristics is given below [4].

(i) Influence of particle size: Particles $> 7\,\mu m$ are larger than the blood capillaries ($\sim 6\,\mu m$) and they become entrapped in the capillary beds of the lungs. Thus, aggregated or flocculated particles tend to accumulate in the lung with fatal consequences. Most of the particles that pass the lung capillary bed become accumulated by the RES of the spleen, liver and bone marrow. The degree of splenic uptake increases with increasing particle size. The splenic removal of particles and liposomes $> 200\,nm$ is due to a non-phagocytic process whereby the splenic architecture acts as a sieve or filter bed. As the particle size is reduced below 200 nm, the extent of splenic uptake decreases and the majority of particles are mostly cleared by the liver. Colloidal particles not cleared by the RES can exit the blood circulation via the sinusoidal fenestrations of the liver and bone marrow provided they are smaller than 150 nm.

(ii) Influence of surface charge: The surface charge determines the electrostatic repulsion between the colloidal nanoparticle and the blood components or a cell surface. However, the range of electrostatic repulsion decreases with increasing ionic strength. The blood has an ionic strength of $\sim 0.15\,mol\,dm^{-3}$ and hence the range of electrostatic repulsion is less than 1 nm. This means the surface charge only influences the protein-protein or particle-macrophage interactions at very short distances. Thus the effect of surface charge on phagocytosis is not due to its effect on electrostatic repulsion, but due to its influence on hydrophobicity of the particles that can determine protein adsorption.

(iii) Influence of surface hydrophobicity: As mentioned above the hydrophobic sites on a nanoparticle determines the adsorption of the serum components. Increasing surface hydrophobicity increases protein adsorption thus increasing the degree of opsonization. It has been shown by in vitro studies that the increased adsorption of proteins on a hydrophobic surface leads to enhanced uptake by phagocytic cells. As will be shown later, the surface modification of nanoparticles by adsorbed or grafted polymers can affect their surface hydrophobicity and hence their ability to be captured by the phagocytic cells.

7.3.1 Surface modified polystyrene latex particles as model drug carriers

The surface of polystyrene latex particles is relatively hydrophobic and can be easily modified by adsorbed nonionic polymers. Poloxamers and poloxamines are composed of a central poly(propylene oxide) (PPO) block and terminal poly(ethylene oxide) (PEO) chains. The general structure of poloxamers and poloxamines is given in Fig. 7.6. The general composition of the molecules that are approved by the FDA is given in Tab. 7.2. The hydrophobic central PPO chain anchors the copolymer to the

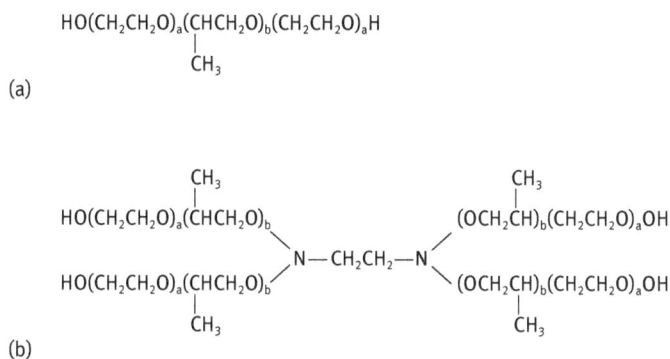

$HO(CH_2CH_2O)_a(CHCH_2O)_b(CH_2CH_2O)_aH$
CH_3
(a)

(b)

Fig. 7.6: General structure of poloxamers (a) and poloxamines (b).

Tab. 7.2: Composition of poloxamers and poloxamines.

Polymer	Average molecular weight / Da	Number of PPO (b) units per chain	Number of units per chain
Poloxamer 108	4 700	16	42
Poloxamer 338	14 600	56	129
Poloxamer 407	12 600	69	98
Poloxamine 908	25 000	17	119

surface of the particle, whereas the hydrophilic PEO blocks provide the required hydrophilic steric barrier. In general, the thickness of the adsorbed layer increases with increasing length of the PEO chains [4].

The coating of polystyrene particles with poloxamers and poloxamines dramatically reduces their sequestration by the liver. The thickness of the PEO layer is crucial to altering the biological fate of the nanoparticles. Poloxamers with short PEO chains, such as poloxamer 108 (M_w of the PEO is 1800 Da), do not provide an effective steric barrier against in vitro phagocytosis. In contrast, poloxamer 338 (M_w of the PEO is 5600 Da) is sufficient to suppress in vitro phagocytosis in the presence of serum and dramatically reduce the liver/spleen uptake of 60 nm polystyrene nanoparticles from 90 % to 45 % following i.v. injection. Coating with poloxamine 908 (M_w of the PEO is 5200 Da) had a more pronounced effect decreasing the amount cleared to less than 25 %.

The presence of a hydrated PEO layer alone does not necessarily prolong the circulatory half-life of all drug carriers. The particle size plays a major role. Particles > 200 nm in diameter with coated poloxamine 908 enhanced spleen uptake and decreased blood levels following i.v. administration to rats. Polystyrene particles with 60 and 150 nm diameters and coated with poloxamer 407 were redirected to the sinusoidal endothelial cells of rabbit bone marrow following i.v. administration. In contrast, poloxamer 407 particles with diameter 250 nm were mostly sequestered by the liver and spleen and only a small portion reached the bone marrow [4].

Polystyrene particles with chemically grafted PEO chains (M_w of the PEO is 2000 Da) were prepared with different surface densities of PEO. In vitro cell interaction studies demonstrated that particle uptake by non-parenchymal rat liver cells (primary Kupffer cells) decreased with increasing PEO surface density until an optimum density is reached. In vivo studies showed that only particles with very low PEO surface density results in considerable liver deposition. However, the results showed that the liver avoidance and blood circulation were not improved above those obtained with the poloxamine 908, even though the surface density of the grafted PEO particles was higher than that of poloxamine 908.

7.3.2 Biodegradable polymeric carriers

As mentioned above, studies using polystyrene nanoparticles as model drug carriers have demonstrated that optimizing the particle size and modifying the surface using a hydrophilic PEO layer (as a steric barrier) can result in an increase in circulation life-time and to some extent selective targeting may be achieved. For practical applications in drug targeting, polymeric nanoparticles constructed from biodegradable and biocompatible materials must be constructed [4]. These polymeric nanoparticles can act as drug carriers by incorporation of the active substance into the core of the nanoparticle. Natural materials such as albumin and gelatine are poorly character-

Tab. 7.3: Biodegradable polymers for drug carriers.

Poly(lactic acid)/Poly(lactic-co-glycolic acid) – PLA/PLGA
Poly(anhydrides)
Poly(caprolactone)
Poly(ortho esters)
Poly(β-maleic acid-co-benzyl malate)
Poly(alkylcyanoacrylate)

---

ized and in some cases can produce an adverse immune response. This led to the use of synthetic, chemically well-defined biodegradable polymers which do not cause any adverse immune response. A list of these biodegradable polymers is given in Tab. 7.3.

The most widely used biodegradable polymers are the aliphatic polyesters based on lactic and glycolic acid which have the following structures:

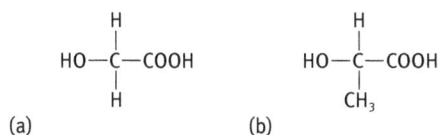

$$HO-\overset{\overset{\displaystyle H}{|}}{\underset{\underset{\displaystyle H}{|}}{C}}-COOH \qquad\qquad HO-\overset{\overset{\displaystyle H}{|}}{\underset{\underset{\displaystyle CH_3}{|}}{C}}-COOH$$

(a)   (b)

Poly(lactic acid) (PLA) and poly(lactic acid-co-glycolic acid) (PLGA) have been used in the production of a wide range of drug carrier nanoparticles. PLA and PLGA degrade by bulk hydrolysis of the ester linkages. The polymers degrade to lactic and glycolic acids which are eliminated in the body, primarily as carbon dioxide and urine.

The preparation of biodegradable nanoparticles with a diameter less than 200 nm (to avoid splenic uptake) remains a technical challenge. Particle formation by in situ emulsion polymerization (that is commonly used for the preparation of polystyrene latex) is not applicable to biodegradable polymers such as polyesters. Instead the biodegradable polymer is directly synthesized by chemical polymerization methods. The polymer is dissolved in a water immiscible solvent such as dichloroethane which is then emulsified into water using a convenient emulsifier such as poly(vinyl alcohol) (PVA). Nanoemulsions can be produced by sonication or homogenization and the organic solvent is then removed by evaporation. Using this procedure, nanoparticles of PLA and PLGA with a diameter ~ 250 nm were produced. Unfortunately, the emulsifier could not be completely removed from the particle surface and hence this procedure was abandoned.

To overcome the above problem nanoparticles were prepared using a surfactant-free method. In this case the polymer is dissolved in a water miscible solvent such as acetone. The acetone solution is carefully added to water while stirring [4]. The polymer precipitates out as nanoparticles which are stabilized against flocculation by electrostatic repulsion (resulting from the presence of COOH groups on the particle surface). Using this procedure, surfactant-free nanoparticles with diameter < 150 nm could be prepared. Later the procedure was modified by incorporation of poloxamers or poloxamines in the aqueous phase. These block copolymers are essential for surface modification of the nanoparticle as discussed below.

Following the encouraging in vivo results using polystyrene latex with surface modification using poloxamer and poloxamine, investigations were carried out using surfactant-free PLGA, ~ 140 nm diameter, which was surface modified using the following block copolymers: water soluble poly(lactic)-poly(ethylene) glycol (PLA-PEG); poloxamers, poloxamines. The results showed that both PLA-PEG 2:5 (M_w of PLA 2000 Da and M_w of PEO is 5000 Da) and poloxamine 908 form an adsorbed layer of 10 nm. The coated PLGA nanoparticles were effectively sterically stabilized towards electrolyte induced flocculation and in vivo studies demonstrated a prolonged systemic circulation and reduced liver/spleen accumulation when compared with the uncoated particles. The main drawback of the polymer adsorption approach is the possibility of desorption in vivo by the blood components. Chemical attachment of the PEG chain to the biodegradable carrier would certainly be beneficial [4].

The best approach is to use block copolymer assemblies as colloidal drug carriers [13–22]. Block copolymers of the B–A and B–A–B types are known to form micelles that can be used as drug carriers. These block copolymers consist of a hydrophobic B block that is insoluble in water and one or two A blocks which are very soluble in water and strongly hydrated by its molecules. In aqueous media the block copolymer will form a core of hydrophobic chains and a shell of hydrophilic chains. These self-assembled structures are referred to as micelles and they are schematically illustrated in Fig. 7.2. The core-shell structure is ideal for drug delivery where the water insoluble drug is incorporated in the core and the hydrophilic shell provides effective steric stabilization, thus minimizing adsorption of the blood plasma components and preventing adhesion to phagocytic cells [4].

The critical micelle concentration (cmc) of block copolymers is much lower than that obtained with surfactants. Typically the cmc is of the order of 10^{-5} g ml^{-1} or less. The aggregation number N (number of copolymer molecules forming a micelle) is typically several tens or even hundreds. This results in assemblies of the order of 10–30 nm which are ideal as drug carriers. The thermodynamic tendency for micellization to occur is significantly higher for block copolymers when compared with low molecular weight surfactants.

The inherent core-shell structure of aqueous block copolymer micelles enhances their potential as a colloidal drug carrier. As mentioned before, the hydrophobic core can be used to solubilize water insoluble substances such as hydrophobic drug molecules. The core acts as a reservoir for the drug which also can be protected against in vivo degradation. Drugs may be incorporated by covalent or noncosolvent binding such as hydrophobic interaction. The hydrophilic shell minimizes the adsorption of blood plasma components. It also prevents the adhesion of phagocytic cells and influences the parakinetics and biodistribution of micelles. The stabilizing chains (PEG) are chemically grafted to the core surface, thus eliminating the possibility of desorption or displacement by serum components. The size of the block copolymer micelles is advantageous for drug delivery [4].

The water solubility of PLA-PEG and PLGA-PEG copolymers depends on the molecular weight of the hydrophobic (PLGA-PEG) and hydrophilic (PEG) blocks. Water soluble PLA-PEG copolymers with relatively low molecular weight PLA blocks self-disperse in water to form block copolymer micelles. For example, water soluble PLA-PEG 2:5 (M_w of PLA is 2000 Da and M_w of PEO is 5000 Da) form spherical micelles ~ 25 nm in diameter. These micelles solubilize model and anticancer drugs by micellar incorporation. However, in vivo, the systemic lifetimes produced were relatively short and the clearance rate was significantly faster when the micelles are administered at low concentration. This suggests micellar dissociation at concentrations below the cmc.

By increasing the PLA/PLGA core molecular weight, the block copolymer becomes insoluble in water and hence it cannot self-disperse to form micelles. In this case the block copolymer is dissolved in a water immiscible solvent such as dichloromethane and the solution is emulsified into water using an emulsifier such as PVA. The solvent is removed by evaporation resulting in the formation of self-assembled nanoparticles with a core-shell structure. Using this procedure, nanoparticles of PLGA-PEG copolymers (M_w of PLGA block of 45 000 Da and M_w of PEO of 5000, 12 000 or 20 000 Da) can be obtained. High loading of drug (up to 45 % by nanoparticle weight) and entrapment efficiencies (more than 95 % of the initial drug used) can be achieved.

The PLGA-PEG nanoparticles shows prolonged blood circulation times and reduced liver deposition when compared with the uncoated PLGA nanoparticles. The adsorption of plasma proteins onto the surfaces of the PEG coated particles is substantially reduced, in comparison with the uncoated PLGA nanoparticles. The qualitative composition of the adsorbed plasma protein is also altered by the presence of the PEG layer. Substantially reduced adsorption of opsinon proteins such as fibrinogen, immunoglobulin G and some apoloproteins is achieved. These results clearly show the importance of the presence of the hydrophilic PEG chain on the surface of the nanoparticles which prevents opsonization [13–22].

The particle size and surface properties are strongly dependent on the emulsification conditions and the choice of emulsifier. By using a water miscible solvent such as acetone, the nanoparticles can be directly precipitated and the solvent is removed by evaporation. Using this procedure one can produce a series of PLA-PEG nanoparticles. The blood circulation of the nanoparticles (e.g. PLA-PEG 30:2) is considerably increased when compared with albumin coated PLA nanoparticles. The albumin molecules are rapidly displaced by the protein in the plasma leading to phagocytosis by Kupffer cells in the liver. The PLA-PEG nanoparticles show a low deposition of proteins on the particle surface.

Functionality is introduced in the core-forming A block in the form of polymers such as poly(L-lysine) or poly(aspartic acid). Both these polymers are biodegradable but not hydrophobic. Hydrophobicity is imparted by covalent or ionic attachment of the drug molecule. In this way potent anticancer drugs can be coupled to the aspartic acid residues of poly(aspartic acid)-poly(ethylene glycol) (P(asp)-PEG) copolymer. In aqueous media the block copolymer-drug conjugate form micelles but some of the drug may become physically entrapped in the core of the micelle. These P(asp)-PEG micelles ($\sim$ 40 nm diameter) remain in the vascular system for prolonged periods, with 68 % of the injected dose remaining 4 hours after intravenous administration. These systems offer a promising route for drug delivery.

7.3.3 The action mechanism of the stabilizing PEG chain

The action mechanism of the hydrophilic PEG chains can be explained in terms of steric interaction that is well known in the theory of steric stabilization (described in detail in Chapter 2). Before considering steric interaction one must know the polymer configuration at the particle/solution interface. The hydrophilic PEG chains can adopt a random coil (mushroom) or an extended (brush) configuration [22]. This depends on the graft density of the PEG chains as will be discussed below. The conformation of the PEG chains on the nanoparticle surface determines the magnitude of steric interaction. This configuration determines the interaction of the plasma proteins with the nanoparticles.

The hydrophilic PEG B chains (buoy blocks) can be regarded as chains terminally attached or grafted to the micellar core (A blocks). If the distance between the grafting points D is much greater than the radius of gyration R_G and the chains will assume a "mushroom" type conformation as illustrated in Fig. 7.7 (a). The extension of the mushroom from the surface will be of the order of $2R_G$ and the volume fraction of the polymer exhibits a maximum away from the surface as illustrated in Fig. 7.8 (a). If the graft density reaches a point whereby $D < R_G$ the chains stretch in solution forming a "brush". A constant segment density throughout the brush with all chains ending a distance Δ (the layer thickness) from the surface and the volume fraction of the polymer shows a step function as is illustrated in Fig. 7.8 (b).

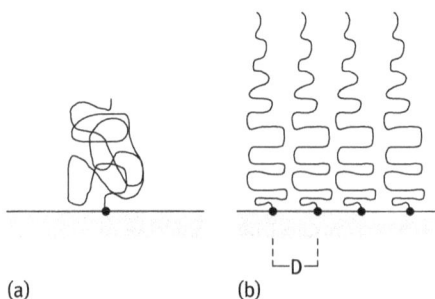

(a)   (b)

Fig. 7.7: Schematic representation of the conformation of terminally attached PEG chains.

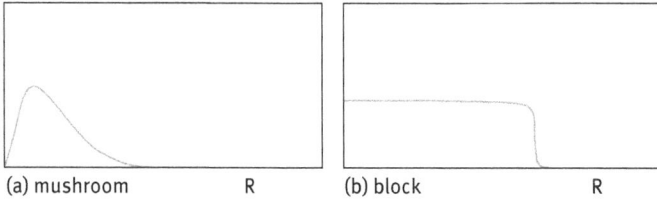

(a) mushroom   R   (b) block   R

Fig. 7.8: Volume fraction profile for (a) mushroom and (b) brush.

The thickness of the block "brush" Δ for a grafted chains of N bonds of length ℓ is given by

$$\frac{\Delta}{\ell} \approx N\left(\frac{\ell}{D}\right)^{2/3}. \tag{7.10}$$

This means that for terminally-attached chains at high graft density (brush) Δ depends linearly on N. This is in contrast to polymer chains in free solutions where $R_G \sim N^{3/5}$ or $R_G \sim N^{1/2}$. In the case of micellar structures, the distance between grafting points D is determined by the aggregation number. Unless high aggregation numbers and hence grafting densities can be achieved, a weaker dependence of Δ on chain length is expected.

For a brush on a flat surface, the attached chain is confined to a cylindrical volume of radius D/2 and height Δ. If the individual chains of the brush are attached to a spherical core (as is the case with nanoparticles), then the volume accessible to each chain increases and the polymer chains have an increased freedom to move laterally resulting in a smaller thickness Δ. This is schematically illustrated in Fig. 7.9 which shows the difference between particles with high surface curvature (Fig. 7.9 (a)) and those with a surface with low surface curvature (Fig. 7.9 (b)). The curvature effect is illustrated for PEO and poloxamer block copolymers using polystyrene latex particles with different sizes. An increase in the layer thickness with increasing particle radius was observed [4].

Most studies with model nonbiodegradable and biodegradable systems showed that the presence of a hydrated PEG steric barrier significantly increases the blood circulation of the nanoparticles following intravenous administration. The hydrophilic PEG layer minimizes the interactions with phagocytic cells and prevents the adsorption of opsinons. Hydrophilicity is necessary but not sufficient for achieving these two effects. This was demonstrated using dextran (which is considerably hydrophilic) coated liposomes which showed shorter circulation times when compared with their PEG counterparts. This clearly showed that chain flexibility is the second prerequisites for inhibiting phagocytic clearance [4].

PEG chains only have a weak tendency to interact hydrophobically with the surrounding proteins. As the protein approaches the stabilizing PEG chains, the configurational entropy of both molecules is reduced. The more mobile the stabilizing PEG chains, the greater the loss in entropy and the more effective the repulsion from the

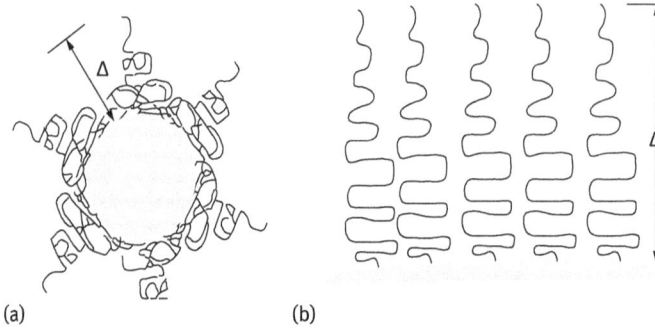

Fig. 7.9: Effect of surface curvature on the adsorbed layer thickness Δ. (a) High surface curvature, (b) low surface curvature.

surface. At sufficiently high surface density, the flexible PEG chains form an impermeable barrier, preventing the interaction of the opsinons with the particle surface. This repulsion is referred to as elastic interaction, G_{el}. With the approach of a second surface to a distance h smaller than the adsorbed layer thickness Δ, a reduction in configurational entropy of the chain occurs [23–26]. The mechanism of elastic interaction was described in detail in Chapter 2.

Four different types of interaction between a protein molecule and hydrophobic substrate can be considered [26]: (i) Hydrophobic attraction between the protein and substrate. (ii) Steric repulsion (osmotic and elastic effects). (iii) Van der Waals attraction between the protein and substrate. (iv) Van der Waals attraction between the protein and PEG chains. These interactions are schematically represented in Fig. 7.10. The interaction of plasma proteins with the PEG steric layer is dependent on the conformation of the chains which is determined by the surface curvature as discussed above.

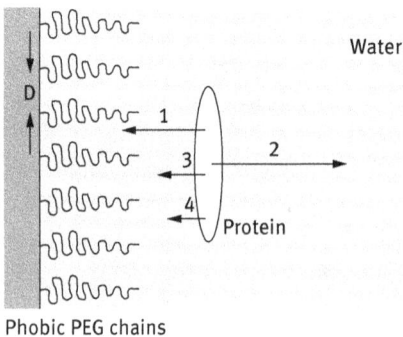

Fig. 7.10: Schematic representation of the various interactions between the PEG layer and a protein molecule.

There is ample evidence to suggest the high surface coverage of long brush-like PEG chains is necessary for preventing serum protein adsorption. However, the precise surface characteristics required for successful PES avoidance are not well established and more research is still required. In vitro phagocytosis of poloxamer coated polystyrene (PS) nanoparticles (60 and 250 nm in diameter) decreases with increasing PEG molecular weight and hence its thickness. However, increasing the PEG molecular weight above 2000 Da did not improve the ability of the coated nanoparticles to avoid phagocytosis. Similar results were obtained in vivo for both coated PS particles and liposomes of phosphatidylamine-PEG. However, results using PLGA-PEG nanoparticles showed an increase in performance when the PEG chain molecular weight was increased from 5000 to 20 000 Da.

7.3.4 Synthesis and characterization of PLA-PEG block copolymers

For systematic investigation of the application of PLA-PEG block copolymers for drug delivery, one needs to prepare a series of block copolymers with various compositions. One can fix the PEG molecular weight to a value that is sufficient to produce good steric stabilization thus prolonging the circulation of a range of colloidal drug carriers. The research carried out using nondegradable (polystyrene particles) and degradable polymers showed that a molecular weight of the PEG chain of 5000 Da is sufficient to provide effective steric stabilization and hence prolonged circulation. In this way a series of PLA-PEG block copolymers could be synthesized keeping the molecular weight of the PEG chain constant at 5000 Da while varying the molecular weight of the PLA chain from 2000 to 100 000 Da. The synthetic route for these block copolymers is briefly described below. This is followed by the methods that can be applied for their characterization.

The preferred method for the synthesis of high molecular weight poly(D,L-lactic acid) is the ring opening polymerization of six membered diester D,L-lactide in the presence of a suitable catalyst such as stannous octanoate $Sn(OCT)_2$. Copolymerization of D,L-lactide with methoxy polyethylene oxide (PEG) yields the A–B block copolymer of PLA-PEG. The polymerization mechanism of $Sn(OCT)_2$ is described as a complexation mechanism in which association of both PEG and D,L-lactide occurs via the sp^3d^2 orbitals [27–29]. A schematic representation of ring opening polymerization is shown in Fig. 7.11.

The reaction will proceed if the temperature is high enough to melt the D,L-lactide (m.p. 126 °C). Higher temperatures are required to produce higher molecular weight polymers by reducing the viscosity of the melt. However, higher temperatures can give rise to secondary reactions such as chain transfer or even depolymerization. Such problems can also occur if the reaction is left to proceed for too long. At 180 °C a maximum of both D,L-lactide conversion and molecular weight is achieved after 4 hours. The concentration of $Sn(Oct)_2$ used should be between 0.005 and 0.05 % w/w.

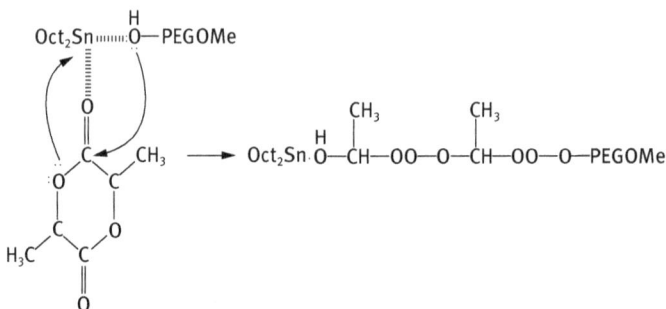

Fig. 7.11: Mechanism of Sn(Oct)$_2$ catalyzed ring opening polymerization of D,L-lactide in the presence of MeOPEG.

A series of AB block copolymers based on a fixed PEG block and a varying PLA segment were prepared using the above mentioned ring opening polymerization method of D,L-lactide in the presence of MeOPEG using stannous octoate as catalyst. This is schematically represented in Fig. 7.12. Pure MeOPEG and D,L-lactide are placed in a dried polymerization tube. 0.008 % w/w stannous octaoate (based on the weight of the reactants) in dried toluene is added. The tube is connected through a vacuum "take-off" adaptor to dry nitrogen and vacuum lines. The tube is placed in an oil bath at 70 °C, purged with nitrogen and the solvent evaporated under reduced pressure. After two hours of drying, the polymerization tube is sealed under vacuum and then transferred to an oven at 170 °C. The polymerization is carried out for 5 hours, removing the tube occasionally to stir the melt [4].

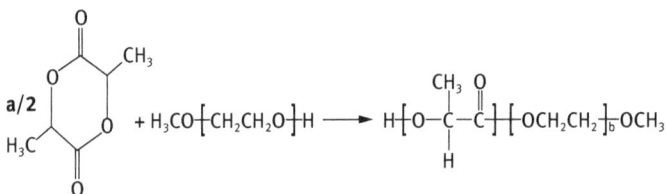

Fig. 7.12: Scheme of synthesis of PLA-PEG block copolymers.

By varying the feed ratio of D,L-lactide to MeOPEG, various block polymers could be synthesized. As mentioned before, the PEG block was fixed at 5000 Da molecular weight while varying the PLA segment molecular weight from 2000–100 000. For example, in a typical polymerization 6 g of D,L-lactide is reacted with 5 g of MeOPEG ($M_w = 5000$) to prepare a copolymer with a PLA to PEG weight ratio of 6 : 5. After completion of the polymerization reaction, the reaction vessel is left to cool slowly and the crude product is dissolved in dichloromethane. The solution of the crude copolymer is then precipitated into an excess of petroleum ether at 40–60 °C, cooled using ice or liquid nitrogen in the case of block copolymers with low molecular weight PLA. For copolymers with a high PLA content (> 50 000 Da), methanol is used as precipitating solvent to remove any low molecular weight PLA-PEG impurities. The resulting copolymer is then dried in a vacuum oven at 70 °C for 24 hours. The copolymers are stored in a vacuum desiccator at room temperature.

^{1}H NMR for PLA-PEG dissolved in deuterated chloroform (CDCl$_3$) is used to characterize the block copolymers [4]. A typical ^{1}H NMR for PLA-PEG dissolved in CDCl$_3$ is shown in Fig. 7.13.

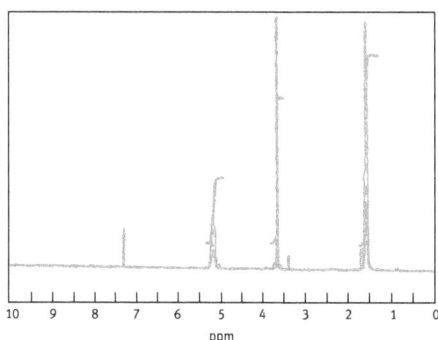

Fig. 7.13: ^{1}H NMR spectrum of PLA-PEG 15 : 5 copolymer dissolved in CDCl$_3$.

The PLA methyl (CH$_3$, δ 5.16 ppm, m) and methylenene (CH$_2$, δ 1.53 ppm, m) groups together with the PEG methylene (CH$_2$CH$_2$, δ 3.63 ppm, m) group are completely resolved. This indicates that the diblock copolymer is dissolved as a homogeneous solution of unimer molecules. The composition of the block copolymer can be obtained from the integration ratio of the peaks corresponding to each group. This is discussed below.

The integration ratios of the peaks corresponding to the CH (PLA) and CH$_2$CH$_2$ (PEG) groups are used to demine the average number of lactic acid segments in the PLA block (N_{PLA}),

$$N_{PLA} = \frac{Integral(CH)_{PLA}}{Integral(CH_2CH_2)_{PEO}/4} \cdot N_{PEG}, \tag{7.11}$$

where the average number of ethylene glycol segments in PEG 5000 block (N_{PEG}) is taken as 114.

The number average molecular weight of the PLA block ($\overline{M}_{nPLA}$) is calculated using the following relationship,

$$\overline{M}_{nPLA} = N_{PLA} \cdot 72 \tag{7.12}$$

The molecular weight results for the various PLA-PEG diblock copolymers are summarized in Tab. 7.4.

Agreement between the theoretical (feed) and NMR determination is obtained with the exception of the block copolymer containing a very high PLA : PEG ratio

Tab. 7.4: Molecular weight data for PLA-PEG diblock copolymers.

PLA-PEG Nomenclature	Copolymer weight ratio ($\overline{M}_{nPLA} : \overline{M}_{nPEG}$)	
	Theoretical (feed)	^{1}HNMR determined
2 : 5	2 000 : 5 000	2 000 : 5 000
3 : 5	3 000 : 5 000	2 600 : 5 000
4 : 5	4 500 : 5 000	3 800 : 5 000
6 : 5	6 000 : 5 000	6 000 : 5 000
9 : 5	10 000 : 5 000	8 700 : 5 000
15 : 5	15 000 : 5 000	14 600 : 5 000
30 : 5	30 000 : 5 000	30 100 : 5 000
45 : 5	45 000 : 5 000	44 500 : 5 000
75 : 5	75 000 : 5 000	75 600 : 5 000
110 : 5	100 000 : 5 000	110 900 : 5 000

(100 000 : 5000) and this could be due to removal of some low molecular impurities during the dissolution precipitation process [4].

7.3.5 Preparation and characterization of PLA-PEG nanoparticles

7.3.5.1 Preparation of PLA-PEG nanoparticles

The preparation of biodegradable nanoparticles for drug delivery requires control of their particle size, surface charge and the role of the hydrophilic layer that provides effective steric stabilization. As discussed before, PLA-PEG copolymers can produce nanoparticles with a core-shell structure. The core that is formed of PLA chains (the anchor chains A), which are insoluble in aqueous media, can solubilize the drug. The shell, consisting of the strongly hydrated PEG chains (the stabilizing buoy B chains), will provide effective steric stabilization. This PEG shell must have sufficient thickness to prevent recognition by the reticuloendothelial system, thus providing prolonged circulation in the blood and preventing phagocytosis. Before describing the procedure for preparation of PLA-PEG nanoparticles, it is essential to understand the theory of steric stabilization which was described in detail in Chapter 2.

With the exception of the water soluble PLA-PEG 2 : 5 copolymer which spontaneously forms micelles on dissolution in water (concentration $\sim 3.3\,\text{mg ml}^{-1}$), all other PLA-PEG nanoparticles could be prepared using the precipitation/solvent evaporation technique. A solution of the polymer consisting of 100 mg of PLA-PEG is dissolved in 10 ml of acetone. This solution is added dropwise to 30 ml deionized water while stirring (using a magnetic stirrer). The stirring is continued overnight to allow the acetone to evaporate. The particle morphology is determined using transmission electron microscopy (TEM). The particle size is determined by dynamic light scattering commonly referred to as photon correlation spectroscopy (PCS). The charge on the particles is determined by measuring the zeta potential [4]. A brief description of these methods is given below.

TEM micrographs of the PLA-PEG nanoparticles are obtained using negative staining with phosphotungestic acid solution (3 % w/v) adjusted to pH 4.7 using KOH. The TEM images clearly show that the nanoparticles precipitated from water miscible solvents such as acetone are spherical in shape. They also clearly show the increase in size of the nanoparticles as the PLA molecular weight is increased while keeping the PEG molecular weight constant at 5000. For example, the 3 : 5 PLA-PEG shows particles with diameters much less than 5 nm. The 30 : 5 PLA-PEG show particles smaller than 10 nm, whereas the 110 : 5 PLA-PEG show larger particles with some greater than 100 nm. As we will see later, the TEM pictures give diameters that are smaller than the hydrodynamic diameter obtained using PCS.

7.3.5.2 Determination of hydrodynamic diameter using dynamic light scattering

Dynamic light scattering (DLS) is a method that measures the time-dependent fluctuation of scattered intensity [30]. It is also referred to as quasi-elastic light scattering (QELS) or photon correlation spectroscopy (PCS). The latter is the most commonly used term for describing the process since most dynamic scattering techniques employ autocorrelation. PCS is a technique that utilizes Brownian motion to measure particle size. As a result of Brownian motion of dispersed particles the intensity of scattered light undergoes fluctuations that are related to the velocity of the particles. Since larger particles move less rapidly than smaller ones, the intensity fluctuation (intensity versus time) pattern depends on particle size as is illustrated in Fig. 7.14. The velocity of the scatterer is measured in order to obtain the diffusion coefficient.

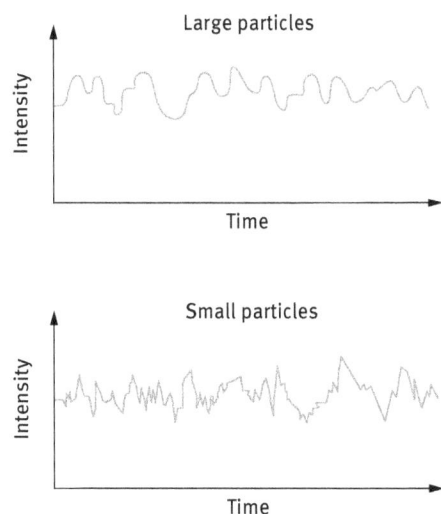

Fig. 7.14: Schematic representation of the intensity fluctuation for large and small particles.

The fluctuations in scattered light are detected by a photomultiplier and are recorded. The data containing information on particle motion are processed by a digital correlator. The latter compares the intensity of scattered light at time t, I(t), to the intensity at a very small time interval τ later, $I(t + \tau)$, and it constructs the second-order autocorrelation function $G_2(\tau)$ of the scattered intensity,

$$G_2(\tau) = \langle I(t)I(t + \tau) \rangle . \tag{7.13}$$

The experimentally measured intensity autocorrelation function $G_2(\tau)$ depends only on the time interval τ, and is independent of t, the time when the measurement started. PCS can be measured in a homodyne where only scattered light is directed to the detector. It can also be measured in heterodyne mode where a reference beam split from the incident beam is superimposed on scattered light. The diverted light beam functions as a reference for the scattered light from each particle.

In the homodyne mode, $G_2(\tau)$ can be related to the normalized field autocorrelation function $g_1(\tau)$ by

$$G_2(\tau) = A + B\,g_1^2(\tau), \tag{7.14}$$

where A is the background term designated as the baseline value and B is an instrument-dependent factor. The ratio B/A is regarded as a quality factor of the measurement or the signal-to-noise ratio and expressed sometimes as the % merit.

The field autocorrelation function $g_1(\tau)$ for a monodisperse suspension decays exponentially with τ,

$$g_1(\tau) = \exp(-\Gamma\tau), \tag{7.15}$$

where Γ is the decay constant (s^{-1}).

Substituting equation (7.15) into equation (7.14) yields the measured autocorrelation function,

$$G_2(\tau) = A + B\exp(-2\Gamma\tau). \tag{7.16}$$

The decay constant Γ is linearly related to the translational diffusion coefficient D_T of the particle,

$$\Gamma = D_T q^2 . \tag{7.17}$$

The modulus q of the scattering vector is given by

$$q = \frac{4\pi n}{\lambda_o} \sin\left(\frac{\theta}{2}\right), \tag{7.18}$$

where n is the refractive index of the dispersion medium, θ is the scattering angle and λ_o is the wavelength of the incident light in vacuum.

PCS determines the diffusion coefficient and the particle radius R is obtained using the Stokes–Einstein equation,

$$D = \frac{kT}{6\pi\eta R}, \tag{7.19}$$

where k is the Boltzmann constant, T is the absolute temperature, and η is the viscosity of the medium.

The Stokes–Einstein equation is limited to noninteracting, spherical and rigid spheres. The effect of particle interaction at relatively low particle concentration c can be taken into account by expanding the diffusion coefficient into a power series of concentration,

$$D = D_0(1 + k_D c), \tag{7.20}$$

where D_0 is the diffusion coefficient at infinite dilution and k_D is the virial coefficient that is related to particle interaction. D_0 can be obtained by measuring D at several particle number concentrations and extrapolating to zero concentration.

For polydisperse particles the first-order autocorrelation function is an intensity-weighted sum of an autocorrelation function of particles contributing to the scattering,

$$g_1(\tau) = \int_0^\infty C(\Gamma) \exp(-\Gamma\tau) \, d\Gamma. \tag{7.21}$$

$C(\Gamma)$ represents the distribution of decay rates.

For narrow particle size distribution the cumulant analysis is usually satisfactory. The cumulant method is based on the assumption that for monodisperse suspensions $g_1(\tau)$ is monoexponential. Hence, the log of $g_1(\tau)$ versus τ yields a straight line with a slope equal to Γ,

$$\ln g_1(\tau) = 0.5 \ln(B) - \Gamma\tau, \tag{7.22}$$

where B is the signal-to-noise ratio.

The cumulant method expands the Laplace transform about an average decay rate,

$$\langle \Gamma \rangle = \int_0^\infty \Gamma C(\Gamma) \, d\Gamma. \tag{7.23}$$

The exponential in equation (7.23) is expanded about an average and integrated term,

$$\ln g_1(\tau) = \langle \Gamma \rangle \tau + (\mu_2 \tau^2)/2! - (\mu_3 \tau^3)/3! + \dots \tag{7.24}$$

An average diffusion coefficient is calculated from $\langle \Gamma \rangle$ and the polydispersity (termed the polydispersity index) is indicated by the relative second moment, $\mu_2/\langle \Gamma \rangle^2$. A constrained regulation method (CONTIN) yields several numerical solutions to the particle size distribution and this is normally included in the software of the PCS machine.

7.3.5.3 Determination of electrophoretic mobility

As mentioned above, the charge on the particles is determined using electrophoretic mobility measurements that can be converted to zeta potential measurements, which give an estimate of particle charge. Electrophoretic mobility is obtained using electrophoretic light scattering (laser Doppler method) or laser velocimetry. As discussed above, laser light scattering can be used to measure the diffusion coefficients of small particles by measuring the Doppler broadening of the frequency of the scattered light due to the velocity of the scattering centres. If an electric field is placed at right angles to the incident light and in the plane defined by the incident and observation beam, the line broadening is unaffected but the centre frequency of the scattered light is shifted to an extent determined by electrophoretic mobility. The shift is very small compared to the incident frequency (~ 100 Hz for an incident frequency of $\sim 6 \times 10^{14}$ Hz) but with a laser source it can be detected by heterodyning (i.e. mixing) the scattered light with the incident beam and detecting the output of the difference frequency.

A homodyne method may be applied, in which case a modulator to generate an apparent Doppler shift at the modulated frequency used. To increase the sensitivity of the laser Doppler method, the electric fields are much higher than those used in conventional electrophoresis. The Joule heating is minimized by pulsing of the electric field in opposite directions. The Brownian motion of the particles also contributes to the Doppler shift and an approximate correction can be made by subtracting the peak width obtained in the absence of an electric field from the electrophoretic spectrum. An He-Ne laser is used as the light source and the output of the laser is split into two coherent beams which are cross-focused in the cell to illuminate the sample. The light scattered by the particle together with the reference beam is detected by a photomultiplier. The output is amplified and analyzed to transform the signals to a frequency distribution spectrum. At the intersection of the beam, interferences of known spacing are formed.

The magnitude of the Doppler shift Δv is used to calculate the electrophoretic mobility u using the following expression,

$$\Delta v = \left(\frac{2n}{\lambda_o}\right) \sin\left(\frac{\theta}{2}\right) uE, \tag{7.25}$$

where n is the refractive index of the medium, λ_o is the incident wavelength in vacuum, θ is the scattering angle and E is the field strength.

Several commercial instruments are available for measuring electrophoretic light scattering: (i) The Coulter DELSA 440SX (Coulter Corporation, USA) is a multiangle laser Doppler system employing heterodyning and autocorrelation signal processing. Measurements are made at four scattering angle (8, 17, 25, and 34°) and the temperature of the cell is controlled by a Peltier device. The instrument reports the electrophoretic mobility, zeta potential, conductivity and particle size distribution. (ii) Malvern (Malvern Instruments, UK) has two instruments: The ZetaSizer 3000 and ZetaSizer 5000. The ZetaSizer 3000 is a laser Doppler system using crossed beam optical configuration and homodyne detection with photon correlation signal processing. The zeta potential is measured using laser Doppler velocimetry and the particle size is measured using photon correlation spectroscopy (PCS). The ZetaSizer 5000 uses PCS to measure both movement of the particles in an electric field for zeta potential determination and random diffusion of particles at different measuring angles for size measurements on the same sample. The manufacturer claims that zeta potential for particles in the range 50 nm to 30 μm can be measured. In both instruments, a Peltier device is used for temperature control.

7.3.5.4 Calculation of zeta potential from particle mobility
Von Smoluchowski (classical) treatment

Von Smoluchowski [31] considered the movement of the liquid adjacent to a flat, charged surface under the influence of an electric field parallel to the surface (i.e. electro-osmotic flow of the liquid). If the surface is negatively charged, there will be a net excess of negative ions in the adjacent liquid and as they move under the influence of the applied field they will draw the liquid along with them. The surface of shear may be taken as a plane parallel to the surface and distant δ from it. The velocity of the liquid in the direction parallel to the wall, v_z, rises from a value of zero at the plane of shear to a maximum value, v_{eo}, at some distance from the wall, after which it remains constant. This is illustrated in Fig. 7.15. v_{eo} is called the electro-osmotic velocity of the liquid. The electrical potential ψ changes from its maximum negative value (ζ) at the shear plane to zero when v_z reaches v_{eo}.

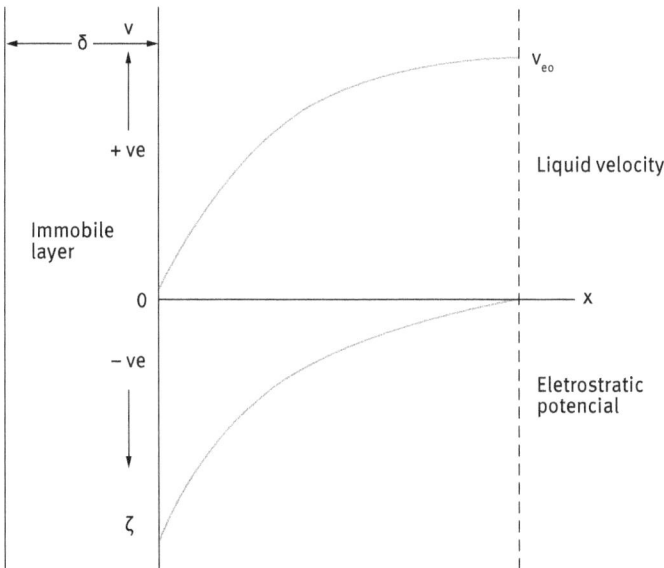

Fig. 7.15: Distribution of electrostatic potential near a charged surface and the resulting electrophoretic velocity.

Using Fig. 7.15, Smoluchowski [31] obtained the following equation for the electrophoretic motion of a large particle with a thin double layer $\kappa R \gg 1$, where $(1/\kappa)$ is the Debye length (thickness of the double layer) and R is the particle radius,

$$u = \frac{\varepsilon_r \varepsilon_0 \zeta}{\eta}.$$ (7.26)

u_E is the electrophoretic mobility (Smoluchowski equation). For water at 25 °C, ε_r is the relative permittivity of the medium, 78.6; ε_0 is the permittivity of free space, 8.85×10^{-12} F m^{-1} and η is the viscosity of the medium; 8.9×10^{-4} Pa s,

$$\zeta = 1.282 \times 10^6 u.$$ (7.27)

u is expressed in m^2 V^{-1} s^{-1} and ζ in V.

Equation (7.26) applies to the case where the particle radius R is much larger than the double layer thickness $(1/\kappa)$, i.e. $\kappa R \gg 1$. This is generally the case for particles that are greater than 0.5 mμ (when the 1:1 electrolyte concentration is lower than 10^{-3} mol dm^{-3}, i.e. $\kappa R > 10$).

The Huckel equation

Soon after the publication by Debye and Huckel of the theory of behaviour of strong electrolytes, Huckel [32] re-examined the electrophoresis problem and obtained a significantly different result from the Smoluchowski equation.

$$u = \frac{2}{3} \frac{\varepsilon_r \varepsilon_0 \zeta}{\eta}.$$ (7.28)

Equation (7.28) applies for small particles (< 100 nm) and thick double layers (low electrolyte concentration), i.e. for the case $\kappa R < 1$.

Equation (7.28) can be simply derived by balancing the electric force on the particle, QE_z, with the frictional force given by Stokes law ($6\pi\eta R v_E$), i.e.,

$$QE_z = 6\pi\eta R v_E,$$ (7.29)

$$u_E = \frac{v_E}{E_z} = \frac{Q}{6\pi\eta R}.$$ (7.30)

The electric charge Q is given by the following equation:

$$Q = 4\pi\varepsilon\varepsilon_0(1 + \kappa R)\zeta.$$ (7.31)

Combining equations (7.30) and (7.31) one obtains,

$$u_E = \frac{2\varepsilon\varepsilon_0\zeta}{3\eta}(1 + \kappa R).$$ (7.32)

When $\kappa R \ll 1$, i.e. small particles with relatively thick double layers, equation (7.32) reduces to equation (7.28).

A more rigorous derivation of equation (7.29) was given by Overbeek and Bijesterbosch [33]. The action of the electric field on the double layer, causing the liquid to
move in, is called electrophoretic retardation because it causes a reduction in the velocity of the migrating particle. Smoluchowski's treatment [32] assumes that this is the
dominant force and that the particle's motion is equal and opposite to the liquid's motion. Huckel [32], on the other hand, also made proper allowance for electrophoretic
retardation in his analysis. However, as mentioned above, equation (7.29) is only valid
for small values of κR when electrophoretic retardation is relatively unimportant and
the main retarding force is the frictional resistance of the medium. The electrophoretic
retardation at small κR remains important in the description of electrolyte conduc-
tion. In this case one must consider the movement of ions of both positive and negative sign and the calculation of the interaction effects for a large number of ions.
In electrophoresis one considers only the particle that is regarded as isolated in an
infinite medium. For large particles with thin double layers, essentially all of the electrophoretic retardation is communicated directly to the particle.

Henry's treatment
Henry [34] accounted for the discrepancy between Smoluchowski's and Huckel's treatment by considering the electric field in the neighbourhood of the particle. Huckel dis-
regarded the deformation of the electric field by the particle, whereas Smoluchowski
assumed the field to be uniform and everywhere parallel to the particle surface. As
shown in Fig. 7.16 these two assumptions are justified in the extreme cases of $\kappa R \ll 1$
and $\kappa R \gg 1$ respectively.

Henry [34] showed that when the external field is superimposed on the local field
around the particle the following expression for the mobility is used:

$$u = \frac{2}{3}\frac{\varepsilon_r\varepsilon_0\zeta}{\eta}f(\kappa R)\,. \tag{7.33}$$

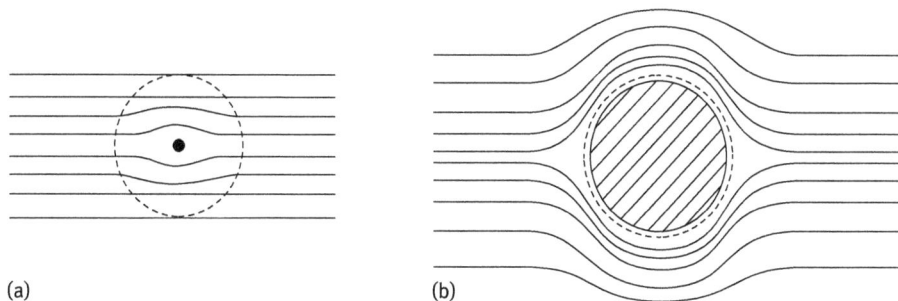

(a)   (b)

Fig. 7.16: Effect of a nonconducting particle on the applied field. (a) $\kappa R \ll 1$; (b) $\kappa R \gg 1$. The broken
line is at a distance $(1/\kappa)$ from the particle surface.

Tab. 7.5: Henry's correction factor f(κR).

κR	0	1	2	3	4	5	10	25	100	∞
f(κR)	1.0	1.027	1.066	1.101	1.133	1.160	1.239	1.370	1.460	1.500

The function f(κR) depends also on the particle shape. Values of f(κR) at increasing values of κR are given in Tab. 7.5.

Henry's calculations are based on the assumption that the external field can be superimposed on the field due to the particle and hence it can only be applied for low potentials ($\zeta < 25$ mV). It also does not take into account the distortion of the field induced by the movement of the particle (relaxation effect).

Wiersema, Loeb, and Overbeek [35] introduced two corrections for Henry's treatment, namely the relaxation and retardation (movement of the liquid with the double layer ions) effects. (i) Distortion of the field induced by the movement of the particles (distortion of the double layer symmetry and its reformation). This is referred to as the relaxation effect. (ii) Movement of the liquid with the double layer ions, which results in reduction of the mobility of the integrating particles. This is referred to as the retardation effect. By considering these two effects, Loeb, Wiersema, and Overbeek [35] derived exact expressions for the relationship between mobility and zeta potential for all κR values and any value of ζ-potential. Numerical tabulation of the relation between mobility and zeta potential has been given by Ottewill and Shaw [36]. Such tables are useful for conversion of u to ζ at all practical values of κR.

7.3.5.5 Investigation of the stability of PLA-PEG nanoparticles

The colloid stability of the PLA-PEG nanoparticles can be assessed by the addition of an electrolyte such as Na_2SO_4 which is known to reduce the solvency of the medium for the PEO chains [4]. This reduction in solvency results in an increase of the Flory–Huggins parameter χ (from its value in water of < 0.5) and when χ reaches 0.5 (the θ condition) flocculation occurs [38]. In this way one can determine the critical flocculation point, CFPT. The CFPT can be easily determined by following the turbidity of the nanoparticle dispersion as a function of Na_2SO_4 concentration. At the CFPT a sharp increase in turbidity is observed. The reversibility of flocculation can be assessed by diluting the flocculated dispersion with water and observing if the flocs can be redispersed by gentle shaking. The effect of the presence of serum protein on nanodispersion stability can also be studied by firstly coating the nanoparticles with the protein and then determining the CFPT using Na_2SO_4.

The hydrodynamic diameter of the nanoparticles and polydispersity index is determined using PCS. The PEG molar mass is fixed at 5000 Da while gradually increasing the PLA molar mass. The nanoparticle composition is expressed as a ratio of PLA : PEG; for example PLA-PEG 2 : 5 refers to a nanoparticle with PLA molar mass of 2000 and PEG molar mass of 5000. For comparison, the hydrodynamic diameter of PLA is also determined at a given molar mass. The zeta potential of the nanoparticles in 1 mM HEPES buffer (adjusted to pH 7.4 by addition of HCl) is also determined.

A summary of the results is given in Tab. 7.6.

Tab. 7.6: Hydrodynamic diameter, polydispersity index and zeta potential of PLA and PLA-PEG nanoparticles.

Polymer	Hydrodynamic diameter, D_{hyd} / nm mean ± SD	Polydispersity index mean ± SD	Zeta potential 1 mM HEPES / mV
PLA (M_w 35 kDa)	124.6 ± 2.5	0.11 ± 0.03	−49.6 ± 0.7
PLA-PEG 2 : 5	26.0 ± 1.6	0.19 ± 0.01	—
PLA-PEG 3 : 5	28.2 ± 0.6	0.14 ± 0.01	—
PLA-PEG 6 : 5	41.1 ± 1.8	0.10 ± 0.04	—
PLA-PEG 15 : 5	50.6 ± 2.0	0.06 ± 0.01	−6.5 ± 0.7
PLA-PEG 30 : 5	63.8 ± 1.8	0.08 ± 0.02	−6.4 ± 1.5
PLA-PEG 45 : 5	80.7 ± 4.8	0.10 ± 0.01	−6.1 ± 0.4
PLA-PEG 75 : 5	118.7 ± 4.9	0.10 ± 0.01	−14.2 ± 0.6
PLA-PEG 110 : 5	156.6 ± 5.0	0.13 ± 0.02	−28.0 ± 0.4

The results of Tab. 7.6 show that the nanoparticles have a relatively low polydispersity index. They suggest that the particle size distribution is monomodal as illustrated in Fig. 7.17. This is confirmed by the absence of subpopulations as shown in Fig. 7.17. Particles prepared from PLA (M_w = 35 000) are significantly larger than those prepared with a near equivalent PLA block (30 : 5). The effect of increasing the polymer concentration in the acetone solution on the particle size is shown in Fig. 7.18 for PLA-PEG copolymers and in Fig. 7.19 for PLA homopolymer. The results of Fig. 7.18 show that up to PLA-PEG 30 : 5, the hydrodynamic diameter is independent on polymer concentration. In contrast, the results of Fig. 7.19 for the PLA homopolymer show a significant increase in particle diameter with increasing polymer concentration. It appears that the PEG block moderates the association of the PLA-PEG copolymer.

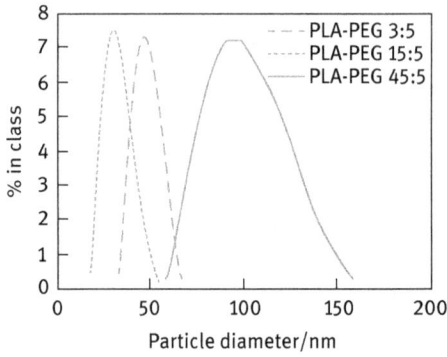

Fig. 7.17: Particle size distribution of PLA-PEG nanoparticles (intensity weighted CONTIN analysis).

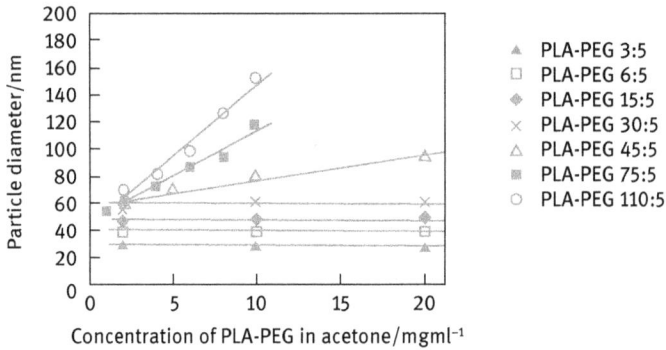

Fig. 7.18: Effect of PLA-PEG concentration on nanoparticle diameter.

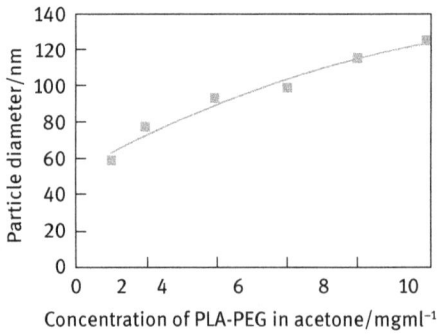

Fig. 7.19: Effect of PLA concentration on nanoparticle diameter.

For a core-shell system, where the core radius is much larger than the shell thickness, i.e. $N_A \gg N_B$ as is the case with the PLA-PEG blocks, scaling theory predicts that [22],

$$R_{hyd} \propto N_A^{2/3} . \tag{7.34}$$

Experimental results (Fig. 7.20) show a linear dependence of R_{hyd} with $N_{PLA}^{1/3}$ which disagrees with the theoretical prediction [4]. This shows that the hydrodynamic radius exhibit a weaker dependence on the length of the PLA block predicted by the large-core mean density model. The results of Fig. 7.18 also show that when the PLA : PEG ratio is increased above 45 : 5 the hydrodynamic diameter increases with increasing polymer concentration, indicating some aggregation of the copolymer.

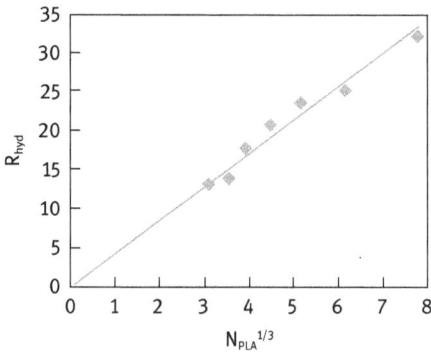

Fig. 7.20: Hydrodynamic radius versus $N_{PLA}^{1/3}$ PLA-PEG nanoparticles.

7.3.5.8 Zeta potential results

The results shown in Tab. 7.6 clearly indicate that the PLA nanoparticles have a high negative zeta potential of 49.6 mV. This is probably due to the presence of ionic carboxylic groups on the nanoparticle surface. At pH 7.4 (which is above the pK_a of COOH groups) and low ionic strength a high negative surface charge is produced. This negative charge provides electrostatic stabilization of the nanoparticles.

The zeta potential of the PLA-PEG nanoparticles is significantly reduced (to ~ -6 mV) up to a PLA : PEG ratio of 45 : 5. This reduction is due to the presence of the PEG layer that causes a significant shift in shear plane and hence reduction of ζ. However, when the PLA : PEG ratio is increased above 45 : 5, ζ starts to increase since the PEG layer thickness become smaller relative to the core of PLA.

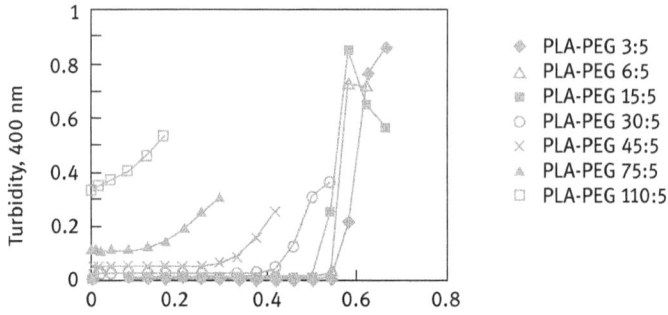

Fig. 7.21: CFPT of PLA-PEG nanoparticles determined by turbidity method.

7.3.5.9 Colloid stability results

Figure 7.21 shows the variation of turbidity with Na_2SO_4 concentration. Above a critical Na_2SO_4 concentration the turbidity shows a rapid increase with a further increase in electrolyte concentration. This critical concentration is defined as the critical flocculation point (CFPT). The CFPT decreases with increasing PLA blocks in the nanoparticle. The results are expected for sterically stabilized dispersion which show that at the CFPT, the medium becomes θ-solvent for the chains at which the Flory–Huggins interaction parameter χ become 0.5 and this is the onset of incipient flocculation. Above the CFPT, χ become higher than 0.5, i.e. the medium becomes worse than a θ-solvent for the chains.

The effect of the PLA core on the CFPT is illustrated in Fig. 7.22 which shows the variation of CFPT with particle diameter.

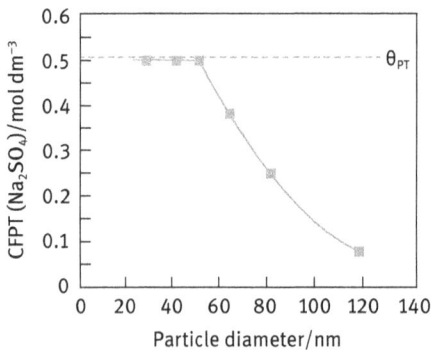

Fig. 7.22: CFPT of PLA-PEG with increasing amounts of PLA (3 : 5 to 75 : 5) as a function of particle diameter.

The results of Fig. 7.22 clearly show that nanoparticles with PLA blocks of $M_w < 15\,000$ give a CFPT close to the θ-point of the PEG chain. When the PLA block M_w is $> 15\,000$ the nanoparticles give a CFPT below the θ-point of the PEG chain. The CFPT decreases with increasing PLA block M_w. This discrepancy between the CFPT and θ-point of the PEG chain may be due to the decrease of surface coverage of the nanoparticles with the PEG chains when the PLA block M_w exceeds a certain value. This reduction in surface coverage leads to lateral movement of the PEG chains which results in a smaller layer thickness. This smaller PEG thickness can result in a deep attractive minimum, causing flocculation under conditions of better than the θ-point of the PEG chain [4].

Increasing the concentration of the PLA-PEG copolymer used during preparation by the solvent/precipitation method results in an increase in nanoparticle size. This is illustrated in Fig. 7.23 for PLA-PEG 45 : 5 which shows the variation of CFPT with particle diameter. The results show a linear increase of the CFPT with increasing particle diameter approaching the θ-point for the PEG chain when the diameter reaches 94 nm (obtained when the concentration of the PLA-PEG reaches 20 mg ml^{-1}). This increase in the stability of the nanoparticle dispersion with increasing particle diameter may be due to the increase surface coverage of the particles with PEG chains as the particle size is increased.

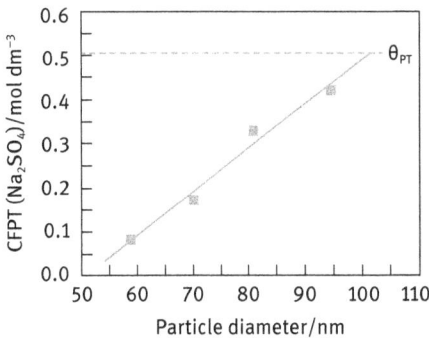

Fig. 7.23: CFPT of PLA-PEG 45 : 5 nanoparticles as a function of particle diameter.

7.3.5.10 Effect of serum protein coating on stability of the nanoparticles

The above results on uncoated nanoparticles showed the larger PLA-PEG copolymers (which are essential for drug solubilization) are flocculated at lower Na_2SO_4 than the θ-point. For example, the PLA-PEG 45 : 5 nanoparticles flocculate at ~ 0.25 mol dm^{-3} Na_2SO_4, whereas the 75 : 5 PLA-PEG flocculate at ~ 0.1 mol dm^{-3} Na_2SO_4. Thus these nanoparticles become unsuitable for drug delivery where the physiological ionic strength is in the region of ~ 0.15 mol dm^{-3}. In addition, at the physiological temperature (37 °C) flocculation will occur at lower Na_2SO_4.

Incubation of the particles in serum enhances colloid stability due to the adsorption of the serum protein on the particle surface [39]. The CFPT of the serum coated nanoparticles is shown in Fig. 7.24 which clearly shows the CFPT approaching the θ-point and its independence on the PLA molecular weight.

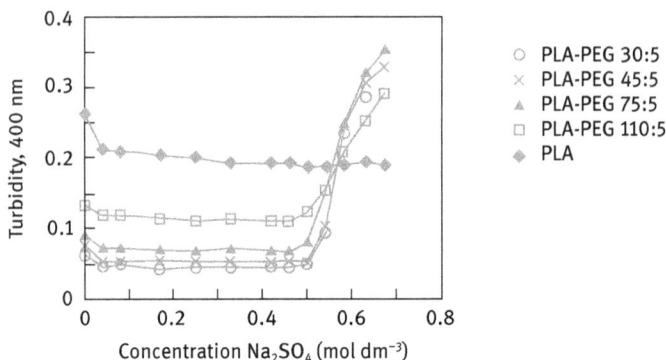

Fig. 7.24: CFPT for serum coated PLA-PEG and PLA nanoparticles.

7.3.5.11 Aggregation number of PLA-PEG nanoparticles

The aggregation number of the PLA-PEG copolymer in aqueous solution determines the properties of the micellar-like structures of the nanoparticles. The aggregation number is the number of copolymer units per micelle and this emphasizes the self-assembly of PLA-PEG of the nanoparticle [40–43]. The process is irreversible with no dynamic equilibrium between the self-assembled structure and unimers. Thus, the process is different from that of surfactant micelles in which a dynamic equilibrium exists between the micelle and the surfactant molecule. The methods that can be applied to determine the aggregation number depend on scattering techniques using radiations of light, X-ray and neutrons. A brief description of static (time-average) light scattering, that is commonly used in practice, is given below.

The structural details of a particle that can be observed depends on the magnitude of the scattering vector Q that is given by

$$Q = \frac{4\pi n_o}{\lambda} \sin\left(\frac{\theta}{2}\right), \tag{7.35}$$

where n_o is the refractive index of the medium, λ is the wavelength of the radiation and θ is the scattering angle.

For light with λ in the range 400–600 nm, Q has a range of 5×10^{-1}–5×10^{-2} nm^{-1} which is suitable for determining the structure of the micelle as a whole, such as its radius of gyration R_G.

Static light scattering can also be applied to obtain the average molar mass of the micelle and hence its aggregation number. For small nanoparticles with $R_G < \lambda/20$ one can apply the simple Rayleigh theory. For larger particles where $R_G > \lambda/20$ one has to use the Rayleigh–Gans–Debye theory.

The intensity of scattered light as a function of the angle between the incident and scattered beams depends very strongly on the weight-average molecular mass M_w of the dispersed particles. Rayleigh showed that small particles ($R_G < \lambda/20$) can be regarded as point masses to the wavelength of the vertically polarized incident light of wavelength λ. In this case the intensity of scattered light is independent of the scattering angle θ. At infinite dilution, the ratio of the scattering intensity of the incident (vertically polarized) beam I_i to that of the scattered light intensity I_p (where $I_p = I_{dispersion} - I_{medium}$) is given by

$$\frac{I_p(\theta)}{I_i(\theta)} = \frac{4\pi^2 n_0^2}{\lambda^4 N_{av} r^2} \left(\frac{dn}{dc}\right)_T^2 c\, M_w ,\qquad (7.36)$$

where N_{av} is Avogadro's number, r is the distance between the sample and the detector, c is the dispersion concentration and (dn/dc) is the specific refractive index increment that is characteristic of the particle solvent system and can be determined using a differential refractometer.

Equation (7.36) can be simplified by using the Rayleigh ratio $R_p(\theta)$,

$$R_p(\theta) = \frac{I_p(\theta) r^2}{I_i(\theta)} .\qquad (7.37)$$

$R_p(\theta)$ can be determined by measuring the scattered light from the particle dispersion and from a scattering standard (such as toluene) R_T (1.9×10^{-7} cm^{-1}),

$$R_p(\theta) = \frac{I_p(\theta)}{I_t(\theta)} R_T \left(\frac{n_p}{n_T}\right)^2 .\qquad (7.38)$$

Combining equations (7.36) and (7.38),

$$\frac{K_1 c}{R_p(\theta)} = \frac{1}{M_w} + 2A_2 c ,\qquad (7.39)$$

where A_2 is the second virial coefficient and K_1 is the optical constant,

$$K_1 = \frac{4\pi^2}{\lambda^4 N_{av}} \frac{n_T^2}{R_T} \left(\frac{dn}{dc}\right)_T^2 .\qquad (7.40)$$

For larger particles ($R_G > \lambda/20$), there will be more than one point in each particle that will scatter light. For any angle $\theta > 0°$, the beams of light scattered by two different points on the same particle reach a reference point perpendicular to the scattering direction out of phase, the phase shift increasing with the angle of observation. The resultant scattered light depends on the angle of observation, but at zero angle the interparticle interferences are never operative as illustrated in Fig. 7.25.

In this case equation (7.38) can be modified to give the Rayleigh–Gans–Debye equation,

$$\frac{Kc}{R_p(\theta)} = \frac{1}{M_w P(\theta)} + 2A_2 c ,\qquad (7.41)$$

where the particle scattering function $P(\theta)$ (the particle form factor) is given by,

$$P^{-1}(\theta) = 1 + \frac{16\pi^2 R_G^2}{3\lambda^2} \sin^2(\theta/2) .\qquad (7.42)$$

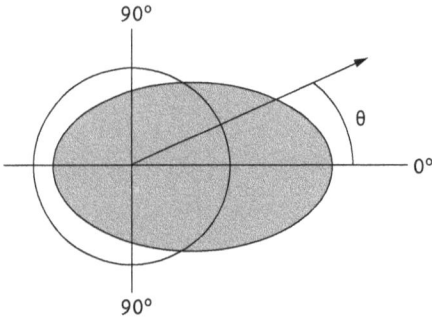

Fig. 7.25: Angular dependence of the intensity of light scattered by a nanoparticle. Unshaded area: "Rayleigh scattering" by small particles, shaded area: "Rayleigh–Gans–Debye scattering" by large particles.

Hence extrapolation of $K_{1c}/R_p(\theta)$ to both infinite dilution and zero angle of observation (Zimm plots) yields $1/M_w$.

Figure 7.26 shows the angular dependence of 2:5 PLA-PEG nanoparticles with $R_G < \lambda/20$ (Rayleigh scatterers). This figure clearly shows the near independence of scattering intensity on the scattering angle θ, especially at low concentrations. Figure 7.27 shows the angular dependence of scattering intensity for the larger 30:5 PLA-PEG nanoparticles, where $R_G > \lambda/20$. The results of Fig. 7.27 clearly show the marked dependence of scattering intensity on θ (Rayleigh–Gans–Debye scatterers). Zimm plots are shown in Fig. 7.28 for the 30:5 PLA-PEG nanoparticles. The slope of the zero angle line can be used to calculate the hydrodynamic radius of the particles. This gave a hydrodynamic radius of the PLA:PEG 30:5 nanoparticles of 28.5 nm which is in good agreement with the result obtained by PCS (31.9 nm).

The weight-average molar mass of the PLA-PEG micellar-like nanoparticles, $M_{w,mic}$, is obtained by extrapolation of $K_1c/R_p(\theta)$ to zero angle and zero concentration using the Zimm plots. The results of $M_{w,mic}$ for various PLA-PEG micellar-like nanoparticles are given in Tab. 7.7.

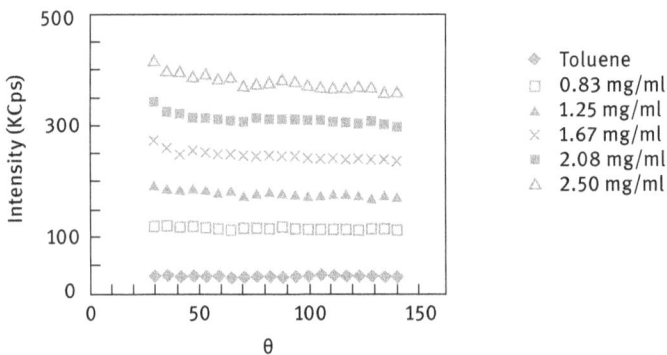

Fig. 7.26: Scattering intensity versus scattering angle θ for 2:5 PLA-PEG micellar-like nanoparticles.

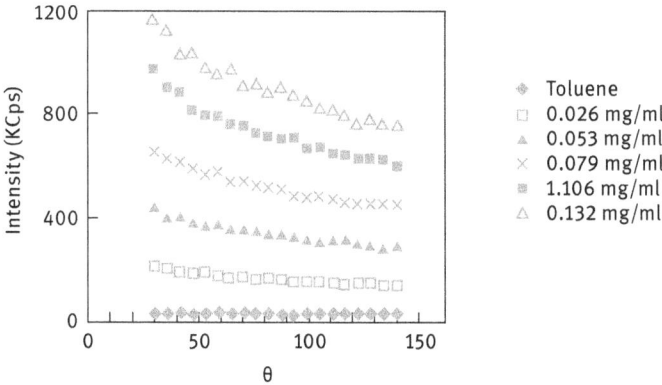

Fig. 7.27: Scattering intensity versus scattering angle θ for 30 : 5 PLA-PEG micellar-like nano-particles.

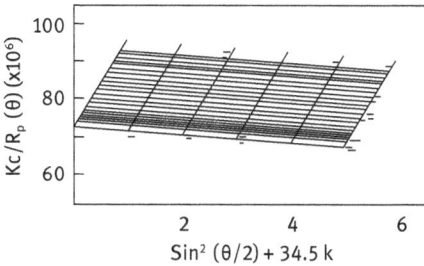

Fig. 7.28: Zimm plots for aqueous dispersions of PLA-PEG 30 : 5 micellar-like nanoparticles.

The micellar aggregation number N_{agg} is then calculated,

$$N_{agg} = \frac{M_{w,mic}}{M_{w,poly}}, \tag{7.43}$$

where $M_{w,poly}$ is taken to be equivalent to the number average molar mass that can be obtained using NMR. The results of N_{agg} for various PLA-PEG micellar-like nanoparticles are given in Tab. 7.7. The same table also gives the surface area available per PEG block at the outer limit of the micelle.

The micellar aggregation number scales with increasing number of monomeric units in the core forming block N_A,

$$N_{agg} \approx N_A^\beta. \tag{7.44}$$

The value of the exponent β depends on the composition of the micelle-forming copolymer. In the large core limit ("crew-cut" micelles, $N_A \gg N_B$) mean density models may be used. The volume fraction of B segments in the corona, ϕ_B, is assumed to be independent of the distance from the core and N_{agg} is predicted to be proportional to N_A (β = 1). If $N_B \gg N_A$ (star model which assumes a concentration profile for ϕ_B) $N_{agg} \sim N_A^{4/5}$.

Tab. 7.7: Molar mass and aggregation numbers of PLA-PEG micellar-like nanoparticles.

PLA-PEG[a]	N_{PLA}	R_{hyd}[b] (nm)	$\overline{M}_{w,mic}$ (Da)	N_{agg}	S_t/N_{agg} (nm^2)
2:5	28	13.0	2.01×10^5	29	73
3:5	42	14.1	3.13×10^5	39	64
4:5	56	17.5	7.02×10^5	78	49
6:5	83	20.6	1.99×10^6	180	30
9:5	125	23.4	2.85×10^6	203	34
15:5	208	25.3	5.57×10^6	278	29
30:5	417	31.9	1.31×10^7	375	34
45:5 (2 mg ml^{-1})	625	30.3	1.19×10^7	238	49
45:5 (10 mg ml^{-1})	625	40.4	3.37×10^7	674	30
45:5 (20 mg ml^{-1})	625	48.0	5.67×10^7	1134	26

a All prepared using 10 mg ml^{-1} solutions of PLA-PEG in acetone unless otherwise stated.
b Results from PCS measurements.

Figure 7.29 shows the variation of aggregation number N_{agg} and hydrodynamic radius R_{hyd} with the number of monomeric units of PLA, N_{PLA}. Figure 7.30 shows log-log plots of N_{agg} versus N_{PLA}. For comparison, results for N_{agg} of PLA : PEG, with a lower molecular weight PEG, namely 1800, are shown in the same figure. For copolymers with relatively low PLA to PEG weight ratio (2:5–6:5) there is a sharp increase in the micellar aggregation number as the molar mass of PLA is increased. This trend can be rationalized in terms of the thermodynamics of micelle formation as discussed below.

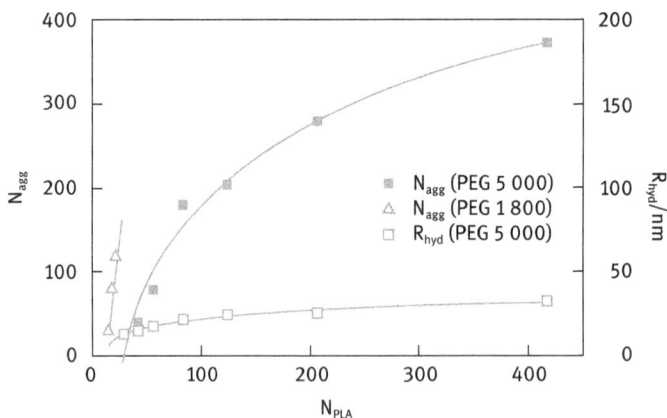

Fig. 7.29: Variation of aggregation number and hydrodynamic radius of PLA-PEG micellar-like nanoparticles with number of monomeric units of PLA.

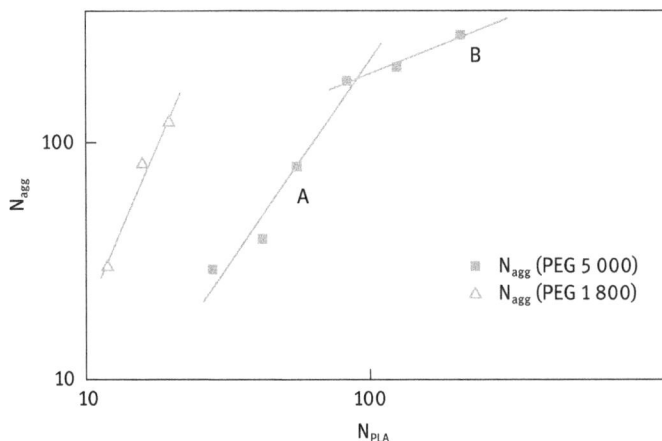

Fig. 7.30: log-log plots of N_{agg} versus N_{PLA}.

In a selective solvent, where the block of the copolymer B is in a good solvent (PEO, $\chi_{BS} < 0.5$) whereas the other segments of A are in a worse than θ-solvent (PLA, $\chi_{AS} > 0.5$) there will be strong attraction between the A segments. These attractive hydrophobic interactions (enthalpic) must overcome the repulsive (entropic) forces between B chains in the corona. PLA-PEG copolymers with very low PLA to PEG ratios (e.g. 400:1800) do not form micelles in aqueous media. The PLA-PEG 2:5 copolymer has the shortest PLA block and although the block copolymer is water soluble with least number of unfavourable interactions between the lactic acid units and the aqueous solvent, yet is still spontaneously forms micellar structures. Since the interactions between the low molecular weight PLA chains within the core of the micelle are weak, the 2:5 PLA:PEG copolymers form a loosely packed micellar assemblies with a lot of free space (solvent) within the corona region. This can be demonstrated by calculating the surface area per copolymer unit (S_t/N_{agg}) at the outer surface of the micelle.

A schematic picture of the PLA:PEG micelle is shown in Fig. 7.31. The external surface area of the 2:5 micelle ($4\pi R_{hyd}^2$) is 2124 nm² and its aggregation number is 29 (see Tab. 7.7). The area per PEG block in the micelle is (2124/29) = 73 nm². This area may be compared with the cross-sectional area of a PEG chain of molar mass of 5 kDa in a good solvent. The radius of gyration R_g of PEG is related to its molar mass by [4],

$$R_g = 0.0215 M_w^{0.583} . \tag{7.45}$$

This gives an R_g of 3.1 nm which in free solution occupies a sphere with maximum cross-sectional area (πR_g^2) of 30 nm². This clearly shows the high area per block copolymer (73 nm²) in the 2:5 PLA:PEG micelle. The loosely packed nature of the micellar-like nanoparticles can be confirmed by ¹H NMR studies on nanoparticles in D_2O.

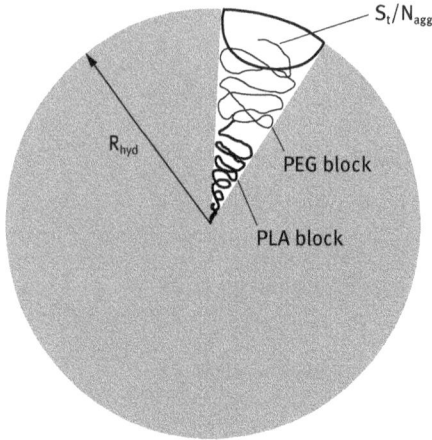

Fig. 7.31: Schematic representation of PLA-PEG micellar-like nanoparticles.

Increasing the length of the hydrophobic PLA block from 2 to 6 kDa increases the number of unfavourable interactions between the lactic acid units of the PLA chains and the aqueous media. This makes the copolymer water insoluble and it is forced to assemble into nanoparticles by precipitation into water from a water miscible solvent (e.g. acetone). The number of attractive hydrophobic interactions between the lactic acid units of the associating PLA chains increases with increasing the length of the chain. This results in an increased packing density of the PLA-PEG subunits and a sharp increase in the micellar aggregation number as illustrated in Fig. 7.29. The surface area per copolymer unit at the outer surface of the micelle (S_t/N_{agg}) falls rapidly as the molar mass of the PLA block is increased from 2 to 6 kDa (Tab. 7.7). (S_t/N_{agg}) appears to tend towards a value that is consistent with the maximum cross-sectional area of a PEG 5 kDa in solution ($\sim$ 30 nm^2). The increasing aggregation number and decreasing surface curvature result in a decrease in the conical volume available to each of the coronal PEG chain [4].

The log-log plots of N_{agg}-NPLA shown in Fig. 7.30 give the following scaling relationship [4],

$$N_{agg} \approx N_{PLA}^{1.74} \ . \tag{7.46}$$

The scaling exponent (1.74) is larger than that predicted theoretically for both "screwcut" and "star" models. This points to a third class of block copolymer micelles, characterized by blocks of different chemical composition or polarity (strongly aggregated). This class predicts an N_A^2 dependence of the micellar aggregation number.

If the micelle core is strongly segregated, then the area and volume of the core are given by

$$4\pi R_c^2 = N_{agg} D^2 \ , \tag{7.47}$$

$$\frac{4}{3}\pi R_c^3 = \frac{N_{agg} N_A}{\rho_{bulk}} \ , \tag{7.48}$$

$$N_{agg} = \frac{36\pi N_A^2}{D^6 \rho_{bulk}}, \tag{7.49}$$

where R_c is the core radius, D^2 is the interfacial area per chain and ρ_{bulk} is the mass density of the core.

The particle size of nanoparticles produced from the PLA-PEG 45 : 5 copolymer depends on the concentration of the polymer dissolved in acetone [4]. Figure 7.32 shows the variation of the aggregation number and hydrodynamic radius with copolymer concentration. Increasing the concentration of the acetonic copolymer solution increases the local concentration of PLA-PEG units available to aggregate at any particle formation site, following precipitation into water. A significant increase in the aggregation number from 238 to 674 occurs when the copolymer concentration is increased from 2 to 10 mg ml^{-1}. At 2 mg ml^{-1} small nanoparticles are produced with a high surface area (S_t/N_{agg}) per PEG block at the outer boundary of the micelle of 49 nm^2 that is similar to that of the smaller PLA-PEG 2 : 5–4 : 5 copolymers. These smaller PLA-PEG 45 : 5 nanoparticles have a loosely packed structure with a lot of solvent in the corona region. The particle radius of these 45 : 5 nanoparticles (30.3 nm) is considerably greater than that produced by the smaller 2 : 5 PLA-PEG block copolymer (13 nm). The largest PLA-PEG 45 : 5 nanoparticles have a low (S_t/N_{agg}) (26 nm^2) comparable to that of the PLA-PEG 6 : 5 micellar-like nanoparticles (30 nm^2). This implies that the largest PLA-PEG 45 : 5 nanoparticles have a high PEG surface coverage which gives them high colloid stability.

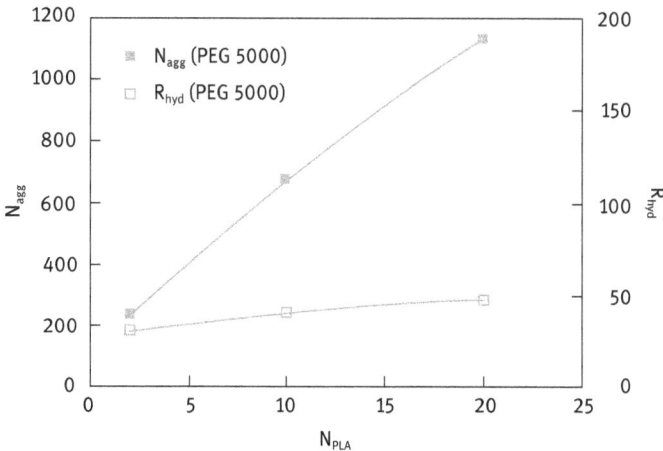

Fig. 7.32: Variation of aggregation number with hydrodynamic radius of PLA-PEG 45 : 5 nanoparticles with block copolymer concentration in the organic phase.

7.3.6 Rheology of PLA-PEG dispersions

As mentioned before, the length and surface density of PEG chains adsorbed grafted to model polymer colloid affects the uptake of the particles by the reticuloendothelial system (RES). To evaluate the in vivo potential of PEG modified nanoparticles it is necessary to determine the thickness of the steric layer Δ. For adsorbed block copolymers Δ can be determined by measuring the hydrodynamic radius of the particle with and without adsorbed layer R_Δ and R ($\Delta = R_\Delta - R$). For PLA-PEG nanoparticles one can only obtain the hydrodynamic radius of the whole particle, for example using PCS which gives the core radius plus the PEG layer thickness. The PEG layer thickness can be obtained using rheological measurements, in particular dynamic (oscillatory) measurements as discussed below.

7.3.6.1 Principles of dynamic (oscillatory) measurements

A strain is applied in a sinusoidal manner, with an amplitude γ_0 and a frequency ν (cycles/s or Hz) or ω (rad s^{-1}). This is usually carried out by moving one of the platens, say the cup (in a concentric cylinder geometry) or the plate (in a cone-and-plate geometry) back and forth in a sinusoidal manner [44–47]. The stress on the other platen, the bob or the cone is simultaneously measured; these platens are usually connected to interchangeable torque bars where the stress can be directly measured. The stress amplitude σ_0 is measured simultaneously. In a viscoelastic system (such as the case with PLA-PEG nanoparticle dispersion), the stress oscillates with the same frequency, but out of phase from the strain.

From the time shift of stress and strain amplitudes (Δt) one can obtain the phase angle shift δ,

$$\delta = \Delta t \omega . \tag{7.50}$$

A schematic representation of the variation of strain and stress with ωt is shown in Fig. 7.33.

From the amplitudes of stress and strain and phase angle shift one can obtain the various viscoelastic parameters,

$$\text{Complex modulus } |G^*| = \sigma_0/\gamma_0 \tag{7.51}$$

$$\text{Storage modulus } \quad G' = |G^*| \cos \delta \tag{7.52}$$

$$\text{Loss modulus } \quad G'' = |G^*| \sin \delta \tag{7.53}$$

$$\tan \delta = G''/G' \tag{7.54}$$

$$\text{Dynamic viscosity } \quad \eta' = G''/\omega \tag{7.55}$$

G' is a measure of the elastic component of the complex modulus – it is a measure of the energy stored in a cycle of oscillation. G'' is a measure of the viscous component of the complex modulus; it is a measure of the energy dissipated as viscous flow in a cycle of oscillation. $\tan \delta$ is a measure of the relative magnitudes of the viscous and elastic components [47]. Clearly the smaller the value of $\tan \delta$, the more elastic the system is and vice versa. η', the dynamic viscosity, shows a decrease with increasing frequency ω. η' reaches a limiting value as $\omega \rightarrow 0$. The value of η' in this limit is identical to the residual (or zero shear) viscosity $\eta(o)$.

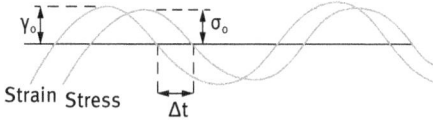

Δt = time shift for sine waves of stress and strain
$\Delta t\, \omega = \delta$ phase angle shift
ω = frequency in radian s^{-1}
$\omega = 2\pi \upsilon$
Perfectly elasic solid $\delta = 0$
Perfectly viscous liquid $\delta = 90°$
Viscoelastic system $0 < \delta < 90°$

Fig. 7.33: Strain and stress amplitudes for a viscoelastic system.

In oscillatory measurements one carries out two sets of experiments: (i) Strain sweep measurements; in this case, the oscillation is fixed (say at 1 Hz) and the viscoelastic parameters are measured as a function of strain amplitude. This allows one to obtain the linear viscoelastic region. In this region all moduli are independent of the applied strain amplitude and become only a function of time or frequency. This is illustrated in Fig. 7.34, which shows a schematic representation of the variation of G^*, G', and G'' with strain amplitude (at a fixed frequency). It can be seen from Fig. 7.34, that G^*, G', and G'' remain virtually constant up to a critical strain value, γ_{cr}; this region is the linear viscoelastic region. Above γ_{cr}, G^* and G' start to fall, whereas G'' starts to increase; this is the nonlinear region. The value of γ_{cr} may be identified with the minimum strain above which the "structure" of the dispersion starts to break down [47].

(ii) Oscillatory sweep: In this case, the strain amplitude is kept constant in the linear viscoelastic region (one usually takes a point far from γ_{cr} but not too low, i.e. in the midpoint of the linear viscoelastic region) and measurements are carried out as a function of frequency. This is shown in Fig. 7.35 for a viscoelastic liquid system. Both G^* and G' increase with increasing frequency and ultimately above a certain frequency they reach a limiting value and show little dependence on frequency. G'' is higher than G' in the low frequency regime; it also increases with increasing frequency and at a certain characteristic frequency ω^* (that depends on the system) it becomes equal to G' (usually referred to as the crossover point), after which it reaches a maximum and then shows a reduction with a further increase in frequency [47].

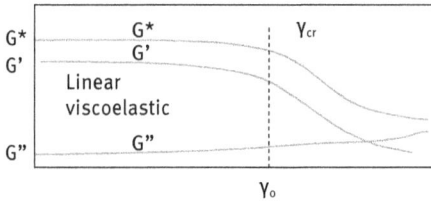

Fig. 7.34: Strain sweep experiments.

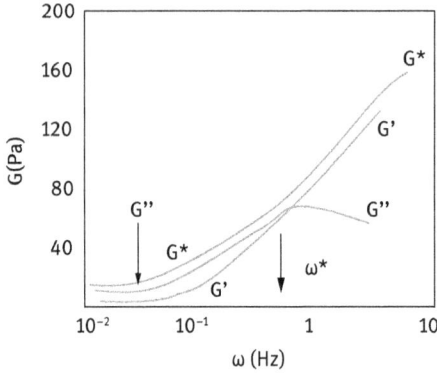

Fig. 7.35: Schematic representation of oscillatory measurements for a viscoelastic liquid.

In the low frequency regime, i.e. below ω^*, $G'' > G'$; this regime corresponds to longer times (remember that the time is reciprocal of frequency) and under these conditions the response is more viscous than elastic. In the high frequency regime, i.e. above ω^*, $G' > G''$; this regime corresponds to short times and under these conditions the response is more elastic than viscous. At sufficiently high frequency, G'' approaches zero and G' becomes nearly equal to G^*; this corresponds to very short time scales whereby the system behaves as a near elastic solid; very little energy dissipation occurs at such high frequency.

The characteristic frequency can be used to calculate the relaxation time of the system t^*,

$$t^* = 1/\omega^* . \tag{7.56}$$

The relaxation time may be used as a guide for the state of the dispersion.

7.3.6.2 Determination of the PEG layer thickness using viscoelastic measurements

The complex, storage and loss modulus (G^*, G', and G'') are measured as a function of the PLA core volume fraction, ϕ_{core}. At relatively low volume fractions the dispersion may be considered dilute and the interaction between the PEG layers of the PLA-PEG nanoparticles are weak (the core-core separation distance h is higher than twice the PEG layer thickness (h > 2Δ)). In this case the viscous component is higher than the elastic component ($G'' > G'$). At high volume fractions, the core-core separation distance h becomes smaller than 2Δ and in this case $G' > G''$. The critical volume fraction $(\phi_{core})^{crit}$ at which the PEG chains just begin to overlap, i.e. at which $G' = G''$, can be used to obtain the layer thickness Δ.

Assuming that the particles are arranged in a random packing fashion, then the overlap should occur at an effective volume fraction ϕ_{eff} of 0.64. ϕ_{eff} can be related to the critical core volume fraction $(\phi_{core})^{crit}$ by the following equation [48, 49]:

$$\phi_{eff} = 0.64 = \phi_{core}^{crit} \left(1 + \frac{\Delta}{R_{hyd} - \Delta} \right)^3 . \tag{7.57}$$

Equation (7.57) can be applied to determine Δ.

As an illustration, Fig. 7.36 shows the strain sweep results at a frequency of 1 Hz of PLA-PEG 6 : 5 dispersions at a PLA core volume fraction ϕ_{core} of 0.094. The plots show a linear viscoelastic region below a critical strain of 0.01. At such a frequency $G' > G''$ indicating steric interaction between the PEG layers. Figure 7.37 shows the frequency sweep results which are typical of a viscoelastic liquid.

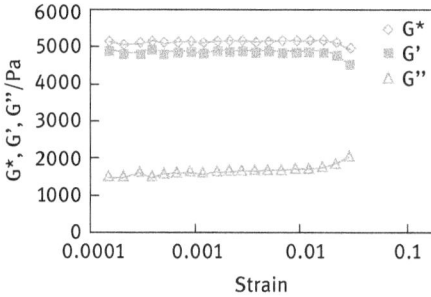

Fig. 7.36: Strain sweep results (at frequency of 1 Hz) for PLA-PEG 6 : 5 nanoparticles with a PLA core volume fraction of 0.094.

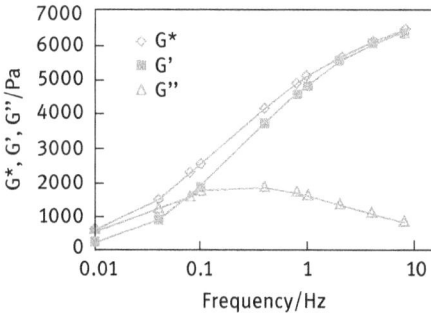

Fig. 7.37: Frequency sweep results (at strain amplitude in the linear region) for PLA-PEG 6 : 5 nanoparticles with a PLA core volume fraction of 0.094.

A characteristic frequency at which $G' = G''$ is obtained at 0.08 Hz. When the frequency is below 0.08 Hz, $G'' > G'$, whereas above 0.08 Hz, $G' > G''$ and at sufficiently high frequency (10 Hz) $G' \sim G^*$, whereas G'' reaches a very low value.

Figure 7.38 shows the frequency sweep results for PLA-PEG 6 : 5 nanoparticles with a PLA core volume fraction of 0.104. At this high core volume fraction $G' \gg G''$ over the whole frequency range. G' shows little dependence on frequency. In this case the dispersion is behaving as nearly "elastic gel". At such high volume fraction the PEG chains undergo significant interpenetration and compression. The high G′ values (> 1000 Pa) may be attributed to the small size of the nanoparticles (with hydrodynamic radius ~ 21 nm) and hence the large number of contact points between the PLA-PEG micelle-like assemblies. Similar strain and frequency sweep results are obtained for the PLA-PEG 3 : 5 dispersions.

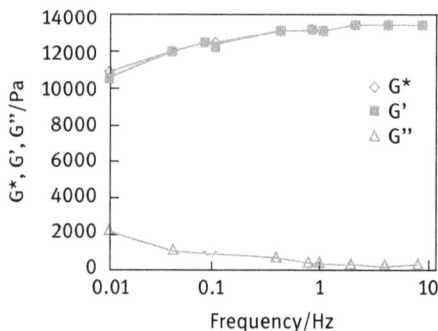

Fig. 7.38: Frequency sweep results (at strain amplitude in the linear region) for PLA-PEG 6 : 5 nanoparticles with a PLA core volume fraction of 0.104.

Figure 7.39 shows plots of G′ and G″ (in the linear viscoelastic region and at a frequency of 1 Hz) as a function of PLA core volume fraction for the PLA-PEG 3 : 5. Figure 7.40 shows the corresponding plots for the PLA-PEG 6 : 5. From these plots one can obtain the critical volume fraction $(\phi_{core})^{crit}$ at which the dispersion changes from predominantly viscous to predominantly elastic. Assuming that that the crossover point at which G′ = G″ represents the onset of interpenetration of the grafted PEG chains, then $(\phi_{core})^{crit}$ can be used to obtain the grafted PEG layer thickness Δ (assuming random packing with ϕ = 0.64) as discussed before. Table 7.8 shows a summary of the results.

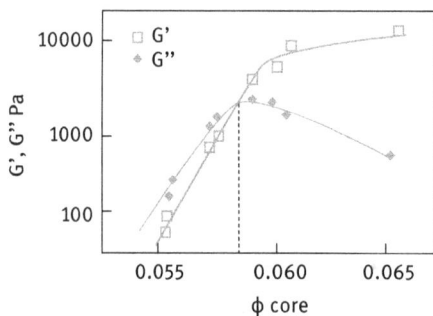

Fig. 7.39: Variation of G′ and G″ with PLA core volume fraction at a frequency of 1 Hz. PLA-PEG 3 : 5.

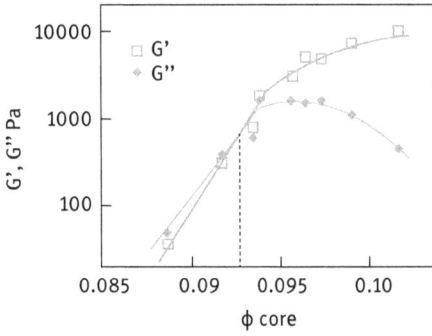

Fig. 7.40: Variation of G′ and G″ with PLA core volume fraction at a frequency of 1 Hz. PLA-PEG 6 : 5.

Tab. 7.8: Hydrodynamic layer thickness of grafted PEG chains Δ and core radius R_c of PLA-PEG micellar-like dispersions.

PLA-PEG	3 : 5	6 : 5
R_{hyd} / nm	11.5	15.2
ϕ_{crit}^{core}	0.059	0.09
Δ / nm	6.3	7.2
R_c / nm	5.2	8.0
Δ/R_c	1.2	0.9

7.3.7 Small angle neutron scattering (SANS) of PLA-PEG nanoparticles

SANS is a simple diffraction technique which can be used to provide direct structural information about PLA-PEG nanoparticles [50]. The shorter wavelength employed in SANS (typically 0.15–2.0 nm) enables shorter length scales (typically 0.5–100 nm) to be probed. One has to use samples with a low hydrogen content to avoid incoherent scatter of neutrons. Selectively deuterated PLA(d)-PEG copolymers with fully deuterated PLA(d) blocks were synthesized [4]. These materials facilitate contrast matching where scattering, and hence structural information, can be obtained from particular parts of a complex system (see below).

7.3.7.1 Basic concepts of SANS

The scattering vector Q describes the relationship between the incident, k_i, and scattered, k_s, wave vectors,

$$Q = |Q| = |k_s - k_i| = \frac{4\pi}{\lambda} \sin\left(\frac{\theta}{2}\right), \tag{7.58}$$

where λ is the neutron wavelength and θ is the scattering angle. Q has dimensions of (length)$^{-1}$.

Substituting equation (7.58) into Bragg's law of diffraction,

$$\lambda = 2d \sin\left(\frac{\theta}{2}\right), \tag{7.59}$$

where d is the molecular-level length scale.

Combining equations (7.58) and (7.59),

$$d = \frac{2\pi}{Q}. \tag{7.60}$$

The differential scattering cross-section $(d\Sigma/d\Omega)(Q)$ is the dependent variable in a SANS experiment. $(d\Sigma/d\Omega)(Q)$ is related to the intensity (flux) of scattered neutrons, I(Q), arriving at the detector by

$$I(Q) = I_i(Q)\, \Delta\Omega\, \eta(\lambda)\, T(\lambda)\, V_s\, \frac{\partial\Sigma}{\partial\Omega}(Q), \tag{7.61}$$

where I_i is the incident flux, $\Delta\Omega$ is the solid angle element defined by the size of the detector pixel, $\eta(\lambda)$ is the detector efficiency, $T(\lambda)$ is the neutron transmission of the sample and V_s is the volume of the sample illuminated by the neutron beam. The first three terms of equation (7.61) are instrument dependent, while the last three terms are sample dependent. In a dilute sample with no interparticle interactions, the differential cross-section is given by

$$\frac{\partial\Sigma}{\partial\Omega}(Q) = N_p V_p (\Delta\rho)^2 P(Q) S(Q) + B, \tag{7.62}$$

where N_p is the number concentration of scattering centres, V_p is the volume of one particle, $(\Delta\rho)^2$ is the contrast (see below), P(Q) is the particle form factor, S(Q) is the interparticle structure factor and B is the background signal. P(Q) is a dimensionless function that describes how $(d\Sigma/d\Omega)(Q)$ is modulated by interference effects between neutrons scattered by different parts of the same scattering centre. S(Q) tends to unity at low concentrations of scattering centres. $(d\Sigma/d\Omega)(Q)$ has dimensions of (length)$^{-1}$ and is normally expressed in units of cm^{-1}.

In SANS, one calculates the difference in neutron scattering length density (that is equivalent to the refractive index difference in light scattering). The scattering length density, ρ, of a molecule is calculated by

$$\rho = \frac{\rho_{bulk} N_A}{M} \sum_i b_i, \tag{7.63}$$

where ρ_{bulk} is the bulk density of the molecule, N_A is Avogadro's number, M is the molar mass and b_i is the coherent neutron scattering length of nuclei i. ρ has dimensions (length)$^{-1}$ and is normally expressed in units of 10^{10} cm^{-2}.

The neutron scattering length densities used for the PLA(d)-PEG nanoparticles are given in Tab. 7.9.

Tab. 7.9: Coherent neutron scattering length densities.

Solvent/Polymer	Bulk density, ρ_{bulk} ($g\,cm^{-3}$)	Scattering length density, ρ ($\times 10^{10}\,cm^{-2}$)
100 % D_2O	1.11	+6.36
80 % D_2O/20 % H_2O	1.09	+5.00
75 % D_2O/25 % H_2O	1.08	+4.65
65 % D_2O/35 % H_2O	1.07	+3.95
100 % H_2O	1.00	−0.56
PEG (hydronated)	1.13 (as solid)	+0.64
PLA(d) (perdeuterated)	1.29 (as solid)	+5.95

The contrast term, $(\Delta\rho)^2$, is simply the square of the difference in neutron scattering length density between the solute (ρ_s) and the surrounding medium (ρ_m),

$$(\Delta\rho)^2 = (\rho_s - \rho_m)^2 . \tag{7.64}$$

If $(\Delta\rho)^2$ is zero, the scattering centres are said to be at "contrast match" and there is no SANS. The scattering from multicomponent systems consists of a contrast weighted summation of the scattering from individual components, plus a contribution from interference terms. The technique of contrast matching can dramatically simplify the scattering pattern, by making certain components "invisible" to neutrons. This is achieved through substituting hydrogen atoms for the deuterium isotope. Deuterium and hydrogen isotopes are characterized by neutron scattering lengths of different signs and magnitude: $b_D = +0.667 \times 10^{-12}\,cm^{-2}$; $b_H = -0.374 \times 10^{-12}\,cm^{-2}$. For aqueous core-shell type systems such as PLA-PEG aqueous dispersions, it is common to deuterate the core. The scattering length density of the medium (ρ_m) is adjusted by the ratio of H_2O/D_2O, such that the core is at match with the medium and the scattering arises from the shell alone. This is schematically represented in Fig. 7.41.

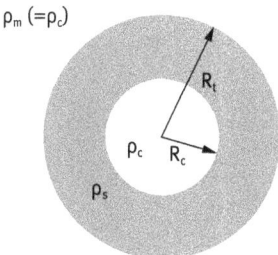

Fig. 7.41: Schematic representation of "contrast" matching for a core-shell system.

7.3.7.2 Analysis of SANS data

Analysis of the scattering data from the PLA(d)-PEG micellar structures assumes a homogeneous core with a fixed scattering length density, ρ_c, and a simple functional form of the scattering length density profile of the shell, $\rho_s(R)$. The overall scattering from such "composite structures" consists of a contrast weighted summation of the scattering from the core and shell components, together with a contribution from interference terms [50]. Equation (7.62) may be written as,

$$\frac{\partial \Sigma}{\partial \Omega}(Q) = K_s P(Q) + B, \tag{7.65}$$

where K_s is a scaling factor and $P(Q)$ is the composite form factor given by

$$P(Q) = \left[(\rho_1 - \rho_2)F(Q, R_1) + (\rho_3 - \rho_2)F(Q, R_2) + \dots\right]^2. \tag{7.66}$$

The form factor for each step or feature, $F(Q, R)$, depends on the scattering length profile.

It is necessary to convolute the composite particle form factor with a suitable particle distribution function, to take into account the polydispersity of the system, e.g. using a Schultz distribution.

Using the above analysis, it is possible to generate a variety of possible scattering length profiles, $\rho(R)$, which constitute the core-shell models [4]. The simplest approach assumes the scattering length density within the shell is uniform and can be represented as a step function for $\rho(R)$ (Fig. 7.42). Alternatively, a diffuse profile for the shell is assumed as represented in Fig. 7.43.

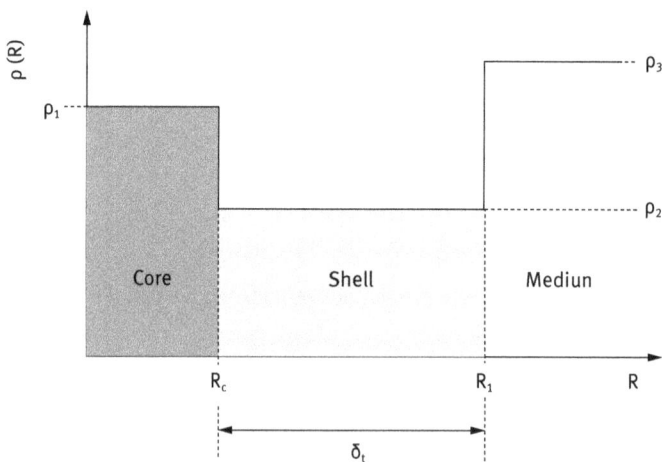

Fig. 7.42: Schematic representation of scattering length profile for a uniform shell model.

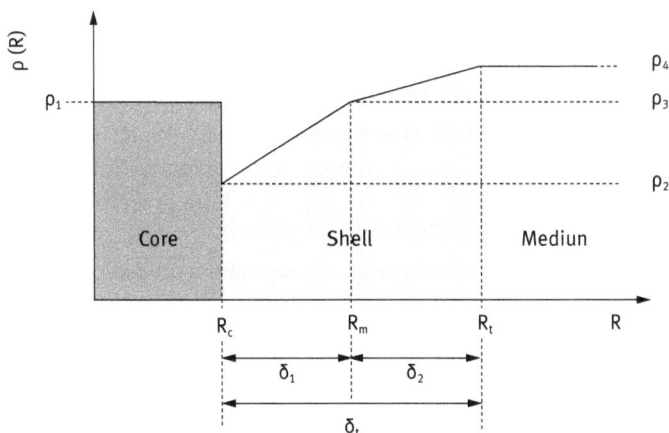

Fig. 7.43: Schematic representation of scattering length profile for a diffuse shell model.

The fully reduced cross-section data in the form of $(d\Sigma/d\Omega)$ versus (Q), for the series of PLA(d)-PEG nanoparticle dispersion, are analyzed using the core-shell model described above. The parameters of the model are adjusted to achieve the "best fit" to the scattering data using a least square programme.

The hydrodynamic diameter, polydispersity index (determined by PCS) and the critical flocculation point (CFPT) of the PLA(d)-PEG systems used in the SANS experiments are given in Tab. 7.10.

Tab. 7.10: Physicochemical characterization of PLA(d)-PEG nanoparticles used in the SANS experiments.

PLA-PEG	3:5	15:5	45:5
Hydrodynamic diameter / nm	32.0	48.4	110.8
Polydispersity index	0.11	0.07	0.13
CFPT (mol dm^{-3} Na$_2$SO$_4$)	0.55	0.50	0.35
Concentration (% w/v)	1.87	2.13	1.71
Volume fraction PLA cores ($\phi_{core}^{expected}$)[a]	0.005	0.012	0.012

[a] Calculated using the literature for the bulk density of PLA.

The colloidal properties of the PLA-PEG systems appear not to be perturbed by isotopic substitution of the PLA block. This implies that the self-assembly of the PLA-PEG copolymers is not affected by deuteration of the PLA block.

The conditions required for the core and the medium to be at a contrast match, can be found by adjusting the D$_2$O/H$_2$O ratio until a minimum scattering intensity is detected.

The main objective of using PLA-PEG nanoparticles for drug delivery is to have long-circulating particulate carriers and minimize opsonization by means of high coverage of brush-like PEG chains. On the basis of this rationale, copolymers with an intermediate PLA to PEG ratio (e.g. PLA-PEG 15 : 5) would appear to form assemblies with optimal protein-resistant surface properties [4].

To test the above hypothesis, PLA and PLA-PEG nanoparticles were radiolabelled for in vivo studies by incorporation of the hydrophobic gamma-emitter, ^{111}IN-oxine (8-hydroxy quinoline). In vitro studies showed that ^{111}IN-oxine is released from PLA and PLA-PEG nanoparticles on incubation with rat serum [4].

For each PLA/PLA-PEG system, a group of three male Wistar rats (150 ±10 g) was injected intravenously via the lateral tail vein with 0.3 ml (equivalent to 1 mg of solid material) of the nanoparticle dispersion. A group of control rats was injected with 10 kBq of unincorporated (free) ^{111}IN-oxine. Blood samples of 20 µl were taken from the contralateral tail vein at various time intervals after administration (5, 15, 30, 60, 120, and 180 minutes). The animals were sacrificed after three hours by intravenous injection of phenobarbitone solution, and the liver, spleen lungs and kidneys removed. The organ- and blood-associated activity was counted using a gamma counter. The carcass-associated radioactivity was determined using a well counter. A total blood volume of 75 % of the body weight was assumed. The results for the blood- and organ-associated activity are expressed as a percentage of the injected dose and are mean values for the three rats ± standard deviation. The data for the lung and kidney are not presented, since the radioactivity associated with these organs was negligible (less than 1 % of the administered dose).

The results for the blood circulation and organ distribution showed some interesting trends [4]. The PLA nanoparticles (uncoated with PEG) were rapidly cleared from the blood circulation, with only 13 % of the injected dose still circulating after 5 minutes. After 3 hours, 70 % of the i.v. administered nanoparticles had been removed by the liver. This is attributed to rapid opsonization of the particle surface and subsequent phagocytosis by the Kupffer cells of the liver. The smallest of the PLA-PEG nanoparticles studied in vivo (PLA-PEG 6 : 5, ~ 40 nm in diameter) was cleared from the circulation, with a high percentage of the radioactivity (~ 70 %) having accumulated in the liver three hours after the i.v. injection. However, the blood clearance rate was significantly slower than for the PLA nanoparticles. An increase in the length of the PLA block produced larger particles which exhibited prolonged circulation times and a reduced liver uptake. For example, in the case of PLA-PEG 110 : 5 nanoparticles (~ 160 nm), 43 % of the injected dose still remained in the systemic circulation after 3 hours, whilst only 23 % of the injected dose accumulated in the liver. Despite avoiding recognition by the Kupffer cells of the liver, 11 % of the injected dose of the PLA-PEG 110 : 5 nanoparticles was found to accumulate in the spleen.

The prolonger circulation times and reduced deposition of the larger PLA-PEG nanoparticles are surprising in view of the low PEG surface coverage of these systems, which are actually stabilized by the presence of adsorbed serum components. It appears that that such low PEG coverage is sufficient for restricting the adsorption of the high molecular weight opsinons. The layer thickness of terminally attached PEG chains with a molecular weight of 5 kDa is approximately 6.2 nm, which may adequately prohibit the adhesion of phagocytic cells.

It is surprising that the smaller micellar-like assemblies prepared from PLA-PEG copolymers with a low molecular weight PLA block were fairly rapidly cleared from the circulation and accumulated in the liver. These nanoparticles are the most colloidally stable of the PLA-PEG assemblies studied and hence the notion that effective steric stabilization is the most crucial effect for achieving blood circulation longevity is now questionable. It appears likely that the short circulation lifetime of the small PLA-PEG micelle-like nanoparticles is partly due to their ability to penetrate deep into the interstitial space of the liver [4].

It seems from the above discussion that the circulatory lifetime of PLA-PEG nanoparticles in vivo does not correlate with their colloid stability in vitro. It seems that the particle size of the PLA assembly is crucial in determining its biological fate. The presence of a low surface coverage of hydrated PEG chains is sufficient to enable relatively large (> 100 nm) PLA-PEG particles to remain in systemic circulation. Regardless of the characteristics of the PEG layer, small nanoparticles (~ 40 nm) are cleared by the liver to a higher degree, with their small size possibly permitting access to all cell types [4].

References

[1] Mills, S. N. and Davis, S. S., "Controlled Drug Delivery", in "Polymers in Controlled Drug Delivery", L. Illum and S. S. Davis (eds.), IOP Publishing, Bristol (1987), pp. 1–14.

[2] Krueter, J., "Colloidal Drug Delivery Systems", Marcel Dekker, New York (1994).

[3] Muller, R. H., "Colloidal Carriers for Controlled Drug Delivery: Modification, Characterisation and in vivo Distribution", Wiss. Verl-Ges., Stuttgart, Germany (1990).

[4] Riley, T., Ph.D. Thesis, Nottingham University (1999).

[5] Kostarelos, K., Ph.D. Thesis, Imperial College, London (1995).

[6] Kostarelos, K., Tadros, T. F. and Luckham, P. F., Langmuir, **15**, 369 (1999).

[7] Israelachvili, J. N., Mitchell, D. J. and Ninham, B. W., J. Chem. Soc., Faraday Trans. II, **72**, 1525 (1976).

[8] Israelachvili, J. N., Marcelja, S. and Horn, R. G., Q. Rev. Biophys, **13(2)**, 121 (1980).

[9] Israelachvili, J. N., "Intermolecular and Surface Forces", Academic Press, San Diego (1991).

[10] Tanford, C., "The Hydrophobic Effect", Wiley, New York (1980).

[11] Tanford, C., Biomembranes, Proc. Int. Sch. Phys. Enrico. Ferm., **90**, 547 (1985).

[12] Israelachvili, J. N. and Mitchell, D. J., Biochim. Biophys. Acta., **389**, 13 (1975).

[13] Mills, S.N and Davis, S. S., in "Polymers in Controlled Drug Delivery", L. Illum and S. S. Davis (eds.), IOP Publishing, Bristol, UK (1987) pp. 1–14.

[14] Kreuter, J., "Colloid Drug Delivery Systems", Marcel Dekker, New York (1994).

[15] Kreuter, J., Nanoparticle based drug delivery systems, J. Control. Rel., **16**, 169–176 (1991).

[16] Illum, L. and Davis, S. S., The organ uptake of intravenously administered colloidal particles can be altered using a non-ionic surfactant (poloxamer 338), FEBS Lett., **167**, 72–82 (1984).

[17] Muir, I. S., Moghimi, S. M., Illum, L., Davis, S. S. and Davies, M. C., The effect of block copolymer on the uptake of model polystyrene microspheres by Kupffer cells – in vitro and in vivo studies, Biochem. Soc. Trans., **19**, 329S (1991).

[18] Illum, L., Davis, S. S., Muller, R. H., Mak, E. and West, P., The organ distribution and circulation life-time of intravenously injected colloidal carriers stabilized with a block copolymer poloxamine 908, Life Sci., **40**, 367–374 (1987).

[19] Chasin, M. and Langer, R. (eds.), "Biodegradable Polymers as Drug Delivery Systems", Marcel Dekker, New York (1990).

[20] Stolnik, S., Dunn, S. E., Davies, M. C., Coombes, A. G. A., Taylor, D. C., Irving, M. P., Purkiss, S. C., Tadros, Th. F., Davis, S. S. and Illum, L., Surface modification of poly(lactide-co-glycolide) nanospheres by biodegradable poly(lactide)-poly(ethylene glycol) copolymers, Pharm. Res., **11**, 1800–1808 (1994).

[21] Kwon, G. S. and Kataoka, K., Block copolymer micelles as long circulating drug vehicles, Advan. Drug Del. Rev., **16**, 295–309 (1995).

[22] de Gennes, P. G., "Scaling Concepts in Polymer Physics", Cornell University Press, Ithaca, London (1979).

[23] Napper, D. H., "Polymeric Stabilization of Colloidal Dispersions", Academic Press, London (1983).

[24] Fleer, G. J., Cohen Stuart, M. A., Scheutjens, J. M. H. M., Cosgrove, T. and Vincent, B., "Polymers at Interfaces", Chapman and Hall, London (1993).

[25] Jeon, S. I., Lee, J. H., Andrade, J. D. and de Gennes, P. G., Protein surface interaction in the presence of polyethylene oxide. I. Simplified theory, J. Colloid Interface Sci., **142**, 149–158 (1991).

[26] Jeon, S. I. and Andrade, J. D., Protein surface interaction in the presence of polyethylene oxide, J. Colloid Interface Sci., **142**, 159–166 (1991).

[27] Hyson, S.-H., Joshed, K. and Ikea, Y., Synthesis of polypeptides with different molecular weights, Biomaterials, **18**, 1503–1508 (1997).

[28] Kohn, F. E., van den Berg, J. W. A., van de Rider, G. and Fijian, J., The ring opening polymerization of D,L-Lactate in the melt initiated with tetraphenyltin, J. Appl. Polym. Sci., **29**, 4265–4277 (1984).

[29] Deng, X. M., Xiong, C. D., Cheng, L. M., Huang, H. H. and Xu, R. P., Studies on the block copolymerization of D,L-Lactide and poly(ethylene glycol) with aluminium complex catalyst, J. Appl. Polym. Sci., **55**, 1193–1196 (1995).

[30] Pusey, P. N., in "Industrial Polymers: Characterisation by Molecular Weights", J. H. S. Green and R. Dietz (eds.), Transcripta Books, London (1973).

[31] von Smoluchowski, M., "Handbuch der Electrizität und des Magnetismus", Vol.II, Barth, Leipzig (1914).

[32] Huckel, E., Phys. Z. **25**, 204 (1924).

[33] Overbeek, J. Th. G. and Bijesterbosch, B. H., in "Electrokinetic Separation Methods", P. G. Righetti, C. J. van Oss and J. W. Vanderhoff Editors, Elsevier/North Holland Biomedical Press (1979).

[34] Henry, D. C., Proc. Royal Soc., London **A133**, 106 (1948).

[35] Wiersema, P. H., Loeb, A. L. and Overbeek, J. Th. G., J. Colloid Interface Sci., **22**, 78 (1967).

[36] Ottewill, R. H. and Shaw, J. N., J. Electroanal. Interfacial Electrochem, **37**, 133 (1972).

[37] Douglas, S. J., Davis, S. S. and Illum, L., Nanoparticles in drug delivery, Crit. Rev. Ther. Drug Carr. Syst., **3**, 233–261 (1987).

[38] Napper, D. H., "Polymeric Stabilization of Colloidal Dispersions", Academic Press, London (1983).

[39] Bazile, D., Prud'homme, C. Bassoulet, M-T. Marlard, M. Spenlehaur G. and Veillard, M. Stealth Me.PEG-PLA nanoparticles avoid uptake by the mononuclear phagocytic system, J. Pharm. Sci., **84**, 493–498 (1995).

[40] Chu, B., "Laser Light Scattering", 2nd edn., Academic Press, New York (1991).

[41] Altinok, H., Yu, G-A., Nixon, K., Gorry, P. A., Attwood, D. and Booth, C., Effect of copolymer architecture on the self-assembly of copolymers of ethylene oxide and propylene oxide in aqueous solution, Langmuir, **13**, 5837–5448 (1997).

[42] Tanodekaew, S., Pannu, R., Heatley, F., Attwood, D. and Booth, C., Association and surface properties of diblock copolymers of ethylene oxide and DL-lactide in aqueous solution, Macromol. Chem. Phys., **198**, 927–944 (1997).

[43] Tuzar, Z. and Kratochvil, P., "Micelles of Block and Graft Copolymers in Solution", in "Surface and Colloid Science", E. Matijevic (ed.), Plenum, New York (1993).

[44] Ferry, J. D., "Viscoelastic Properties of Polymers", John Wiley & Sons, NY (1980).

[45] Whorlow, R. W., "Rheological Techniques", Ellis Horwood, Chister (1980)

[46] Goodwin, J. W. and Hughes, R., "Rheology for Chemists", Royal Society of Chemistry Publication, Cambridge (2000).

[47] Tadros, Th. F. "Rheology of Dispersions", Wiley-VCH, Germany (2010).

[48] Prestidge, C. and Tadros, Th. F., J. Colloid Interface Sci., **124**, 660–665 (1988).

[49] Hagen, S. A., Davis, S. S., Illum, L., Davies, M. C., Garnett, M. C., Taylor, D. C., Irving, M. P. and Tadros, Th. F., Langmuir, **11**, 1482–1485 (1995).

[50] King, S. M., Griffiths, P. C. and Cosgrove, T., "Using SANS to Study Adsorbed Layers in Colloidal Dispersions", in "Applications of Neutrons in Soft Condensed", by B. J. Gabrys (ed.), Gordon and Breach, Amsterdam (1998).

8 Preparation of nanoemulsion using high pressure homogenizers

8.1 Introduction

The production of small droplets (submicron) requires application of high energy; the process of emulsification is generally inefficient. Simple calculations show that the mechanical energy required for emulsification exceeds the interfacial energy by several orders of magnitude. For example, to produce an emulsion at $\phi = 0.1$ with a volume to surface diameter (Sauter diameter) $d_{32} = 0.6\,\mu m$, using a surfactant that gives an interfacial tension $\gamma = 10\,mN\,m^{-1}$, the net increase in surface free energy is $A\gamma = 6\phi\gamma/d_{32} = 10^4\,J\,m^{-3}$. The mechanical energy required in a homogenizer is $10^7\,J\,m^{-3}$, i.e. an efficiency of 0.1%. The rest of the energy (99.9%) is dissipated as heat [1–5].

Before describing the methods that can be applied to prepare submicron droplets (nanoemulsions), it is essential to consider the thermodynamics of emulsion formation and breakdown, the role of the emulsifier in preventing coalescence during emulsification and the procedures that can be applied for selecting the emulsifier.

8.2 Thermodynamics of emulsion formation and breakdown

Consider a system in which an oil is represented by a large drop 2 of area A_1 immersed in a liquid 2, which is now subdivided into a large number of smaller droplets with total area A_2 ($A_2 \gg A_1$) as shown in Fig. 8.1. The interfacial tension γ_{12} is the same for the large and smaller droplets since the latter are generally in the region of 0.1 to few μm.

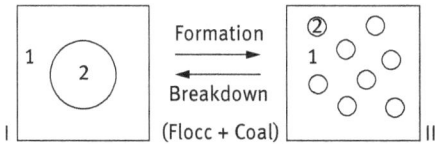

Fig. 8.1: Schematic representation of emulsion formation and breakdown.

The change in free energy in going from state I to state II is made up from two contributions: A surface energy term (that is positive) that is equal to $\Delta A\gamma_{12}$ (where $\Delta A = A_2 - A_1$) and an entropy of dispersions term which is also positive (since producing a large number of droplets is accompanied by an increase in configurational entropy) which is equal to $T\Delta S^{conf}$.

From the second law of thermodynamics,

$$\Delta G^{form} = \Delta A \gamma_{12} - T \Delta S^{conf}. \tag{8.1}$$

In most cases $\Delta A \gamma_{12} \gg T \Delta S^{conf}$, which means that ΔG_{form} is positive, i.e. the formation of emulsions is nonspontaneous and the system is thermodynamically unstable. In the absence of any stabilization mechanism, the emulsion will break down by flocculation, coalescence, Ostwald ripening or a combination of all these processes. This is illustrated in Fig. 8.2 which shows several paths for emulsion breakdown processes [1–5].

In the presence of a stabilizer (surfactant and/or polymer), an energy barrier is created between the droplets and therefore the reversal from state II to state I becomes noncontinuous as a result of the presence of these energy barriers. This is illustrated in Fig. 8.3. In the presence of the above energy barriers, the system becomes kinetically stable. The energy barrier is produced by electrostatic or steric stabilization as described in detail in Chapter 2.

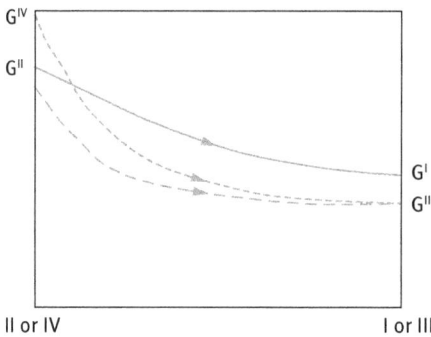

Fig. 8.2: Free energy path in emulsion breakdown: ——, flocc. + coal.; – – –, flocc. + coal. + sed.; ······, flocc. + coal. + sed. + Ostwald ripening.

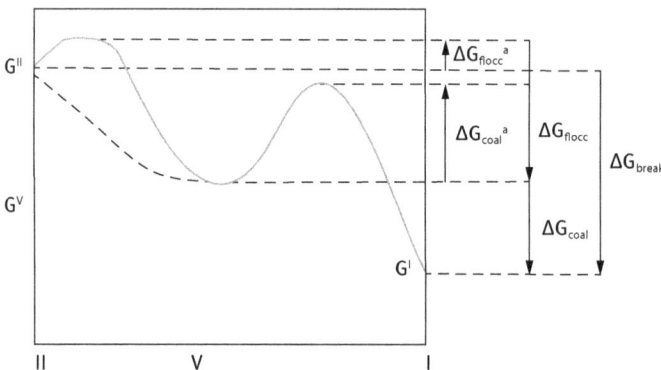

Fig. 8.3: Schematic representation of free energy path for breakdown (flocculation and coalescence) for systems containing an energy barrier.

8.3 Adsorption of surfactants at the liquid/liquid interface

Surfactants accumulate at interfaces, a process described as adsorption. The simplest interfaces are air/water (A/W) and oil/water (O/W). The surfactant molecule orients itself at the interface with the hydrophobic portion orienting towards the hydrophobic phase (air or oil) and the hydrophilic portion orienting at the hydrophilic phase (water). This is schematically illustrated in Fig. 8.4. As a result of adsorption, the surface tension of water is reduced from its value of $72\,\mathrm{mN\,m^{-1}}$ before adsorption to ~ 30–$40\,\mathrm{mN\,m^{-1}}$ and the interfacial tension for the O/W system decreases from its value of $50\,\mathrm{mN\,m^{-1}}$ (for an alkane oil/water) before adsorption to a value of 1–$10\,\mathrm{mN\,m^{-1}}$ depending on the nature of the surfactant.

Two approaches can be applied to treat surfactant adsorption at the A/L and L/L interface: The Gibbs approach, which treats the process as an equilibrium phenomenon [3]. In this case one can apply the second law of thermodynamics. Secondly, the equation of state approach in which the surfactant film is treated as a "two-dimensional" layer with a surface pressure π. The Gibbs approach allows one to obtain surfactant adsorption from surface tension measurements. The equation of state approach allows one to study the surfactant orientation at the interface [3]. In this section, only the Gibbs approach will be described.

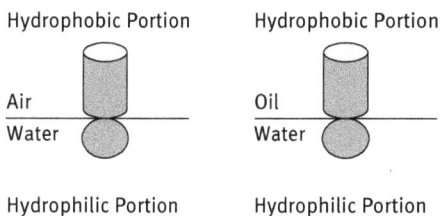

Hydrophobic Portion   Hydrophobic Portion

Air   Oil

Water   Water

Hydrophilic Portion   Hydrophilic Portion

Fig. 8.4: Schematic representation of orientation of surfactant molecules.

Gibbs derived a thermodynamic relationship between the variation of surface or interfacial tension with concentration and the amount of surfactant adsorbed Γ (moles per unit area), referred to as the surface excess. At equilibrium, the Gibbs free energy $dG^\sigma = 0$ and the Gibbs–Deuhem equation becomes

$$dG^\sigma = -S^\sigma\,dT + A\,d\gamma + \sum_i n_i^\sigma\,d\mu_i = 0. \tag{8.2}$$

At constant temperature,

$$A\,d\gamma = -\sum_i n_i^\sigma\,d\mu_i \tag{8.3}$$

or

$$d\gamma = -\sum \frac{n_i^\sigma}{A}\,d\mu_i = -\sum \Gamma_i^\sigma\,d\mu_i. \tag{8.4}$$

For a surfactant (component 2) adsorbed at the surface of a solvent (component 1),

$$-d\gamma = \Gamma_1^\sigma\,d\mu_1 + \Gamma_2^\sigma\,d\mu_2. \tag{8.5}$$

If the Gibbs dividing surface is used and the assumption $\Gamma_1^\sigma = 0$ is made,

$$- d\gamma = \Gamma_{2,1}^\sigma \, d\mu_2 . \tag{8.6}$$

The chemical potential of the surfactant μ_2 is given by the expression

$$\mu_2 = \mu_2^0 + RT \ln a_2^L , \tag{8.7}$$

where μ_2^0 is the standard chemical potential, a_2^L is the activity of surfactant that is equal to $C_2 f_2 \sim x_2 f_2$ where C_2 is the concentration in mol dm^{-3} and x_2 is the mole fraction that is equal to $C_2/(C_2 + 55.5)$ for a dilute solution and f_2 is the activity coefficient that is also ~ 1 in dilute solutions.

Differentiating equation (8.7) one obtains,

$$d\mu_2 = RT \, d \ln a_2^L . \tag{8.8}$$

Combining equations (8.6) and (8.8),

$$- d\gamma = \Gamma_{2,1}^\sigma RT \, d \ln a_2^L \tag{8.9}$$

or

$$\frac{d\gamma}{d \ln a_2^L} = -RT \, \Gamma_{2,1}^L . \tag{8.10}$$

In dilute solutions, $f_2 \sim 1$ and

$$\frac{d\gamma}{d \ln C_2} = -\Gamma_2 RT . \tag{8.11}$$

Equations (8.10) and (8.11) are referred to as the Gibbs adsorption equations which show that Γ_2 can be determined from the experimental results of variation of γ with log C_2 as illustrated in Fig. 8.5 for the A/W and O/W interfaces.

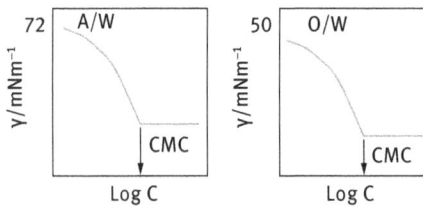

Fig. 8.5: Surface or interfacial tension-log C curves.

Γ_2 can be calculated from the linear portion of the γ–log C curve just before the critical micelle concentration (cmc):

$$\text{slope} = -\frac{d\gamma}{d \log C_2} = -2.303 \Gamma_2 RT . \tag{8.12}$$

From Γ_2 the area per molecule of surfactant (or ion) can be calculated,

$$\text{Area/molecule} = \frac{1}{\Gamma_2 N_{av}} \ (m^2) = \frac{10^{18}}{\Gamma_2 N_{av}} \ (nm^2). \tag{8.13}$$

N_{av} is Avogadro's constant that is equal to 6.023×10^{23}.

The area per surfactant ion or molecule gives information on the orientation of the ion or molecule at the interface. The area depends on whether the molecules lie flat or vertical at the interface. It also depends on the length of the alkyl chain (if the molecules lie flat) or the cross-sectional area of the head group (if the molecules lie vertical). For example, for an ionic surfactant such as sodium dodecyl sulphate (SDS), the area per molecule depends on the orientation. If the molecule lies flat, the area is determined by the area occupied by the alkyl chain and by the sulphate head group. In this case the area per molecule increases with increasing alkyl chain length and will be in the range 1–$2\,nm^2$. In contrast, for vertical orientation, the area per molecule is determined by the cross-sectional area of the sulphate group which is $\sim 0.4\,nm^2$ and virtually independent of the alkyl chain length. Addition of electrolytes screens the charge on the head group and hence the area per molecule decreases. For nonionic surfactants such as alcohol ethoxylates, the area per molecule for flat orientation is determined by the length of the alkyl chain and the number of ethylene oxide (EO) units. For vertical orientation, the area per molecule is determined by the cross-sectional area of the polyethylene oxide chain and this increases with an increasing number of EO units.

At concentrations just before the breakpoint, the slope of the γ-log C curve is constant,

$$\left(\frac{\partial \gamma}{\partial \log C_2} \right) = \text{const.} \tag{8.14}$$

This indicates that saturation of the interface occurs just below the cmc.

Above the breakpoint (C > cmc), the slope is zero

$$\left(\frac{\partial \gamma}{\partial \log C_2} \right) = 0 \tag{8.15}$$

or

$$\gamma = \text{const.} \cdot \log C_2 \tag{8.16}$$

Since γ remains constant above the cmc, then C_2 or a_2 of the monomer must remain constant.

Addition of surfactant molecules above the cmc must result in association to form micelles which have low activity and hence a_2 remains virtually constant.

The hydrophilic head group of the surfactant molecule can also affect its adsorption. These head groups can be unionized, e.g. alcohol or poly(ethylene oxide) (PEO), weakly ionized, e.g. COOH, or strongly ionized, e.g. sulphates $-O-SO_3^-$, sulphonates $-SO_3^-$ or ammonium salts $-N^+(CH_3)_3^-$. The adsorption of the different surfactants at

the A/W and O/W interface depends on the nature of the head group. With nonionic surfactants repulsion between the head groups is smaller than with ionic head
groups and adsorption occurs from dilute solutions; the cmc is low, typically 10^{-5}–
10^{-4} mol dm^{-3}. Nonionic surfactants with medium PEO form closely packed layers at
C < cmc. Adsorption is slightly affected by moderate addition of electrolytes or change
in the pH. Nonionic surfactant adsorption is relatively simple and can be described
by the Gibbs adsorption equation.

With ionic surfactants, adsorption is more complicated depending on the repulsion between the head groups and addition of indifferent electrolyte. The Gibbs adsorption equation has to be solved to take into account the adsorption of the counterions and any indifferent electrolyte ions.

For a strong surfactant electrolyte such as R–O–SO$_3^-$Na$^+$ (R–Na$^+$)

$$\Gamma_2 = -\frac{1}{2RT}\left(\frac{\partial\gamma}{\partial\ln a_\pm}\right).$$ (8.17)

The factor 2 in equation (8.17) arises because both surfactant ion and counterion must
be adsorbed to maintain neutrality. ($\partial\gamma/$ d ln a$\pm$) is twice as large for an unionized surfactant molecule.

For a nonadsorbed electrolyte such as NaCl, any increase in Na$^+$R$^-$ concentration
produces a negligible increase in Na$^+$ concentration (dμ Na$^+$ is negligible, dμ Cl$^-$ is
also negligible).

$$\Gamma_2 = -\frac{1}{RT}\left(\frac{\partial\gamma}{\partial\ln C_{NaR}}\right),$$ (8.18)

which is identical to the case of nonionics.

The above analysis shows that many ionic surfactants may behave like nonionics
in the presence of a large concentration of an indifferent electrolyte such as NaCl.

8.4 Mechanism of emulsification

As mentioned before, to prepare an emulsion oil, water, surfactant and energy are
needed. This process can be analyzed from a consideration of the energy required to
expand the interface, $\Delta A\gamma$ (where ΔA is the increase in interfacial area when the bulk
oil with area A$_1$ produces a large number of droplets with area A$_2$; A$_2 \gg$ A$_1$, γ is the
interfacial tension). Since γ is positive, the energy to expand the interface is large and
positive; this energy term cannot be compensated by the small entropy of dispersion
TΔS (which is also positive) and the total free energy of formation of an emulsion, ΔG
given by equation (8.1), is positive. Thus, emulsion formation is nonspontaneous and
energy is required to produce the droplets.

The formation of large droplets (few μm), as is the case for macroemulsions, is
fairly easy and hence high speed stirrers such as the Ultra-Turrax or Silverson Mixer
are sufficient to produce the emulsion [6]. In contrast, the formation of small drops

(submicron as is the case with nanoemulsions) is difficult and this requires a large amount of surfactant and/or energy. The high energy required for formation of nanoemulsions can be understood from a consideration of the Laplace pressure Δp (the difference in pressure between inside and outside the droplet),

$$\Delta p = \frac{2\gamma}{r} . \tag{8.19}$$

To break up a drop into smaller ones, it must be strongly deformed and this deformation increases Δp. This is illustrated in Fig. 8.6 which shows the situation when a spherical drop deforms into a prolate ellipsoid.

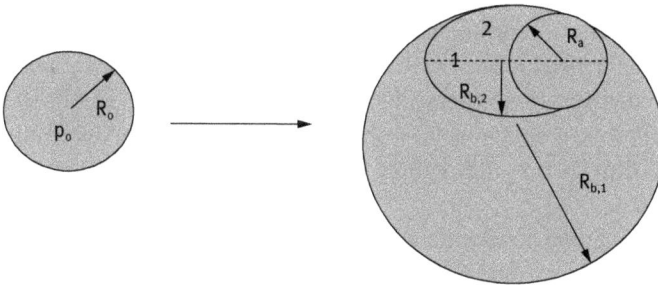

Fig. 8.6: Illustration of increasing Laplace pressure when a spherical drop is deformed to a prolate ellipsoid.

Near 1 there is only one radius of curvature R_a, whereas near 2 there are two radii of curvature $R_{b,1}$ and $R_{b,2}$. Consequently, the stress needed to deform the drop is higher for a smaller drop. Since the stress is generally transmitted by the surrounding liquid via agitation, higher stresses need more vigorous agitation, and hence more energy is needed to produce smaller drops [6].

Surfactants play major roles in the formation of emulsions: By lowering the interfacial tension, Δp is reduced and hence the stress needed to break up a drop is reduced. Surfactants also prevent coalescence of newly formed drops

To describe emulsion formation one has to consider two main factors: hydrodynamics and interfacial science. In hydrodynamics one has to consider the type of flow: laminar flow or turbulent flow. This depends on the Reynolds number as will be discussed later.

To assess emulsion formation, one usually measures the droplet size distribution using for example laser diffraction techniques. A useful average diameter d is

$$d_{nm} = \left(\frac{S_m}{S_n} \right)^{1/(n-m)} . \tag{8.20}$$

In most cases d_{32} (the volume/surface average or Sauter mean) is used. The width of the size distribution can be given as the variation coefficient c_m which is the standard deviation of the distribution weighted with d_m divided by the corresponding average d. Generally, C_2 will be used, which corresponds to d_{32}.

An alternative way to describe the emulsion quality is to use the specific surface area A (surface area of all emulsion droplets per unit volume of emulsion),

$$A = \pi s^2 = \frac{6\phi}{d_{32}} . \tag{8.21}$$

8.5 Methods of emulsification

Several procedures may be applied for emulsion preparation [6], these range from simple pipe flow (low agitation energy, L), static mixers and general stirrers (low to medium energy, L–M), high speed mixers such as the Ultra-Turrax (M), colloid mills and high pressure homogenizers (high energy, H), ultrasound generators (M–H). The method of preparation can be continuous (C) or batchwise (B): pipe flow and static mixers – C; stirrers and Ultra-Turrax – B,C; colloid mill and high pressure homogenizers – C; ultrasound – B,C.

In all methods, there is liquid flow; unbounded and strongly confined flow. In the unbounded flow any droplet is surrounded by a large amount of flowing liquid (the confining walls of the apparatus are far away from most of the droplets). The forces can be frictional (mostly viscous) or inertial. Viscous forces cause shear stresses to act on the interface between the droplets and the continuous phase (primarily in the direction of the interface). The shear stresses can be generated by laminar flow (LV) or turbulent flow (TV); this depends on the Reynolds number Re,

$$Re = \frac{vl\rho}{\eta} , \tag{8.22}$$

where v is the linear liquid velocity, ρ is the liquid density and η is its viscosity. l is a characteristic length that is given by the diameter of flow through a cylindrical tube and by twice the slit width in a narrow slit.

For laminar flow, Re $\leq$ 1000, whereas for turbulent flow Re $\geq$ 2000. Thus, whether the regime is linear or turbulent depends on the scale of the apparatus, the flow rate and the liquid viscosity [7–10].

If the turbulent eddies are much larger than the droplets, they exert shear stresses on the droplets. If the turbulent eddies are much smaller than the droplets, inertial forces will cause disruption (TI).

In bounded flow other relations hold. If the smallest dimension of the part of the apparatus in which the droplets are disrupted (say a slit) is comparable to droplet size, other relations hold (the flow is always laminar). A different regime prevails if the droplets are directly injected through a narrow capillary into the continuous phase (Injection regime), i.e. membrane emulsification.

Within each regime, an essential variable is the intensity of the forces acting; the viscous stress during laminar flow $\sigma_{viscous}$ is given by

$$\sigma_{viscous} = \eta G, \qquad (8.23)$$

where G is the velocity gradient.

The intensity in turbulent flow is expressed by the power density ε (the amount of energy dissipated per unit volume per unit time); for turbulent flow [6],

$$\varepsilon = \eta G^2. \qquad (8.24)$$

The most important regimes are: laminar/viscous (LV) – turbulent/viscous (TV) – turbulent/inertial (TI). For water as the continuous phase, the regime is always TI. For higher viscosity of the continuous phase ($\eta_C = 0.1\,Pa\,s$), the regime is TV. For still higher viscosity or a small apparatus (small l), the regime is LV. For a very small apparatus (as is the case with most laboratory homogenizers), the regime is nearly always LV.

For the above regimes, a semi-quantitative theory is available that can give the timescale and magnitude of the local stress σ_{ext}, the droplet diameter d, timescale of droplets deformation τ_{def}, timescale of surfactant adsorption, τ_{ads} and mutual collision of droplets.

An important parameter that describes droplet deformation is the Weber number We (which gives the ratio of the external stress over the Laplace pressure),

$$We = \frac{G \eta_C R}{2\gamma}. \qquad (8.25)$$

The viscosity of the oil plays an important role in the break-up of droplets; the higher the viscosity, the longer it will take to deform a drop. The deformation time τ_{def} is given by the ratio of oil viscosity to the external stress acting on the drop,

$$\tau_{def} = \frac{\eta_D}{\sigma_{ext}}. \qquad (8.26)$$

The viscosity of the continuous phase η_C plays an important role in some regimes: For a turbulent inertial regime, η_C has no effect on droplets size. For a turbulent viscous regime, larger η_C leads to smaller droplets. For a laminar viscous regime the effect is even stronger.

8.6 Role of surfactants in emulsion formation

Surfactants lower the interfacial tension γ and this causes a reduction in droplet size. The latter decreases with decreasing γ. For laminar flow, the droplet diameter is proportional to γ; for a turbulent inertial regime, the droplet diameter is proportional to $\gamma^{3/5}$.

The effect of reducing γ on the droplet size is illustrated in Fig. 8.7 which shows a plot of the droplet surface area A and mean drop size d_{32} as a function of surfactant concentration m for various systems.

The amount of surfactant required to produce the smallest drop size will depend on its activity a (concentration) in the bulk which determines the reduction in γ, as given by the Gibbs adsorption equation,

$$- d\gamma = RT\Gamma\, d\ln a \,, \tag{8.27}$$

where R is the gas constant, T is the absolute temperature and Γ is the surface excess (number of moles adsorbed per unit area of the interface).

Γ increases with increasing surfactant concentration and eventually it reaches a plateau value (saturation adsorption). This is illustrated in Fig. 8.8 for various emulsifiers.

The value of γ obtained depends on the nature of the oil and surfactant used; small molecules such as nonionic surfactants lower γ more than polymeric surfactants such as PVA.

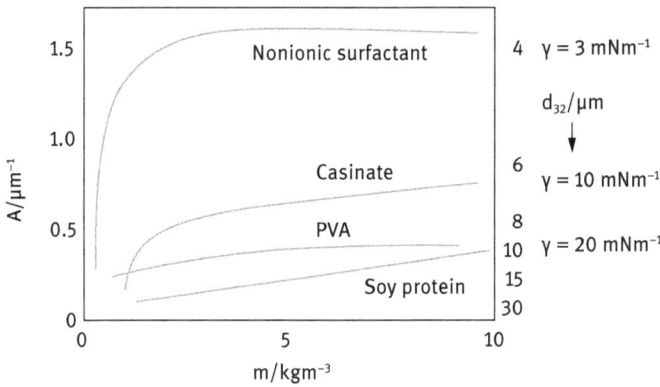

Fig. 8.7: Variation of A and d_{32} with m for various surfactant systems.

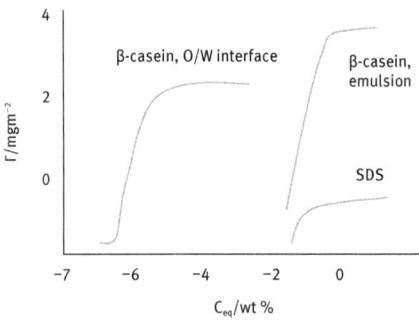

Fig. 8.8: Variation of Γ (mg m^{-2}) with log C_{eq} / % wt. The oils are β-casein (O/W interface) toluene, β-casein (emulsions) soybean, and SDS benzene.

Another important role of the surfactant is its effect on the interfacial dilational modulus ε,

$$\varepsilon = \frac{d\gamma}{d \ln A} . \tag{8.28}$$

During emulsification an increase in the interfacial area A takes place and this causes a reduction in Γ. The equilibrium is restored by adsorption of surfactant from the bulk, but this takes time (shorter times occur at higher surfactant activity). Thus ε is small at small a and also at large a. Because of the lack or slowness of equilibrium with polymeric surfactants, ε will not be the same for expansion and compression of the interface.

In practice, surfactant mixtures are used and these have pronounced effects on γ and ε. Some specific surfactant mixtures give lower γ values than either of the two individual components. The presence of more than one surfactant molecule at the interface tends to increase ε at high surfactant concentrations. The various components vary in surface activity. Those with the lowest γ tend to predominate at the interface, but if present at low concentrations, it may take a long time before reaching the lowest value. Polymer-surfactant mixtures may show some synergetic surface activity.

8.7 Role of surfactants in droplet deformation

Apart from their effect on reducing γ, surfactants play major roles in deformation and break-up of droplets – this is summarized as follows. Surfactants allow the existence of interfacial tension gradients which are crucial for formation of stable droplets. In the absence of surfactants (clean interface), the interface cannot withstand a tangential stress; liquid motion will be continuous (Fig. 8.9 (a)).

If a liquid flows along the interface with surfactants, the latter will be swept downstream causing an interfacial tension gradient (Fig. 8.9 (b)). A balance of forces will be established,

$$\eta \left[\frac{dV_x}{dy} \right]_{y=0} = -\frac{dy}{dx} . \tag{8.29}$$

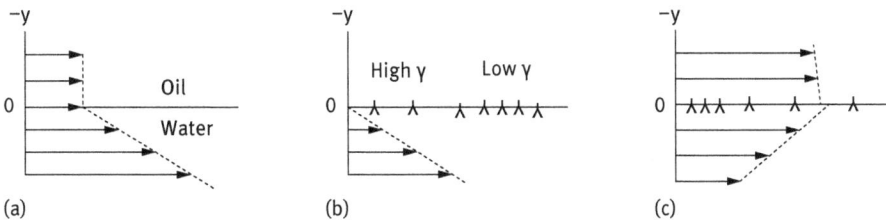

Fig. 8.9: Interfacial tension gradients and flow near an oil/water interface: (a) no surfactant; (b) velocity gradient causes an interfacial tension gradient; (c) interfacial tension gradient causes flow (Marangoni effect).

If the y-gradient can become large enough, it will arrest the interface. If the surfactant is applied at one site on the interface, a γ-gradient is formed that will cause the interface to move roughly at a velocity given by

$$v = 1.2[\eta \rho z]^{-1/3} |\Delta \gamma|^{2/3} . \tag{8.30}$$

The interface will then drag some of the bordering liquid with it (Fig. 8.9 (c)).

Interfacial tension gradients are very important in stabilizing the thin liquid film between the droplets which is very important during the beginning of emulsification (films of the continuous phase may be drawn through the disperse phase and collision is very large). The magnitude of the γ-gradients and of the Marangoni effect depends on the surface dilational modulus ε, which for a plane interface with one surfactant-containing phase is given by the expression

$$\varepsilon = \frac{-d\gamma/d\ln\Gamma}{(1 + 2\xi + 2\xi^2)^{1/2}}, \tag{8.31}$$

$$\xi = \frac{dm_C}{d\Gamma} \left(\frac{D}{2\omega} \right)^{1/2}, \tag{8.32}$$

$$\omega = \frac{d\ln A}{dt}, \tag{8.33}$$

where D is the diffusion coefficient of the surfactant and ω represents a timescale (time needed for doubling the surface area) that is roughly equal to τ_{def}.

During emulsification, ε is dominated by the magnitude of the denominator in equation (8.31) because ζ remains small. The value of $dm_C/d\Gamma$ tends to go to very high values when Γ reaches its plateau value; ε goes to a maximum when m_C is increased.

For conditions that prevail during emulsification, ε increases with m_C and it is given by the relationship

$$\varepsilon = \frac{d\pi}{d\ln\Gamma}, \tag{8.34}$$

where π is the surface pressure (π = γ_0 − γ). Figure 8.10 shows the variation of π with ln Γ; ε is given by the slope of the line.

The SDS shows a much higher ε value when compared with β-casein and lysozome – this is because the value of Γ is higher for SDS. The two proteins show differences in their ε values which may be attributed to the conformational change that occurs upon adsorption.

The presence of a surfactant means that during emulsification the interfacial tension need not be the same everywhere (see Fig. 8.9). This has two consequences: (i) the equilibrium shape of the drop is affected; (ii) any γ-gradient formed will slow down the motion of the liquid inside the drop (this diminishes the amount of energy needed to deform and break up the drop).

Another important role of the emulsifier is to prevent coalescence during emulsification. This is certainly not due to the strong repulsion between the droplets, since the pressure at which two drops are pressed together is much greater than the repulsive

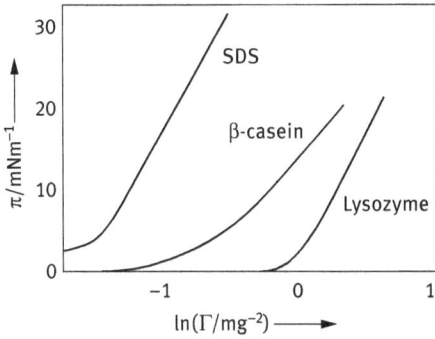

Fig. 8.10: π versus ln Γ for various emulsifiers.

stresses. The counteracting stress must be due to the formation of γ-gradients. When two drops are pushed together, liquid will flow out from the thin layer between them, and the flow will induce a γ-gradient. This was shown in Fig. 8.9 (c). This produces a counteracting stress given by

$$\tau_{\Delta\gamma} \approx \frac{2|\Delta\gamma|}{(1/2)d}. \tag{8.35}$$

The factor 2 follows from the fact that two interfaces are involved. Taking a value of $\Delta\gamma = 10\,\mathrm{mN\,m^{-1}}$, the stress amounts to 40 kPa (which is of the same order of magnitude as the external stress).

The Gibbs–Marangoni effect [11–14], schematically represented in Fig. 8.11, is closely related to the above mechanism. The depletion of surfactant in the thin film between approaching drops results in γ-gradient without liquid flow being involved. This results in an inward flow of liquid that tends to drive the drops apart.

The Gibbs–Marangoni effect also explains the Bancroft rule which states that the phase in which the surfactant is most soluble forms the continuous phase. If the surfactant is in the droplets, a γ-gradient cannot develop and the drops would be prone

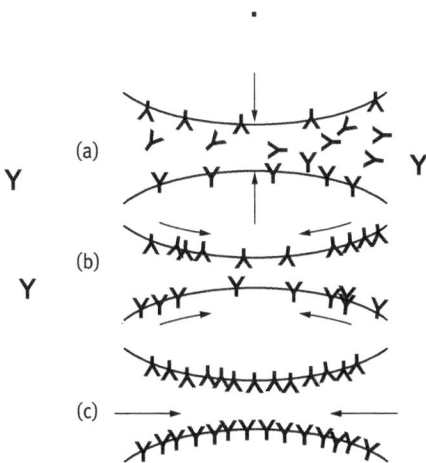

Fig. 8.11: Schematic representation of the Gibbs–Marangoni effect for two approaching drops.

to coalescence. Thus, surfactants with HLB > 7 tend to form O/W emulsions and those with HLB < 7 tend to form W/O emulsions.

The Gibbs–Marangoni effect also explains the difference between surfactants and polymers for emulsification – polymers give larger drops when compared with surfactants and polymers give a smaller value of ε at small concentrations when compared to surfactants (Fig. 8.10).

Various other factors should also be considered for emulsification: The disperse phase volume fraction ϕ. An increase in ϕ leads to an increase in droplet collision and hence coalescence during emulsification. With increasing ϕ, the viscosity of the emulsion increases and could change the flow from being turbulent to being laminar (LV regime).

The presence of many particles results in a local increase in velocity gradients. This means that G increases. In turbulent flow, increasing ϕ will induce turbulence depression. This will result in larger droplets. Turbulence depression by added polymers tends to remove the small eddies, resulting in the formation of larger droplets.

If the mass ratio of surfactant to continuous phase is kept constant, increasing ϕ results in a decrease in surfactant concentration and hence an increase in γ_{eq} resulting in larger droplets. If the mass ratio of surfactant to disperse phase is kept constant, the above changes are reversed.

General conclusions cannot be drawn since several of the above mentioned mechanism may come into play. Experiments using a high pressure homogenizer at various ϕ values at constant initial m_C (regime TI changing to TV at higher ϕ) showed that with increasing ϕ (> 0.1) the resulting droplet diameter increased and the dependence on energy consumption became weaker. Figure 8.12 shows a comparison of the average droplet diameter versus power consumption using different emulsifying machines.

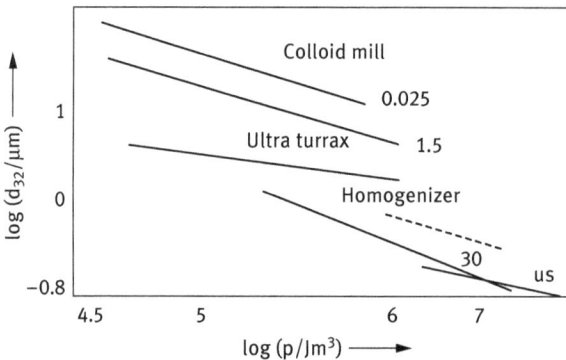

Fig. 8.12: Average droplet diameters obtained in various emulsifying machines as a function of energy consumption p. The number near the curves denote the viscosity ratio λ – the results for the homogenizer are for $\phi = 0.04$ (solid line) and $\phi = 0.3$ (broken line) – us means ultrasonic generator.

It can be seen that the smallest droplet diameters were obtained by using the high pressure homogenizers.

8.8 Selection of emulsifiers

8.8.1 The hydrophilic-lipophile balance (HLB) concept.

The selection of different surfactants for preparing either O/W or W/O emulsions is often still made on an empirical basis. A semi-empirical scale for selecting surfactants is the hydrophilic-lipophilic balance (HLB number) developed by Griffin [15]. This scale is based on the relative percentage of hydrophilic to lipophilic (hydrophobic) groups in the surfactant molecule(s). For an O/W emulsion droplet, the hydrophobic chain resides in the oil phase whereas the hydrophilic head group resides in the aqueous phase. For a W/O emulsion droplet, the hydrophilic group(s) reside in the water droplet, whereas the lipophilic groups reside in the hydrocarbon phase.

Table 8.1 gives a guide to the selection of surfactants for a particular application. The HLB number depends on the nature of the oil. As an illustration, Tab. 8.2 gives the required HLB numbers to emulsify various oils.

The relative importance of the hydrophilic and lipophilic groups was first recognized when using mixtures of surfactants containing varying proportions of a low and high HLB number.

The efficiency of any combination (as judged by phase separation) was found to pass a maximum when the blend contained a particular proportion of the surfactant

Tab. 8.1: Summary of HLB ranges and their applications.

HLB Range	Application
3–6	W/O Emulsifier
7–9	Wetting agent
8–18	O/W Emulsifier
13–15	Detergent
15–18	Solubilizer

Tab. 8.2: Required HLB numbers to emulsify various oils.

Oil	W/O Emulsion	O/W Emulsion
Paraffin oil	4	10
Beeswax	5	9
Linolin, anhydrous	8	12
Cyclohexane	—	15
Toluene	—	15

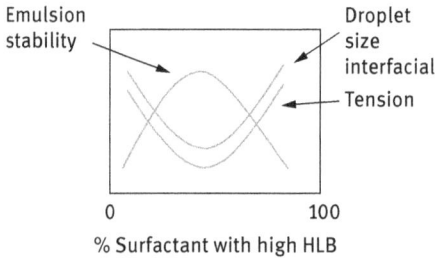

Fig. 8.13: Variation of emulsion stability, droplet size and interfacial tension with % surfactant with high HLB number.

with the higher HLB number. This is illustrated in Fig. 8.13 which shows the variation of emulsion stability, droplet size and interfacial tension with % surfactant with high HLB number.

The average HLB number may be calculated from additivity,

$$HLB = x_1 HLB_1 + x_2 HLB_2 . \tag{8.36}$$

x_1 and x_2 are the weight fractions of the two surfactants with HLB_1 and HLB_2.

Griffin [15] developed simple equations for calculating the HLB number of relatively simple nonionic surfactants. For a polyhydroxy fatty acid ester,

$$HLB = 20 \left(1 - \frac{S}{A} \right) . \tag{8.37}$$

S is the saponification number of the ester and A is the acid number. For a glyceryl monostearate, S = 161 and A = 198; the HLB is 3.8 (suitable for W/O emulsion).

For a simple alcohol ethoxylate, the HLB number can be calculated from the weight percent of ethylene oxide (E) and polyhydric alcohol (P),

$$HLB = \frac{E + P}{5} . \tag{8.38}$$

If the surfactant contains PEO as the only hydrophilic group, the contribution from one OH group is neglected,

$$HLB = \frac{E}{5} . \tag{8.39}$$

For a nonionic surfactant $C_{12}H_{25}-O-(CH_2-CH_2-O)_6$, the HLB is 12 (suitable for O/W emulsion).

The above simple equations cannot be used for surfactants containing propylene oxide or butylene oxide. They also cannot be applied for ionic surfactants. Davies [16, 17] devised a method for calculating the HLB number for surfactants from their chemical formulae, using empirically determined group numbers. A group number is assigned to various component groups. A summary of the group numbers for some surfactants is given in Tab. 8.3.

Tab. 8.3: HLB group numbers.

	Group Number
Hydrophilic	
$-SO_4Na^+$	38.7
$-COO-$	21.2
$-COONa$	19.1
N (tertiary amine)	9.4
Ester (sorbitan ring)	6.8
$-O-$	1.3
CH– (sorbitan ring)	0.5
Lipophilic	
$(-CH-), (-CH_2-), CH_3$	0.475
Derived	
$-CH_2-CH_2-O$	0.33
$-CH_2-CH_2-CH_2-O-$	-0.15

The HLB is given by the following empirical equation:

$$HLB = 7 + \sum(\text{hydrophilic group numbers}) - \sum(\text{lipohilic group numbers}). \quad (8.40)$$

Davies has shown that the agreement between HLB numbers calculated from the above equation and those determined experimentally is quite satisfactory.

Various other procedures were developed to obtain a rough estimate of the HLB number. Griffin found good correlation between the cloud point of 5 % solution of various ethoxylated surfactants and their HLB number.

Davies [16, 17] attempted to relate the HLB values to the selective coalescence rates of emulsions. Such correlations were not realized since it was found that the emulsion stability and even its type depend to a large extent on the method of dispersing the oil into the water and vice versa. At best, the HLB number can only be used as a guide for selecting optimum compositions of emulsifying agents.

One may take any pair of emulsifying agents that fall at opposite ends of the HLB scale, e.g. Tween 80 (sorbitan monooleate with 20 mol EO, HLB = 15) and Span 80 (sorbitan monooleate, HLB = 5) using them in various proportions to cover a wide range of HLB numbers. The emulsions should be prepared in the same way, with a few percent of the emulsifying blend. The stability of the emulsions is then assessed at each HLB number from the rate of coalescence or qualitatively by measuring the rate of oil separation. In this way one may be able to find the optimum HLB number for a given oil. Having found the most effective HLB value, various other surfactant pairs are compared at this HLB value, to find the most effective pair.

Shinoda and co-workers [18, 19] found that many O/W emulsions stabilized with non-ionic surfactants undergo a process of inversion at a critical temperature (PIT). The PIT can be determined by following the emulsion conductivity (small amount of electrolyte is added to increase the sensitivity) as a function of temperature. The conductivity of the O/W emulsion increases with increasing temperature until the PIT is reached, above which there will be a rapid reduction in conductivity (W/O emulsion is formed). Shinoda and co-workers found that the PIT is influenced by the HLB number of the surfactant. The size of the emulsion droplets was found to depend on the temperature and HLB number of the emulsifiers. The droplets are less stable towards coalescence close to the PIT. However, by rapid cooling of the emulsion a stable system may be produced. Relatively stable O/W emulsions were obtained when the PIT of the system was 20–65 °C higher than the storage temperature. Emulsions prepared at a temperature just below the PIT followed by rapid cooling generally have smaller droplet sizes. This can be understood if one considers the change of interfacial tension with temperature as illustrated in Fig. 8.14. The interfacial tension decreases with increasing temperature reaching a minimum close to the PIT, after which it increases.

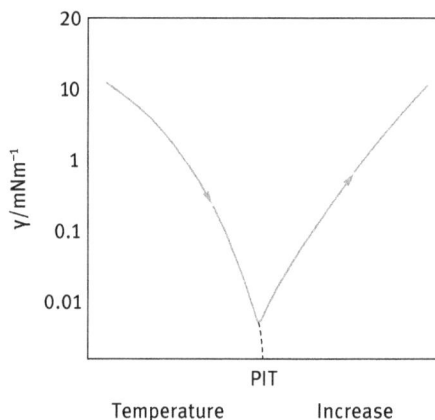

Fig. 8.14: Variation of interfacial tension with temperature increase for an O/W emulsion.

Thus, the droplets prepared close to the PIT are smaller than those prepared at lower temperatures. These droplets are relatively unstable towards coalescence near the PIT, but by rapid cooling of the emulsion one can retain the smaller size. This procedure may be applied to prepare mini(nano)emulsions.

The optimum stability of the emulsion was found to be relatively insensitive to changes in the HLB value or the PIT of the emulsifier, but instability was very sensitive to the PIT of the system.

It is essential, therefore, to measure the PIT of the emulsion as a whole (with all other ingredients).

At a given HLB value, stability of the emulsion against coalescence increases markedly as the molar mass of both the hydrophilic and lipophilic components increases. The enhanced stability using high molecular weight surfactants (polymeric surfactants) can be understood from a consideration of the steric repulsion which produces more stable films. Films produced using macromolecular surfactants resist thinning and disruption, thus reducing the possibility of coalescence. The emulsions showed maximum stability when the distribution of the PEO chains was broad. The cloud point is lower, but the PIT is higher than in the corresponding case for narrow size distributions. The PIT and HLB number are directly related parameters.

Addition of electrolytes reduces the PIT and hence an emulsifier with a higher PIT value is required when preparing emulsions in the presence of electrolytes. Electrolytes cause dehydration of the PEO chains and in effect this reduces the cloud point of the nonionic surfactant. One needs to compensate for this effect by using a surfactant with higher HLB. The optimum PIT of the emulsifier is fixed if the storage temperature is fixed.

In view of the above correlation between PIT and HLB and the possible dependence of the kinetics of droplet coalescence on the HLB number, Sherman and coworkers suggested the use of PIT measurements as a rapid method for assessing emulsion stability. However, one should be careful in using such methods for assessment of the long-term stability since the correlations were based on a very limited number of surfactants and oils.

Measuring the PIT can at best be used as a guide for preparing stable emulsions. An assessment of the stability should be evaluated by following the droplet size distribution as a function of time using a Coulter counter or light diffraction techniques. Following the rheology of the emulsion as a function of time and temperature may also be used for assessing stability against coalescence. Care should be taken in analyzing the rheological results. Coalescence results in an increase in droplet size and this is usually followed by a reduction in the viscosity of the emulsion. This trend is only observed if the coalescence is not accompanied by flocculation of the emulsion droplets (which results in an increase in the viscosity). Ostwald ripening can also complicate the analysis of the rheological data.

8.8.3 The cohesive energy ratio (CER) concept

Beerbower and Hills [20] considered the dispersing tendency on the oil and water interfaces of the surfactant or emulsifier in terms of the ratio of the cohesive energies of the mixtures of oil with the lipophilic portion of the surfactant and the water with the hydrophilic portion. They used the Winsor R_0 concept which is the ratio of the intermolecular attraction of oil molecules (O) and lipophilic portion of surfactant (L), C_{LO}, to that of water (W) and hydrophilic portion (H), C_{HW},

$$R_0 = \frac{C_{LO}}{C_{HW}}. \tag{8.41}$$

C_{LL}, C_{OO}, C_{LO} (at oil side)

C_{HH}, C_{WW}, C_{HW} (at water side)

C_{LW}, C_{HO}, C_{LH} (at the interface)

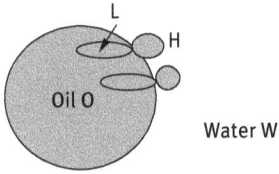

Fig. 8.15: The cohesive energy ratio concept.

Several interaction parameters may be identified at the oil and water sides of the interface. One can identify at least nine interaction parameters as schematically represented in Fig. 8.15.

In the absence of emulsifier, there will be only three interaction parameters: C_{OO}, C_{WW}, C_{OW}; if $C_{OW} \ll C_{WW}$, the emulsion breaks.

The above interaction parameters may be related to the Hildebrand [21] solubility parameter δ (at the oil side of the interface) and the Hansen [22] nonpolar, hydrogen bonding and polar contributions to δ at the water side of the interface. The solubility parameter of any component is related to its heat of vaporization ΔH by the expression

$$\delta^2 = \frac{\Delta H - RT}{V_m}, \tag{8.42}$$

where V_m is the molar volume.

Hansen considered δ (at the water side of the interface) to consist of three main contributions, a dispersion contribution, δ_d, a polar contribution, δ_p, and a hydrogen bonding contribution, δ_h. These contributions have different weighting factors,

$$\delta^2 = \delta_d^2 + 0.25\delta_p^2 + 0.25\delta_h^2. \tag{8.43}$$

Beerbower and Hills [20] used the following expression for the HLB number:

$$HLB = 20\left(\frac{M_H}{M_L + M_H}\right) = 20\left(\frac{V_H\rho_H}{V_L\rho_L + V_H\rho_H}\right), \tag{8.44}$$

where M_H and M_L are the molecular weights of the hydrophilic and lipophilic portions of the surfactants. V_L and V_H are their corresponding molar volumes whereas ρ_H and ρ_L are the densities respectively.

The cohesive energy ratio was originally defined by Winsor, equation (8.41).

When $C_{LO} > C_{HW}$, $R > 1$ and a W/O emulsion forms. If $C_{LO} < C_{HW}$, $R < 1$ and an O/W emulsion forms. If $C_{LO} = C_{HW}$, $R = 1$ and a planer system results – this denotes the inversion point.

R_o can be related to V_L, δ_L and V_H, δ_H by the expression,

$$R_o = \frac{V_L\delta_L^2}{V_H\delta_H^2}. \tag{8.45}$$

Using equation (8.43),

$$R_0 = \frac{V_L(\delta_d^2 + 0.25\delta_p^2 + 0.25\delta_h^2)_L}{V_H(\delta_d^2 + 0.25\delta_p^2 + 0.25\delta_h^2)_H}.$$ (8.46)

Combining equations (8.45) and (8.46), one obtains the following general expression for the cohesive energy ratio:

$$R_0 = \left(\frac{20}{HLB} - 1\right)\frac{\rho_H(\delta_d^2 + 025\delta_p^2 + 0.25\delta_h^2)_L}{\rho_L(\delta_d^2 + 0.25\delta_p^2 + 0.25\delta_p^2)_L}.$$ (8.47)

For an O/W system, HLB = 12–15 and R_0 = 0.58–0.29 ($R_0 < 1$). For a W/O system, HLB = 5–6 and R_0 = 2.3–1.9 ($R_0 > 1$). For a planer system, HLB = 8–10 and R_0 = 1.25–0.85 ($R_0 \sim 1$).

The R_0 equation combines both the HLB and cohesive energy densities – it gives a more quantitative estimate of emulsifier selection. R_0 considers HLB, molar volume and chemical match. The success of this approach depends on the availability of data on the solubility parameters of the various surfactant portions. Some values are tabulated in the book by Barton [23].

8.8.4 The critical packing parameter (CPP) for emulsion selection

The critical packing parameter (CPP) is a geometric expression relating the hydrocarbon chain volume (v) and length (l) and the interfacial area occupied by the head group (a) [24],

$$CPP = \frac{v}{l_c a_o}.$$ (8.48)

a_o is the optimal surface are per head group and l_c is the critical chain length.

Regardless of the shape of any aggregated structure (spherical or cylindrical micelle or a bilayer), no point within the structure can be farther from the hydrocarbon-water surface than l_c. The critical chain length, l_c, is roughly equal but less than the fully extended length of the alkyl chain.

The above concept can be applied to predict the shape of an aggregated structure. Consider a spherical micelle with radius r and aggregation number n; the volume of the micelle is given by

$$\left(\frac{4}{3}\right)\pi r^3 = nv,$$ (8.49)

where v is the volume of a surfactant molecule.

The area of the micelle is given by

$$4\pi r^2 = na_o,$$ (8.50)

where a_o is the area per surfactant head group.

Combining equations (8.49) and (8.50),

$$a_o = \frac{3v}{r}. \tag{8.51}$$

The cross-sectional area of the hydrocarbon chain a is given by the ratio of its volume to its extended length l_c

$$a = \frac{v}{l_c}. \tag{8.52}$$

From equations (8.51) and (8.52),

$$P = \frac{a}{a_o} = \left(\frac{1}{3}\right)\left(\frac{r}{l_c}\right). \tag{8.53}$$

Since $r < l_c$, then CPP $\leq \frac{1}{3}$.

For a cylindrical micelle with length d and radius r,

$$\text{Volume of the micelle} = \pi r^2 d = nv, \tag{8.54}$$

$$\text{Area of the micelle} = 2\pi rd = na_o. \tag{8.55}$$

Combining equations (8.54) and (8.55),

$$a_o = \frac{2v}{r}, \tag{8.56}$$

$$a = \frac{v}{l_c}, \tag{8.57}$$

$$P = \frac{a}{a_o} = \left(\frac{1}{2}\right)\left(\frac{r}{l_c}\right). \tag{8.58}$$

Since $r < l_c$, then $\frac{1}{3} < $ CPP $\leq \frac{1}{2}$.

For vesicles (liposomes) $1 > $ CPP $\geq \frac{2}{3}$ and for lamellar micelles P ~ 1. For inverse micelles CPP > 1. A summary of the various shapes of micelles and their CPP is given in Tab. 8.4.

Surfactants that make spherical micelles with the above packing constraints, i.e. CPP $\leq \frac{1}{3}$, are more suitable for O/W emulsions. Surfactants with CPP > 1, i.e. forming inverted micelles, are suitable for forming W/O emulsions.

8.9 Preparation of nanoemulsions using high energy methods

As mentioned above, emulsification combines the creation of fine droplets and their stabilization against coalescence. The emulsion droplets are created by premixing the lipophilic and hydrophilic phases. The coarse droplets are then finely dispersed in the μm range or even smaller by deforming and disrupting them at high specific energy. These droplets have to be stabilized against coalescence by using an efficient emulsifier. The latter must adsorb quickly at the oil/water interface to prevent droplet

Tab. 8.4: CPP and shape of micelles.

Lipid	Critical packing parameter v/a_0l_c	Critical packing shape	Structures formed
Single-chained lipids (surfactants) with large head-group areas: – SDS in low salt	< 1/3	Cone	Spherical micelles
Single-chained lipids with small head-group areas: – SDS and CTAB in high salt – nonionic lipids	1/3–1/2	Truncated cone	Cylindrical micelles
Double-chained lipids with large head-group areas, fluid chains: – Phosphatidyl choline (lecithin) – phosphatidyl serine – phosphatidyl glycerol – phosphatidyl inositol – phosphatidic acid – sphingomyelin, DGDG[a] – dihexadecyl phosphate – dialkyl dimethyl ammonium – salts	1/2–1	Truncated one	Flexible bilayers, vesicles
Double-chained lipids with small head-group areas, anionic lipids in high salt, saturated frozen chains: – phosphatidyl ethanaiamine – phosphatidyl serine + Ca^{2+}	~ 1	Cylinder	Planar bilayers
Double-chained lipids with small head-group areas, nonionic lipids, poly(cis) unsaturated chains, high T: – unsat. phosphatidyl ethanolamine – cardiolipin + Ca^{2+} – phosphatidic acid + Ca^{2+} – cholesterol, MGDG[b]	> 1	Inverted truncated cone or wedge	Inverted micelles

a DGDG: digalactosyl diglyceride, diglucosyldiglyceride
b MGDG: monogalactosyl diglyceride, monoglucosyl diglyceride

coalescence during emulsification. In most cases, a synergestic mixture of emulsifiers is used.

In most cases the nanoemulsions is produced in two stages, firstly by using a rotor-stator mixer (such as an Ultra-Turrax or Silverson) that can produce droplets in the μm range, followed by high pressure homogenization (reaching 3000 bar) to produce droplets in the nanometer size (as low as 50 nm).

The rotor-stator mixer consists of a rotating and a fixed machine part [25]. Different geometries are available with various sizes and gaps between the rotor and stator. The simplest rotor-stator machine is a vessel with a stirrer, which is used to produce the emulsion batchwise or quasi-continuously. The power density is relatively low and broadly distributed. Therefore, small mean droplet diameter (< 1 μm) can rarely be produced. In addition, a long residence time and emulsification for several minutes are required, often resulting in broad droplet size distribution. Some of these problems can be overcome by reducing the disruption zone that enhances the power density, e.g. using colloid mills or toothed-disc dispersing machines.

To produce submicron droplets, high pressure homogenization is commonly used [25]. These homogenizers are operated continuously and throughputs up to several thousand liters per hour can be achieved. The homogenizer consists essentially of a high pressure pump and a homogenization nozzle. The pump creates the pressure which is then transferred within the nozzle to kinetic energy that is responsible for droplet disintegration. The design of the homogenization nozzle influences the flow pattern of the emulsion in the nozzle and hence droplet disruption. A good example of efficient homogenization nozzles are opposing jets that operate in the Microfluidizer. Other examples are the jet disperser (designed by Bayer) and the simple orifice valve. Droplet disruption in high pressure homogenizers is predominantly due to inertial forces in turbulent flow, shear forces in laminar elongational flow, as well as cavitation.

Droplets can also be disrupted by means of ultrasonic waves (frequency > 18 kHz) which cause cavitation that induces microjets and zones of high microturbulence [25]. A batchwise operation at small scale has been applied in the laboratory, especially for low viscosity systems. Continuous application requires the use of a flow chamber of special design into which the ultrasound waves are introduced. Due to the limited power of sound inducers, there are technical limits for high throughput.

Another method that can be applied for droplet disintegration is the use of microchannel systems (membrane emulsification). This can be realized by pressing the disperse phase through microporous membrane pores [25]. Droplets are formed at the membrane surface and detached from it by wall shear stress of the continuous phase. In addition to tubular membranes made from ceramics like aluminium oxide, special porous glasses and polymers like polypropylene, polyterafluoroethylene (PTFE), nylon and silicon have been used. The membrane's surface wetting behaviour is of major influence; if the membrane is wetted by the continuous phase only, emulsions of very narrow droplet size distribution are produced with mean droplet sizes in the range of three times the mean droplet diameter of the pore. The pressure to be applied should ideally be a little above the capillary pressure. Membrane emulsification reduces the shear forces acting in droplet formation.

8.10 Emulsification process functions

The droplets are disrupted if they are deformed over a period of time t_{def} that is longer than a critical deformation time $t_{defcrit}$ and if the deformation described by the Weber number We, equation (8.25), exceeds a critical value We_{cr}. The droplet-deforming tensions are supplied by the continuous phase.

In turbulent flow, the droplets are disrupted mostly by inertial forces that are generated by energy dissipating small eddies. Due to internal viscous forces the droplets try to regain their initial form and size. Two dimensionless numbers, the turbulent Weber number We_{turb} and the Ohnesorge number Oh, characterize the tensions working on droplets in deformation and break-up [25].

$$We_{turb} = \frac{C^2 P_v^{2/3} \rho_c^{1/3} x^{5/3}}{\gamma} \tag{8.59}$$

$$Oh = \frac{\eta_d}{(\gamma \rho_d x)^{1/2}} \tag{8.60}$$

C is a constant, P_v is the volumetric power density, ρ_c the continuous phase viscosity, ρ_d the droplet density and x is the droplet diameter.

Droplet disruption in laminar shear flow is restricted to a narrow range of viscosity ratio between the disperse phase and continuous phase (η_d/η_c) for single droplet disruption, or between the disperse phase and emulsion (η_d/η_e) for emulsions. For laminar shear flow,

$$x_{3,2} \propto E_v^{-1} f(\eta_d/\eta_e) . \tag{8.61}$$

And for laminar elongational flow,

$$x_{3,2} \propto E_v^{-1} . \tag{8.62}$$

where E_v is the volumetric energy density or specific disruption energy.

Laminar elongational flow is successfully applied in innovative high pressure homogenization valves, where it adds to the effect of turbulent droplet disruption by predeforming the droplets. Thus, the droplet disruption efficiency of high pressure homogenization can be significantly increased, especially for droplets with high viscosities.

8.11 Enhancing of the process of forming nanoemulsions

The intensity of the process or the effectiveness in making small droplets is often governed by the net power density ($\varepsilon(t)$).

$$p = \varepsilon(t)\,dt\,, \tag{8.63}$$

where t is the time during which emulsification occurs.

Break-up of droplets will only occur at high ε values, which means that the energy dissipated at low ε levels is wasted. Batch processes are generally less efficient than continuous processes. This shows why with a stirrer in a large vessel, most of the energy applied at low intensity is dissipated as heat. In a homogenizer, p is simply equal to the homogenizer pressure [4, 5].

Several procedures may be applied to enhance the efficiency of emulsification when producing nanoemulsions: One should optimize the efficiency of agitation by increasing ε and decreasing dissipation time. The emulsion is preferably prepared at high volume fraction ϕ of the disperse phase and diluted afterwards. However, very high ϕ values may result in coalescence during emulsification. Alternatively, more surfactant could be added which would create a smaller γ_{eff} and possibly diminish recoalescence. Also a surfactant mixture could be used that shows more reduction in γ of the individual components. If possible, dissolve the surfactant in the disperse phase rather than the continuous phase; this often leads to smaller droplets. It may be useful to emulsify in steps of increasing intensity, particularly with emulsions having highly viscous disperse phase.

References

[1] Tadros, Th. F. and Vincent, B., in "Encyclopedia of Emulsion Technology", P. Becher (ed.), Marcel Dekker, New York (1983).
[2] Binks, B. P. Editor, "Modern Aspects of Emulsion Science", The Royal Society of Chemistry Publication (1998).
[3] Tadros, Th. F., "Applied Surfactants" Wiley-VCH, Germany (2005).
[4] Tadros, Th. F. (ed.), "Emulsion Science and Technology", Wiley-VCH, Germany (2009).
[5] Tadros, Th. F. (ed.), "Emulsion Formation and Stability", Wiley-VCH, Germany (2013).
[6] Walstra, P. and Smolders, P. E. A., in "Modern Aspects of Emulsions", B. P. Binks (ed.), The Royal Society of Chemistry, Cambridge (1998).
[7] Stone, H. A., Ann. Rev. Fluid Mech., **226**, 95 (1994).
[8] Wierenga, J. A., ven Dieren, F., Janssen, J. J. M. and Agterof, W. G. M., Trans. Inst. Chem. Eng., **74-A**, 554 (1996).
[9] Levich, V. G., "Physicochemical Hydrodynamics", Prentice-Hall, Englewood Cliffs (1962).
[10] Davis, J. T., "Turbulent Phenomena", Academic Press, London (1972).
[11] Lucassen-Reynders, E. H., in "Encyclopedia of Emulsion Technology", P. Becher (ed.), Marcel Dekker, New York (1996).
[12] Lucassen-Reynders, E. H., Colloids and Surfaces, **A91**, 79 (1994).

[13] Lucassen, J., in "Anionic Surfactants", E. H. Lucassen-Reynders (ed.), Marcel Dekker, New York (1981).

[14] van den Tempel, M., Proc. Int. Congr. Surf. Act., **2**, 573 (1960).

[15] Griffin, W. C., J. Cosmet. Chemists, **1**, 311 (1949); **5**, 249 (1954).

[16] Davies, J. T., Proc. Int. Congr. Surface Activity, Vol. 1, p. 426 (1959).

[17] Davies, J. T. and Rideal, E. K., "Interfacial Phenomena", Academic Press, New York (1961).

[18] Shinoda, K., J. Colloid Interface Sci., **25**, 396 (1967).

[19] Shinoda, K. and Saito, H., J. Colloid Interface Sci., **30**, 258 (1969).

[20] Beerbower, A. and Hill, M. W., Amer. Cosmet. Perfum., **87**, 85 (1972).

[21] Hildebrand, J. H., "Solubility of Non-Electrolytes", 2nd Ed., Reinhold, New York (1936).

[22] Hansen, C. M., J. Paint Technol., **39**, 505 (1967).

[23] Barton, A. F. M., "Handbook of Solubility Parameters and Other Cohesive Parameters", CRC Press, New York (1983).

[24] Israelachvili, J. N., Mitchell, J. N. and Ninham, B. W., J. Chem. Soc., Faraday Trans. II, **72**, 1525 (1976).

[25] Schuchmann, H. P., "Emulsification Techniques for the Formulation of Emulsions and Suspensions", in "Product Design and Engineering", U. Brockel, W. Meier and G. Wagner (eds.), Vol. 1, Ch. 4, Wiley-VCH, Germany (2007).

9 Low energy methods for preparation of nanoemulsions and practical examples of nanoemulsions

9.1 Introduction

The low energy methods for preparing nanoemulsions are of particular interest, being more economical to produce and offering the possibility of producing narrow droplet distribution nanoemulsions. In these methods, the chemical energy of the components is the key factor for the emulsification. The most well-known low energy emulsification methods are direct or self-emulsification [1–5] and phase inversion methods [6–8]. Generally, emulsification by low energy methods allows obtaining smaller and more uniform droplets.

In the so-called direct or self-emulsification methods, emulsification is achieved by a dilution process at a constant temperature, without any phase transitions (no change in the spontaneous curvature of the surfactant) taking place in the system during emulsification [1–5]. In this case, oil-in-water nanoemulsions (O/W) are obtained by the addition of water over a direct microemulsion phase, whereas water-in-oil nanoemulsions (W/O) are obtained by the addition of oil over an indirect microemulsion phase. This method is described in detail below. This self-emulsification method uses the chemical energy of dissolution in the continuous phase of the solvent present in the initial system (which is going to constitute the disperse phase). When the intended continuous phase and the intended disperse phase are mixed, the solvent present in the later phase is dissolved into the continuous phase, dragging and dispersing the micelles of the initial system, thus giving rise to the nanoemulsion droplets.

Phase inversion methods make use of the chemical energy released during the emulsification process as a consequence of a change in the spontaneous curvature of surfactant molecules, from negative to positive (obtaining oil-in-water, O/W, nanoemulsions) or from positive to negative (obtaining water-in-oil, W/O, nanoemulsions). This change of surfactant curvature can be achieved by a change in composition keeping the temperature constant (phase inversion composition method, PIC) [6, 7], or by a rapid change in temperature with no variation in composition (phase inversion temperature method, PIT) [8]. The PIT method can only be applied to systems with surfactants sensitive to changes in temperature, i.e. the POE-type surfactants, in which changes in temperature induce a change in the hydration of the poly(oxyethylene) chains, and thus, a change of curvature [8, 9]. In the PIC method, the change in curvature is induced by the progressive addition of the intended continuous phase, which may be pure water or oil [6, 7] over the mixture of the intended disperse phase (oil or water and surfactant/s).

Studies on surfactant phase behaviour are important when the low energy emulsification methods are used, since the phases involved during emulsification are crucial
in order to obtain nanoemulsions with small droplet size and low polydispersity. In
contrast, if shear methods are used, only phases present at the final composition are
important.

9.2 Phase inversion composition (PIC) Principle

A study of the phase behaviour of water/oil/surfactant systems demonstrated that
emulsification can be achieved by three different low energy emulsification methods,
as shown schematically in Fig. 9.1: (A) Stepwise addition of oil to a water surfactant
mixture. (B) Stepwise addition of water to a solution of the surfactant in oil. (C) Mixing all the components in the final composition, pre-equilibrating the samples prior to
emulsification. In these studies, the system water/Brij 30 (polyoxyethlene lauryl ether
with an average of 4 mol of ethylene oxide)/decane was chosen as a model to obtain
O/W emulsions. The results showed that nanoemulsions with droplet sizes of the order of 50 nm were formed only when water was added to mixtures of surfactant and oil
(method (B)) whereby inversion from W/O emulsion to O/W nanoemulsion occurred.

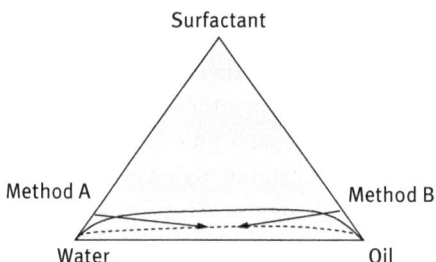

Fig. 9.1: Schematic representation of the experimental path in two emulsification methods.
Method (A): addition of decane to water/
surfactant mixture. Method (B): addition of
water to decane/Brij 30 solutions.

9.3 Phase inversion temperature (PIT) Principle

Phase inversion in emulsions can be one of two types: Transitional inversion induced
by changing factors which affect the HLB of the system, e.g. temperature and/or electrolyte concentration, or catastrophic inversion which is induced by increasing the
volume fraction of the disperse phase.

Transitional inversion can also be induced by changing the HLB number of the
surfactant at constant temperature using surfactant mixtures. This is illustrated in
Fig. 9.2 which shows the average droplet diameter and rate constant for attaining constant droplet size as a function of the HLB number. It can be seen that the diameter
decreases and the rate constant increases as inversion is approached.

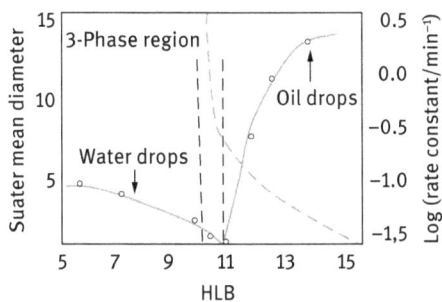

Fig. 9.2: Emulsion droplet diameters (circles) and rate constant for attaining steady size (squares) as a function of HLB – cyclohexane/nonylphenol ethoxylate.

For application of the phase inversion principle one uses the transitional inversion method which has been demonstrated by Shinoda and co-workers [8, 9] when using nonionic surfactants of the ethoxylate type. These surfactants are highly dependent on temperature, becoming lipophilic with increasing temperature due to the dehydration of the polyethyleneoxide chain. When an O/W emulsion is prepared using a nonionic surfactant of the ethoxylate type and is heated, then at a critical temperature (the PIT), the emulsion inverts to a W/O emulsion. At the PIT the droplet size reaches a minimum and the interfacial tension also reaches a minimum. However, the small droplets are unstable and they coalesce very rapidly. By rapid cooling of the emulsion that is prepared at a temperature near the PIT, very stable and small emulsion droplets could be produced.

A clear demonstration of the phase inversion that occurs on heating an emulsion is illustrated from a study of the phase behaviour of emulsions as a function of temperature. This is illustrated in Fig. 9.3 which shows schematically what happens when the temperature is increased [10, 11]. At low temperature, over the Winsor I region, O/W macroemulsions can be formed and are quite stable. On increasing the temperature, the O/W emulsion stability decreases and the macroemulsion finally resolves when the system reaches the Winsor III phase region (both O/W and W/O emulsions are unstable). At higher temperature, over the Winsor II region, W/O emulsions become stable.

Near the HLB temperature, the interfacial tension reaches a minimum. This is illustrated in Fig. 9.4. Thus, by preparing the emulsion at a temperature 2–4 °C below the PIT (near the minimum in γ) followed by rapid cooling of the system, nanoemulsions may be produced. The minimum in γ can be explained in terms of the change in curvature H of the interfacial region, as the system changes from O/W to W/O. For an O/W system and normal micelles, the monolayer curves towards the oil and H is given a positive value. For a W/O emulsion and inverse micelles, the monolayer curves towards the water and H is assigned a negative value. At the inversion point (HLB temperature) H becomes zero and γ reaches a minimum.

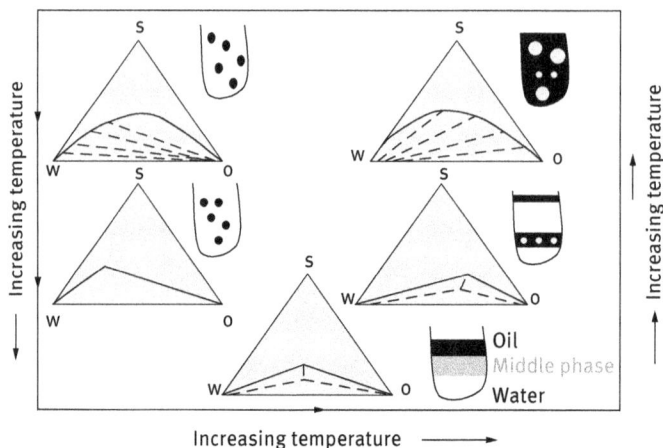

Fig. 9.3: The PIT concept.

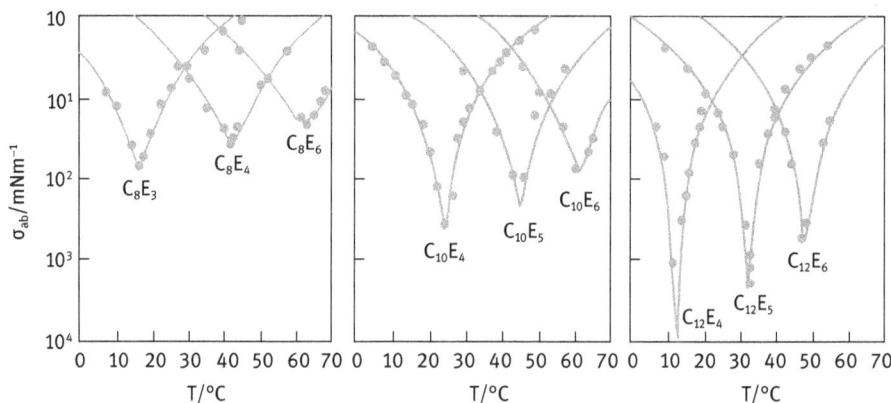

Fig. 9.4: Interfacial tension of n-octane against water in the presence of various C_nE_m surfactants above the cmc as a function of temperature.

9.4 Preparation of nanoemulsions by dilution of microemulsions

A common way to prepare nanoemulsions by self-emulsification is to dilute with water an O/W microemulsion. When diluting a microemulsion with water, part of the surfactant and/or cosurfactant diffuses to the aqueous phase. The droplets are no longer thermodynamically stable, since the surfactant concentration is not high enough to maintain the ultra-low interfacial tension ($< 10^{-4}$ mN m^{-1}) for thermodynamic stability. The system becomes unstable and the droplets show a tendency to grow by coalescence and/or Ostwald ripening forming a nanoemulsion. This is illustrated in Fig. 9.5 which shows the phase diagram of the system water/SDS-hexanol (ratio of 1 : 1.76)/dodecane.

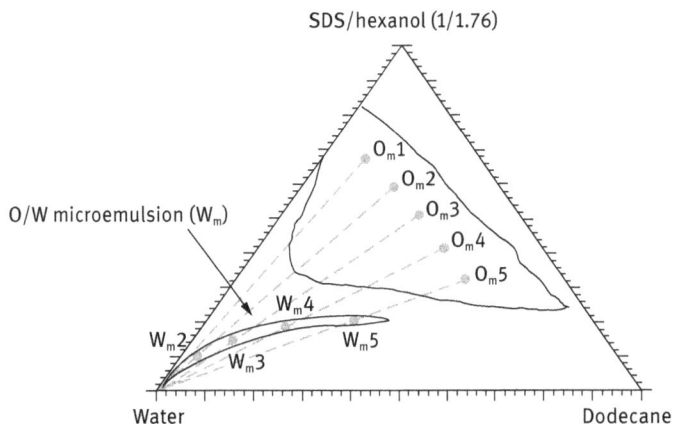

Fig. 9.5: Pseudoternary phase diagram of water/SDS/hexanol/dodecane with SDS : hexanol ratio of 1 : 1.76. Solid and dashed lines indicate the emulsification paths followed starting from both O/W (W_m) and W/O (O_m) microemulsion domains.

Nanoemulsions can be prepared starting from microemulsions located in the inverse microemulsion domain, O_m, and in the direct microemulsion domain, W_m, at different oil : surfactant ratios ranging from 12 : 88 to 40 : 60, and coincident for both types of microemulsions. The water concentration is fixed at 20 % for microemulsions in the O_m domain labelled as O_m1, O_m2, O_m3, O_m4, O_m5. The microemulsions in the W_m region are accordingly W_m2, W_m3, W_m4, W_m5 and their water content decreases from W_m2 to W_m5.

Several emulsification methods can be applied: (a) addition of microemulsion into water in one step; (b) addition of microemulsion into water stepwise; (c) addition of water into microemulsion in one step; (d) addition of water into microemulsion stepwise. The final water content is kept constant at 98 wt %.

Starting emulsification from W_m microemulsions, low-polydispersed nanoemulsions with droplet sizes within the range 20–40 nm are obtained regardless of the emulsification method used. When starting from O_m microemulsions, the nanoemulsion formation and properties depend on the emulsification method. From O_m1 microemulsion, a turbid emulsion with rapid creaming is obtained whatever method is used. In this case the direct microemulsion region W_m is not crossed. Starting from O_m2 to O_m5 and using emulsification method (d) in which water is gradually added to the microemulsion, the nanoemulsion droplet sizes coincide with those obtained starting from microemulsions in the W_m domain for the corresponding O : S ratio. Methods (a), (b), and (c) produce coarse emulsions.

9.5 Steric stabilization and the role of the adsorbed layer thickness

Since most nanoemulsions are prepared using nonionic and/or polymeric surfactants, it is necessary to consider the interaction forces between droplets containing adsorbed layers (steric stabilization). This was described in detail in Chapter 2 and here only the importance of the ratio of adsorbed layer thickness to droplet radius is emphasized. The energy G_T–distance h curve for a sterically stabilized system shows a minimum G_{min} at a separation distance h ~ 2δ and when h < 2δ, G_T increases very sharply with decreasing h. The magnitude of G_{min} depends on the particle radius R, the Hamaker constant A and the adsorbed layer thickness δ. As an illustration, Fig. 9.6 shows the variation of G_T with h at various ratios of δ/R. The depth of the minimum decreases with increasing δ/R. This is the basis of the high kinetic stability of nanoemulsions. With nanoemulsions having a radius in the region of 50 nm and an adsorbed layer thickness of say 10 nm, the value of δ/R is 0.2. This high value (when compared with the situation with macroemulsions where δ/R is at least an order of magnitude lower) results in a very shallow minimum (which could be less than kT). This above situation results in very high stability with no flocculation (weak or strong). In addition, the very small size of the droplets and the dense adsorbed layers ensure lack of deformation of the interface, lack of thinning and disruption of the liquid film between the droplets and hence coalescence is also prevented.

9.6 Ostwald ripening in nanoemulsions

The only instability problem with nanoemulsions is Ostwald ripening which was discussed in detail in Chapter 3. Several methods may be applied to reduce Ostwald ripening [12–14]: (i) Addition of a second disperse phase component which is insoluble in the continuous phase (e.g. squalane). In this case significant partitioning between different droplets occurs, with the component having low solubility in the continuous phase expected to be concentrated in the smaller droplets. During Ostwald ripening in a two component disperse phase system, equilibrium is established when the difference in chemical potential between differently sized droplets (which results from curvature effects) is balanced by the difference in chemical potential resulting from partitioning of the two components. If the secondary component has zero solubility in the continuous phase, the size distribution will not deviate from the initial one (the growth rate is equal to zero). In the case of limited solubility of the secondary component, the distribution is the same, i.e. a mixture growth rate is obtained which is still lower than that of the more soluble component. (ii) Modification of the interfacial film at the O/W interface: Reduction in γ results in a reduction in Ostwald ripening. However, this alone is not sufficient since one has to reduce γ by several orders of magnitude. Walstra [14] suggested that by using surfactants which are strongly adsorbed at the O/W interface (i.e. polymeric surfactants) and which do not desorb during ripening, the rate could be significantly reduced. An increase in the surface dilational modulus and a decrease in γ would be observed for the shrinking drops. The difference in γ between the droplets would balance the difference in capillary pressure (i.e. curvature effects). To achieve this effect it is useful to use A–B–A block copolymers that are soluble in the oil phase and insoluble in the continuous phase. The polymeric surfactant should enhance the lowering of γ by the emulsifier. In other words, the emulsifier and the polymeric surfactant should show synergy in lowering γ.

9.7 Practical examples of nanoemulsions

Several experiments have been carried out to investigate methods of preparing nanoemulsions and their stability [15]. The first method applied the PIT principle for preparation of nanoemulsions. Experiments were carried out using hexadecane and isohexadecane (Arlamol HD) as the oil phase and Brij 30 ($C_{12}EO_4$) as the nonionic emulsifier. The phase diagrams of the ternary system water–$C_{12}EO_4$–hexadecane and water–$C_{12}EO_4$–isohexadecane are shown in Figures 9.6 and 9.7. The main features of the pseudoternary system are as follows: (i) O_m isotropic liquid transparent phase, which extends along the hexadecane–$C_{12}EO_4$ or isohexadecane–$C_{12}EO_4$ axis, corresponding to inverse micelles or W/O microemulsions; (ii) L_α lamellar liquid crystalline phase extending from the W–$C_{12}EO_4$ axis toward the oil vertex; (iii) the rest of the phase diagram consists of two- or three-phase regions: ($W_m + O$) two-liquid-phase region, which appears along the water-oil axis; ($W_m + L_\alpha + O$) three-phase region, which consists of a bluish liquid phase (O/W microemulsion), a lamellar liquid crystalline phase (L_α) and a transparent oil phase; ($L_\alpha + O_m$) two-phase region consisting of an oil and liquid crystalline region; MLC a multiphase region containing a lamellar liquid crystalline phase (L_α). The HLB temperature was determined using conductivity measurements, whereby 10^{-2} mol dm^{-3} NaCl was added to the aqueous phase (to increase the sensitivity of the conductibility measurements). The concentration of NaCl was low and hence it had little effect on the phase behaviour.

Figure 9.8 shows the variation of conductivity versus temperature for 20 % O/W emulsions at different surfactant concentrations. It can be seen that there is a sharp decrease in conductivity at the PIT or HLB temperature of the system.

The HLB temperature decreases with increasing surfactant concentration – this could be due to the excess nonionic surfactant remaining in the continuous phase.

However, at a concentration of surfactant higher than 5 %, the conductivity plots showed a second maximum (Fig. 9.8). This was attributed to the presence of L_α phase and bicontinuous L3 or D′ phases [16].

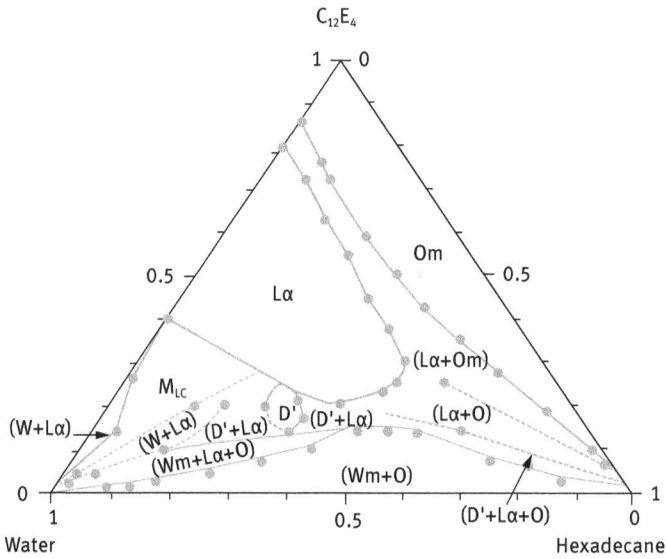

Fig. 9.6: Pseudoternary phase diagram at 25 °C of the system water–$C_{12}EO_4$–hexadecane.

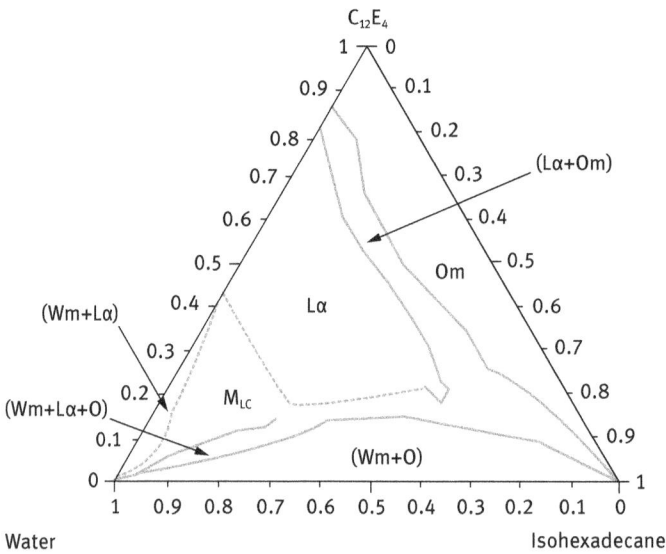

Fig. 9.7: Pseudoternary phase diagram at 25 °C of the system water–$C_{12}EO_4$–isohexadecane.

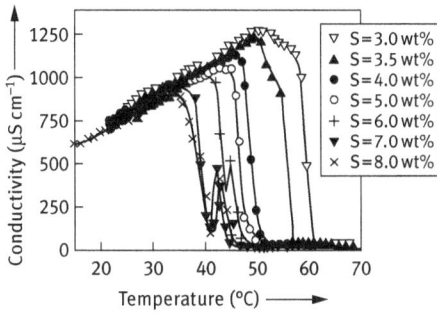

Fig. 9.8: Conductivity versus temperature for a 20 : 80 hexadecane : water emulsion at various $C_{12}EO_4$ concentrations.

Nanoemulsions were prepared by rapid cooling of the system to 25 °C. The droplet diameter was determined using photon correlation spectroscopy (PCS). The results are summarized in Tab. 9.1, which shows the exact composition of the emulsions, HLB temperature, z-average radius and polydispersity index.

O/W nanoemulsions with droplet radii in the range 26–66 nm could be obtained at surfactant concentrations between 4 and 8 %. The nanoemulsion droplet size and polydispersity index decreases with increasing surfactant concentration. The decrease in droplet size with increasing surfactant concentration is due to the increase in surfactant interfacial area and the decrease in interfacial tension, γ.

Tab. 9.1: Composition, HLB temperature (T_{HLB}), droplet radius r and polydispersity index (pol.) for the system water–$C_{12}EO_4$–hexadecane at 25 °C.

Surfactant (wt %)	Water (wt %)	Oil/Water	T_{HLB} (°C)	r (nm)	pol.
2.0	78.0	20.4/79.6	—	320	1.00
3.0	77.0	20.6/79.4	57.0	82	0.41
3.5	76.5	20.7/79.3	54.0	69	0.30
4.0	76.0	20.8/79.2	49.0	66	0.17
5.0	75.0	21.2/78.9	46.8	48	0.09
6.0	74.0	21.3/78.7	45.6	34	0.12
7.0	73.0	21.5/78.5	40.9	30	0.07
8.0	72.0	21.7/78.3	40.8	26	0.08

As mentioned above, γ reaches a minimum at the HLB temperature. Therefore, the minimum in interfacial tension occurs at a lower temperature as the surfactant concentration increases. This temperature becomes closer to the cooling temperature as the surfactant concentration increases and this results in smaller droplet sizes.

All nanoemulsions showed an increase in droplet size with time, as a result of Ostwald ripening. Figure 9.9 shows plots of r^3 versus time for all the nanoemulsions studied. The slope of the lines gives the rate of Ostwald ripening ω ($m^3\,s^{-1}$) and this showed an increase from 2×10^{-27} to $39.7 \times 10^{-27}\ m^3\,s^{-1}$ as the surfactant concentration is increased from 4 to 8 wt %. This increase could be due to a number of factors: (i) Decreasing droplet size increases Brownian diffusion and this enhances the rate. (ii) Presence of micelles, which increases with increasing surfactant concentration. This has the effect of increasing the solubilization of the oil into the core of the micelles. This results in an increase of the flux J of diffusion of oil molecules from different size droplets. Although the diffusion of micelles is slower than the diffusion of oil molecules, the concentration gradient $(\partial C/\partial X)$ can be increased by orders of magnitude as a result of solubilization. The overall effect will be an increase in J and this may enhance Ostwald ripening. (iii) Partition of surfactant molecules between the oil and aqueous phases. With higher surfactant concentrations, the molecules with shorter EO chains (lower HLB number) may preferentially accumulate at the O/W interface and this may result in reduction of the Gibbs elasticity, which in turn results in an increase in the Ostwald ripening rate.

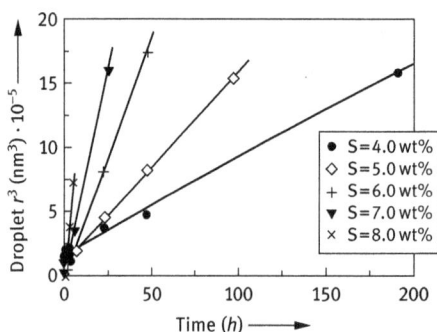

Fig. 9.9: r^3 versus time at 25 °C for nanoemulsions prepared using the system water–$C_{12}EO_4$–hexadecane.

The results with isohexadecane are summarized in Tab. 9.2. As with the hexadecane system, droplet size and the polydispersity index decreased with increasing surfactant concentration. Nanoemulsions with droplet radii of 25–80 nm were obtained at 3–8 % surfactant concentration. It should be noted, however, that nanoemulsions could be produced at a lower surfactant concentration when using isohexadecane, when compared with the results obtained with hexadecane. This could be attributed to the higher solubility of the isohexadecane (a branched hydrocarbon), the lower HLB temperature and the lower interfacial tension.

The stability of the nanoemulsions prepared using isohexadecane was assessed by following the droplet size as a function of time. Plots of r^3 versus time for four surfactant concentrations (3, 4, 5, and 6 wt %) are shown in Fig. 9.10. The results show an increase in the Ostwald ripening rate as the surfactant concentration is increased from 3 to 6 % (the rate increased from 4.1×10^{-27} to 50.7×10^{-27} m^3 s^{-1}). The nanoemulsions prepared using 7 wt % surfactant were so unstable that they showed significant creaming after 8 hours. However, when the surfactant concentration was increased to 8 wt %, a very stable nanoemulsion could be produced with no apparent increase in

Tab. 9.2: Composition, HLB temperature (T_{HLB}), droplet radius r and polydispersity index (pol.) at 25 °C for emulsions in the system water–$C_{12}EO_4$–isohexadecane.

Surfactant (wt %)	Water (wt %)	Oil/Water	T_{HLB} (°C)	r (nm)	pol.
2.0	78.0	20.4/79.6	–	97	0.50
3.0	77.0	20.6/79.4	51.3	80	0.13
4.0	76.0	20.8/79.2	43.0	65	0.06
5.0	75.0	21.2/78.9	38.8	43	0.07
6.0	74.0	21.3/78.7	36.7	33	0.05
7.0	73.0	21.5/78.5	33.4	29	0.06
8.0	72.0	21.7/78.3	32.7	27	0.12

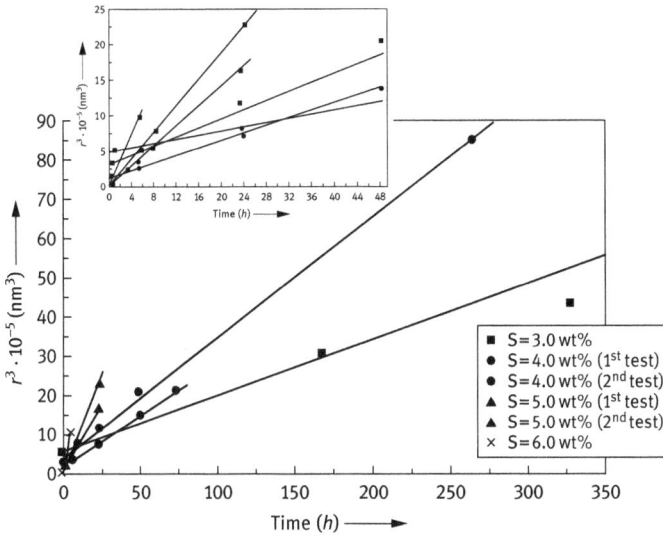

Fig. 9.10: r^3 versus time at 25 °C for the system water–$C_{12}EO_4$–isohexadecane at various surfactant concentrations; O/W ratio 20/80.

droplet size over several months. This unexpected stability was attributed to the phase behaviour at such surfactant concentrations. The sample containing 8 wt % surfactant showed birefringence to shear when observed under polarized light. It seems that the ratio between the phases ($W_m + L_\alpha + O$) may play a key factor in nanoemulsion stability. Attempts were made to prepare nanoemulsions at higher O/W ratios (hexadecane being the oil phase), while keeping the surfactant concentration constant at 4 wt %. When the oil content was increased to 40 and 50 %, the droplet radius increased to 188 and 297 nm respectively. In addition, the polydispersity index also increased to 0.95. These systems become so unstable that they showed creaming within a few hours. This is not surprising, since the surfactant concentration is not sufficient to produce the nanoemulsion droplets with high surface area. Similar results were obtained with isohexadecane. However, nanoemulsions could be produced using 30/70 O/W ratio (droplet size being 81 nm), but with high polydispersity index (0.28). The nanoemulsions showed significant Ostwald ripening.

The effect of changing the alkyl chain length and branching was investigated using decane, dodecane, tetradecane, hexadecane and isohexadecane. Plots of r^3 versus time are shown in Fig. 9.11 for 20/80 O/W ratio and surfactant concentration of 4 wt %. As expected, by reducing the oil solubility from decane to hexadecane, the rate of Ostwald ripening decreases. The branched oil isohexadecane also shows a higher Ostwald ripening rate when compared with hexadecane. A summary of the results is shown in Tab. 9.3 which also shows the solubility of the oil $C(\infty)$.

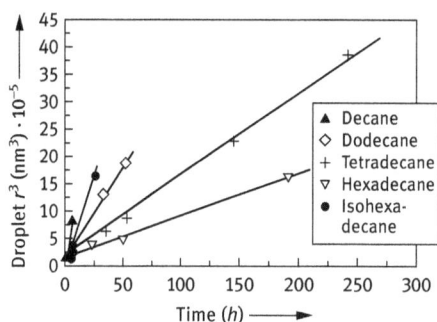

Fig. 9.11: r^3 versus time at 25 °C for nanoemulsions (O/W ratio 20/80) with hydrocarbons of various alkyl chain lengths. System water–$C_{12}EO_4$–hydrocarbon (4 wt % surfactant).

Tab. 9.3: HLB temperature (T_{HLB}), droplet radius r, Ostwald ripening rate ω and oil solubility for nanoemulsions prepared using hydrocarbons with different alkyl chain length.

Oil	T_{HLB} (°C)	r (nm)	ω (10^{27} m^3 s^{-1})	$C(\infty)$ (ml ml^{-1})
Decane	38.5	59	20.9	710.0
Dodecane	45.5	62	9.3	52.0
Tetradecane	49.5	64	4.0	3.7
Hexadecane	49.8	66	2.3	0.3
Isohexadecane	43.0	60	8.0	–

As expected from the Ostwald ripening theory (see Chapter 3), the rate of Ostwald ripening decreases as the oil solubility decreases. Isohexadecane has a rate of Ostwald ripening similar to that of dodecane.

As discussed before, one would expect that the Ostwald ripening of any given oil should decrease on incorporation of a second oil with much lower solubility. To test this hypothesis, nanoemulsions were made using hexadecane or isohexadecane to which various proportions of a less soluble oil, namely squalane, was added. The results using hexadecane did significantly decrease in stability on addition of 10 % squalane. This was thought to be due to coalescence rather than an increasing Ostwald ripening rate. In some cases addition of a hydrocarbon with a long alkyl chain can induce instability as a result of change in the adsorption and conformation of the surfactant at the O/W interface.

In contrast to the results obtained with hexadecane, addition of squalane to the O/W nanoemulsion system based on isohexadecane showed a systematic decrease in the Ostwald ripening rate as the squalane content was increased. The results are shown in Fig. 9.12 as plots of r^3 versus time for nanoemulsions containing varying amounts of squalane. Addition of squalane up to 20 % based on the oil phase showed a systematic reduction in the rate (from 8.0×10^{-27} to 4.1×10^{-27} m^3 s^{-1}). It should be noted that when squalane alone was used as the oil phase, the system was very unstable and it showed creaming within 1 hour. This shows that the surfactant used is not suitable for emulsification of squalane.

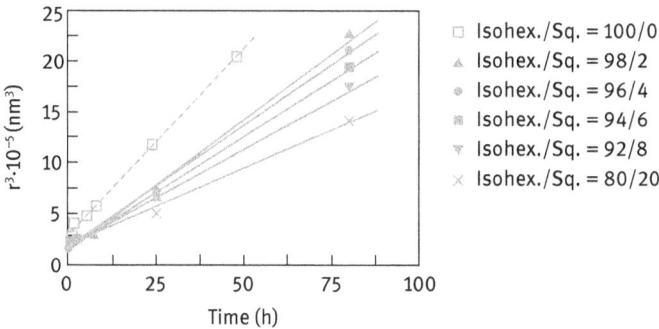

Fig. 9.12: r^3 versus time at 25 °C for the system water–C$_{12}$EO$_4$–isohexadecane–squalane (20/80 O/W and 4 wt % surfactant).

The effect of HLB number on nanoemulsion formation and stability was investigated by using mixtures of C$_{12}$EO$_4$ (HLB = 9.7) and C$_{12}$EO$_6$ (HLB = 11.7). Two surfactant concentrations (4 and 8 wt %) were used and the O/W ratio was kept at 20/80. Figure 9.13 shows the variation of droplet radius with HLB number. This figure shows that the droplet radius remain virtually constant in the HLB range 9.7–11.0, after which there is a gradual increase in droplet radius with increasing HLB number of the surfactant mixture. All nanoemulsions showed an increase in droplet radius with time, except for the sample prepared at 8 wt % surfactant with an HLB number of 9.7 (100 % C$_{12}$EO$_4$). Figure 9.14 shows the variation of Ostwald ripening rate constant ω with HLB number of surfactant. The rate seems to decrease with increasing surfactant HLB number and when the latter is > 10.5, the rate reaches a low value (< 4×10^{-27} m^3 s^{-1}).

As discussed above, one would expect that incorporating an oil soluble polymeric surfactant that adsorbs strongly at the O/W interface would reduce the Ostwald ripening rate. To test this hypothesis, an A–B–A block copolymer of polyhydroxystearic acid (PHS, the A chains) and polyethylene oxide (PEO, the B chain) PHS–PEO–PHS (Arlacel P135) was incorporated in the oil phase at low concentrations (the ratio of

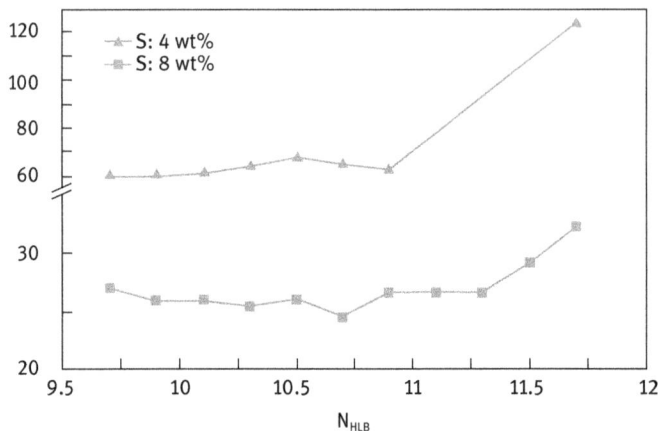

Fig. 9.13: r versus HLB number at two different surfactant concentrations (O/W ratio 20/80).

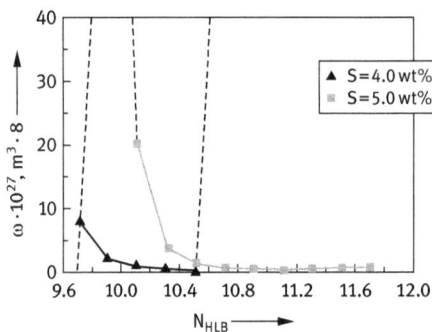

Fig. 9.14: ω versus HLB number in the systems water–$C_{12}EO_4$–$C_{12}EO_6$–isohexadecane at two surfactant concentrations.

surfactant to Arlacel was varied between 99 : 1 to 92 : 8). For the hexadecane system, the Ostwald ripening rate showed a decrease with the addition of Arlacel P135 surfactant at ratios lower than 94 : 6. Similar results were obtained using isohexadecane. However, at higher polymeric surfactant concentrations, the nanoemulsion became unstable.

As mentioned above, nanoemulsions prepared using the PIT method are relatively polydisperse and they generally give higher Ostwald ripening rates when compared to nanoemulsions prepared using high pressure homogenization techniques. To test this hypothesis, several nanoemulsions were prepared using a Microfluidizer (that can apply pressures in the range 5000–15 000 psi or 350–1000 bar). Using an oil : surfactant ratio of 4 : 8 and O/W ratios of 20 : 80 and 50 : 50, emulsions were prepared first using the Ultra-Turrax followed by high pressure homogenization (ranging from 1500 to 15 000 psi). The best results were obtained using a pressure of 15 000 psi (one cycle of homogenization). The droplet radius was plotted versus the oil : surfactant ratio, R(O/S) as shown in Fig. 9.15.

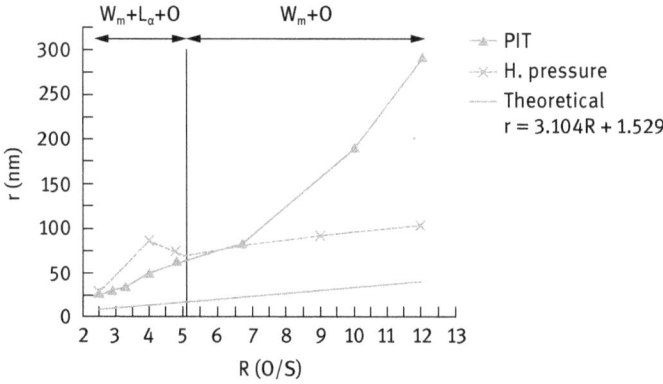

Fig. 9.15: r versus R(O/S) at 25 °C for the system water–$C_{12}EO_4$–hexadecane. W_m = micellar solution or O/W microemulsion, L_α = lamellar liquid crystalline phase; O = oil phase.

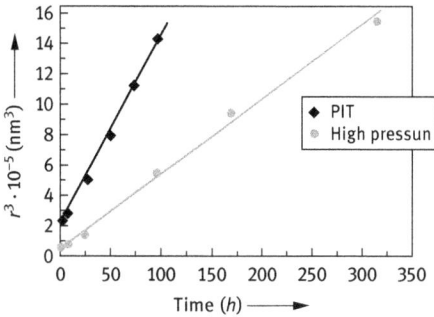

Fig. 9.16: r^3 versus time for nanoemulsion systems prepared using the PIT method and the Microfluidizer. 20 : 80 O/W and 4 wt % surfactant.

For comparison, the theoretical radii values calculated by assuming that all surfactant molecules are at the interface was calculated using the Nakajima equation [16, 17],

$$ r = \left(\frac{3M_b}{AN\rho_a} \right) R + \left(\frac{3\alpha M_b}{AN\rho_b} \right) + d, \qquad (9.1) $$

where M_b is the molecular weight of the surfactant, A is the area occupied by a single molecule, N is Avogadro's number, ρ_a is the oil density, ρ_b is the density of the surfactant alkyl chain, α is the alkyl chain weight fraction and d is the thickness of the hydrated layer of PEO.

In all cases, there is an increase in nanoemulsion radius with increasing R(O/S). However, when using the high pressure homogenizer, the droplet size can be maintained to values below 100 nm at high R(O/S) values. With the PIT method, there is a rapid increase in r with increasing R(O/S) when the latter exceeds 7.

As expected, the nanoemulsions prepared using high pressure homogenization showed a lower Ostwald ripening rate when compared to the systems prepared using the PIT method. This is illustrated in Fig. 9.16 which shows plots of r^3 versus time for the two systems.

9.8 Nanoemulsions based on polymeric surfactants

The use of polymeric surfactants for preparing nanoemulsions is expected to significantly reduce Ostwald ripening due to the high interfacial elasticity produced by the adsorbed polymeric surfactant molecules [18]. To test this hypothesis, several nanoemulsions were formulated using a graft copolymer of hydrophobically modified inulin. The inulin backbone consists of polyfructose with a degree of polymerization >23. This hydrophilic backbone is hydrophobically modified by attachment of several C_{12} alkyl chains [19]. The polymeric surfactant (with a trade name of INUTEC® SP1) adsorbs with several alkyl chains that can be soluble in the oil phase or strongly attached to the oil surface, leaving the strongly hydrated hydrophilic polyfructose loops and tails "dangling" in the aqueous phase. These hydrated loops and tails (with a hydrodynamic thickness > 5 nm) provide effective steric stabilization.

Oil/water (O/W) nanoemulsions were prepared by two step emulsification processes. In the first step, an O/W emulsion was prepared using a high speed stirrer, namely an Ultra-Turrax [20]. The resulting coarse emulsion was subjected to high pressure homogenization using a Microfluidizer (Microfluidics, USA). In all cases, the pressure used was 700 bar and homogenization was carried out for 1 minute. The z-average droplet diameter was determined using PCS measurements as discussed before.

Figure 9.17 shows plots of r^3 versus t for nanoemulsions of the hydrocarbon oils that were stored at 50 °C. It can be seen that both paraffinum liquidum with low and high viscosity give almost a zero slope, indicating absence of Ostwald-ripening in this case. This is not surprising since both oils have very low solubility and the hy-

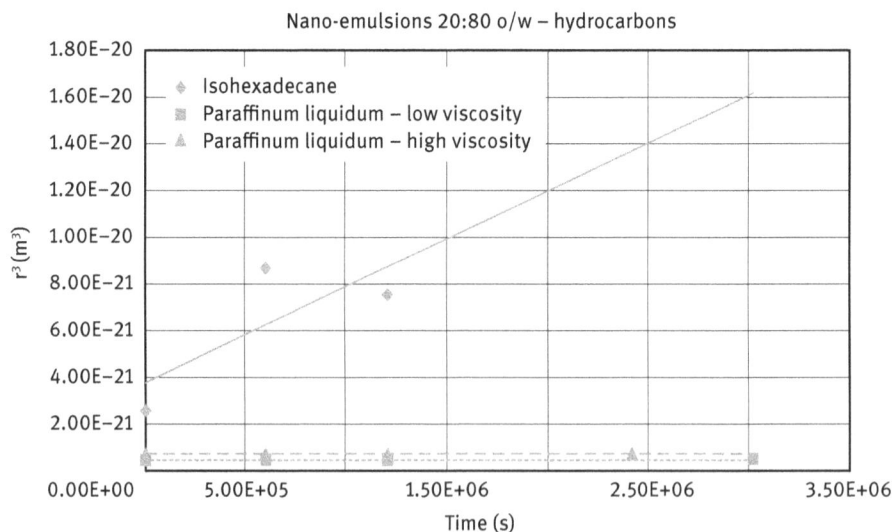

Fig. 9.17: r^3 versus time for nanoemulsions based on hydrocarbon oils.

drophobically modified inulin, INUTEC® SP1, strongly adsorbs at the interface giving high elasticity that reduces both Ostwald ripening and coalescence. However, with the more soluble hydrocarbon oils, namely isohexadecane, there is an increase in r^3 with time, giving a rate of Ostwald ripening of 4.1×10^{-27} m^3 s^{-1}. The rate for this oil is almost three orders of magnitude lower than that obtained with a nonionic surfactant, namely laureth-4 (C$_{12}$-alkylchain with 4 mol ethylene-oxide) when stored at 50 °C. This clearly shows the effectiveness of INUTEC® SP1 in reducing Ostwald ripening. This reduction can be attributed to the enhancement of the Gibbs dilational elasticity [18] which results from the multi point attachment of the polymeric surfactant with several alkyl groups to the oil droplets. This results in a reduction of the molecular diffusion of the oil from the smaller to the larger droplets.

Figure 9.18 shows the results for the isopropylalkylate O/W nanoemulsions. As with the hydrocarbon oils, there is a significant reduction in the Ostwald ripening rate with increasing alkyl chain length of the oil. The rate constants are 1.8×10^{-27}, 1.7×10^{-27} and 4.8×10^{-28} m^3 s^{-1} respectively.

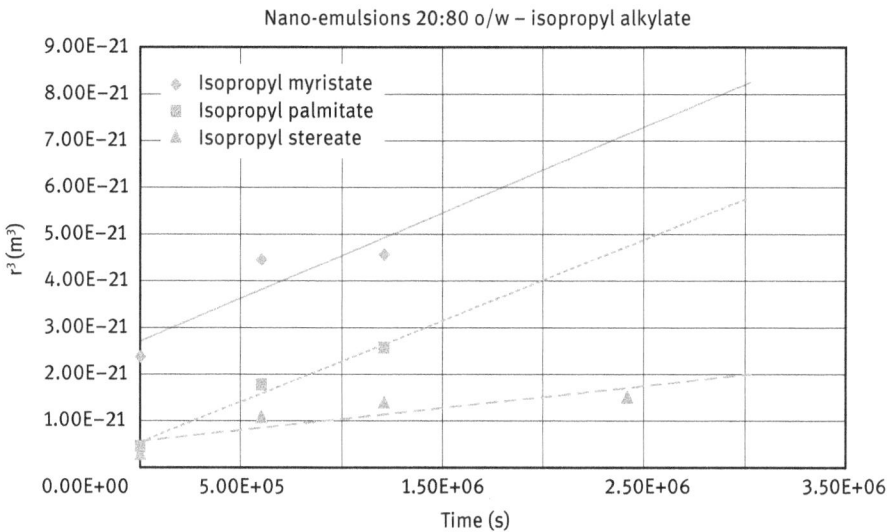

Fig. 9.18: r^3 versus time for nanoemulsions based on isopropylalkylate.

Figure 9.19 shows the r^3–t plots for nanoemulsions based on natural oils. In all cases, the Ostwald ripening rate is very low. However, a comparison between squalene and squalane shows that the rate is relatively higher for squalene (unsaturated oil) when compared with squalane (with lower solubility). The Ostwald ripening rate for these natural oils is given in Tab. 9.4.

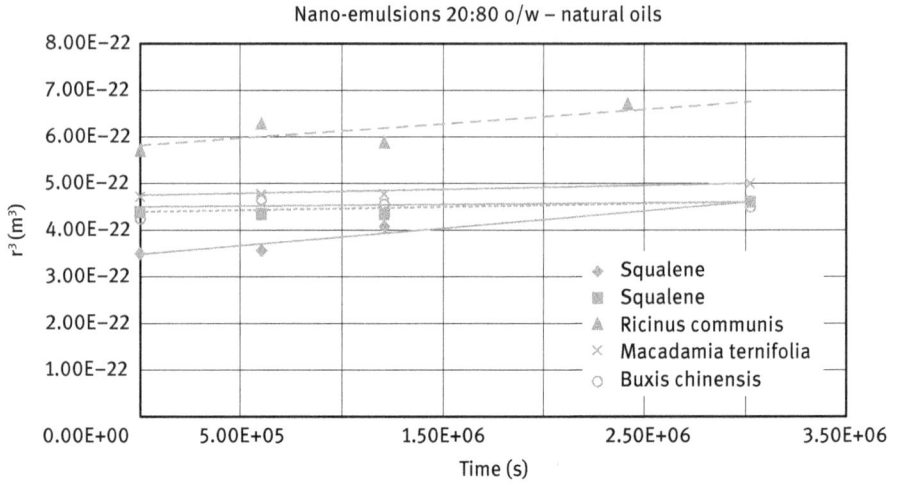

Fig. 9.19: r^3 versus time for nanoemulsions based on natural oils.

Tab. 9.4: Ostwald ripening rates for nanoemulsions based on natural oils.

Oil	Ostwald ripening rate ($m^3\ s^{-1}$)
Squalene	2.9×10^{-28}
Squalane	5.2×10^{-30}
Ricinus Communis	3.0×10^{-29}
Macadamia Ternifolia	4.4×10^{-30}
Buxis Chinensis	~ 0

Figure 9.20 shows the results based on silicone oils. Both dimethicone and phenyl trimethicone give an Ostwald ripening rate close to zero, whereas cyclopentasiloxane gives a rate of $5.6 \times 10^{-28}\ m^3\ s^{-1}$.

Figure 9.21 shows the results for nanoemulsions based on esters and the Ostwald ripening rates are given in Tab. 9.5. C_{12-15} alkylbenzoate seems to give the highest rate.

Tab. 9.5: Ostwald ripening rates for nanoemulsions based on esters.

Oil	Ostwald ripening rate ($m^3\ s^{-1}$)
Butyl stearate	1.8×10^{-28}
Caprylic capric triglyceride	4.9×10^{-29}
Cetearyl ethylhexanoate	1.9×10^{-29}
Ethylhexyl palmitate	5.1×10^{-29}
Cetearyl isononanoate	1.8×10^{-29}
C_{12-15} alkyl benzoate	6.6×10^{-28}

Nano-emulsions 20:80 o/w – silicon oils

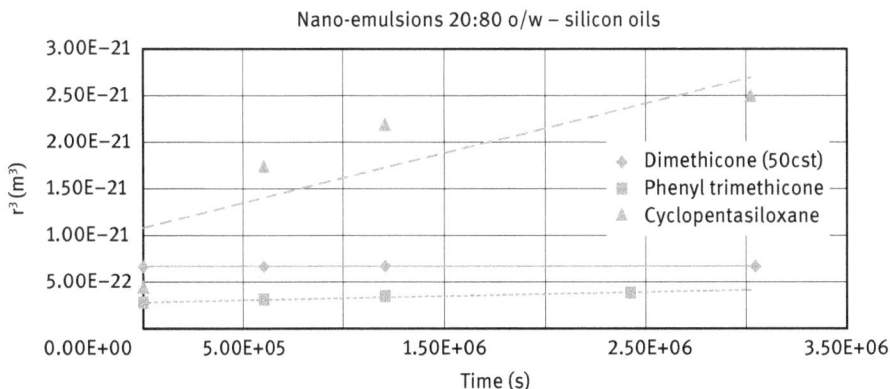

Fig. 9.20: r^3 versus time for nanoemulsions based on silicon oils.

Nano-emulsions 20:80 o/w – natural oils

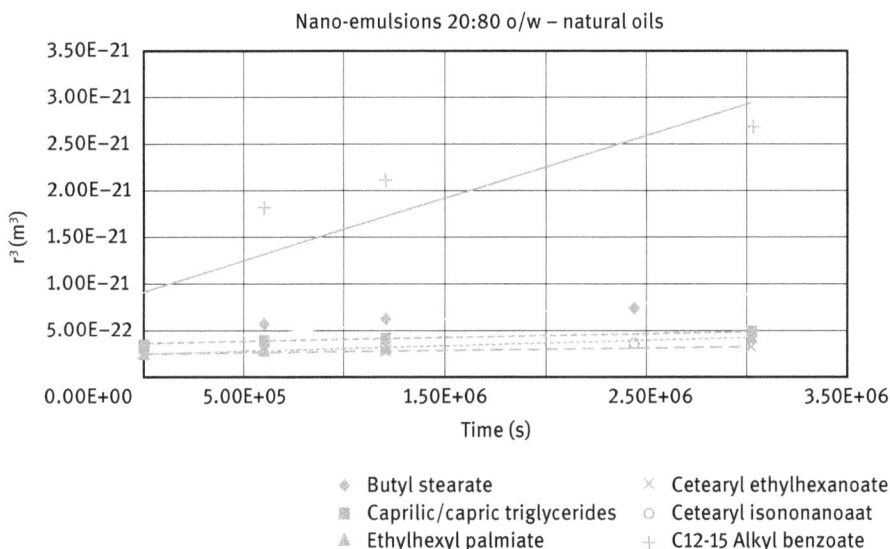

Fig. 9.21: r^3 versus time for nanoemulsions based on esters.

Figure 9.22 gives a comparison for two nanoemulsions based on polydecene, a highly insoluble nonpolar oil and PPG-15 stearyl ether which is relatively more polar. Polydecene gives a low Ostwald ripening rate of 6.4×10^{-30} m^3 s^{-1} which is one order of magnitude lower than that of PPG-15 stearyl ether (5.5×10^{-29} m^3 s^{-1}).

The influence of adding glycerol (which is sometimes added to personal care formulations as a humectant) which can be used to prepare transparent nanoemulsions (by matching the refractive index of the oil and the aqueous phase) on the Ostwald ripening rate is shown in Fig. 9.23. With the more insoluble silicone oil, addition of

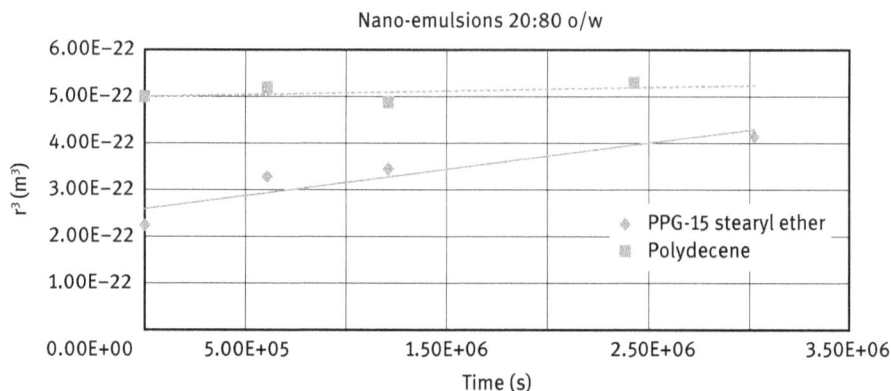

Fig. 9.22: r^3 versus time for nanoemulsions based on PPG-15 stearyl ether and polydecene.

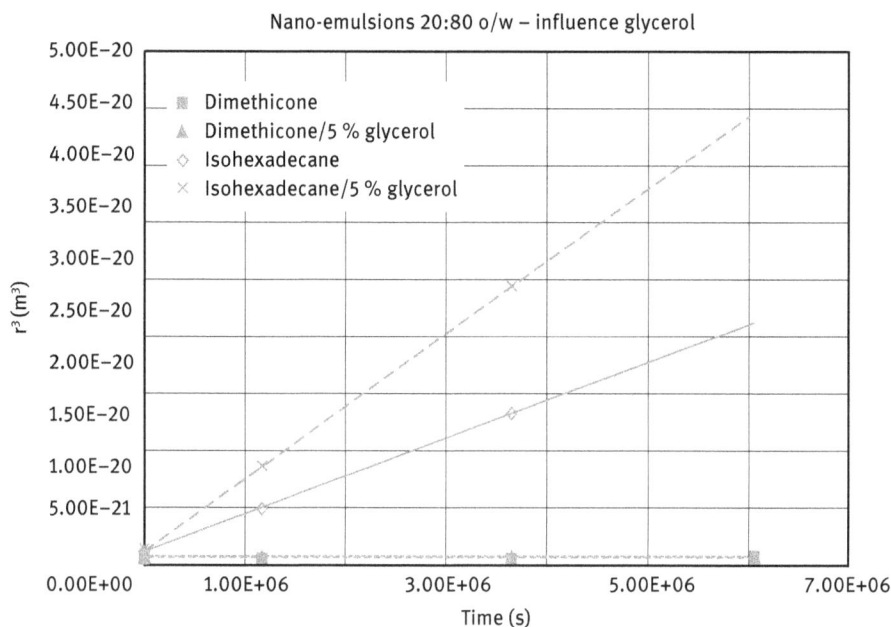

Fig. 9.23: Influence of glycerol on the Ostwald ripening rate of nanoemulsions.

5% glycerol does not show an increase in the Ostwald ripening rate, whereas for the more soluble isohexadecane oil, glycerol increases the rate.

It can be seen that hydrophobically modified inulin, HMI (INUTEC® SP1), reduces the Ostwald ripening rate of nanoemulsions when compared with nonionic surfactants such as laureth-4. This is due to the strong adsorption of INUTEC® SP1 at the oil-water interface (by multi-point attachment) and enhancement of the Gibbs dilational elasticity, both reducing the diffusion of oil molecules from the smaller to the larger droplets [20, 21]. The present study also showed a large influence of the nature of the oil phase with the more soluble and more polar oils giving the highest Ostwald ripening rate. However, in all cases, when using INUTEC® SP1, the rates are reasonably low allowing one to use this polymeric surfactant in formulation of nanoemulsions for personal care applications.

References

[1] Solans, C., Izquierdo, P., Nolla, J., Azemar, N. and Garcia-Celma, M. J., Nanoemulsions. Current opinion, Colloid & Interface Science, **10**, (3–4), 102–110 (2005).
[2] Ganachaud, F. and Katz J. L., Nanoparticles and nanocapsules created using the ouzo effect: Spontaneous emulsification as an alternative to ultrasonic and high-shear devices, Chem. Phys. Chem., **6**, 209–216 (2005).
[3] Bouchemal, K., Briançon, S., Perrier E. and Fessi, H., Nanoemulsion formulation using spontaneous emulsification: Solvent, oil and surfactant optimization, Int. J. Pharm., **280**, 241–251 (2004).
[4] Vitale, S. A. and Katz, J. L., Liquid droplet dispersions formed by homogeneous liquid-liquid nucleation: "The ouzo effect", Langmuir, **19**, 4105–4110 (2003).
[5] Tadros, T., Izquierdo, P., Esquena, J. and Solans, C., Formation and stability of nano-emulsions, Adv. Colloid Interface Sci., **108–109**, 303–318 (2004).
[6] Forgiarini, A., Esquena, J., Gonzalez, C. and Solans, C., Formation of nano-emulsions by low-energy emulsification methods at constant temperature, Langmuir, **17**, (7), 2076–2083 (2001).
[7] Izquierdo, P., Esquena, J., Tadros, T. F., Dederen, C., Garcia, M. J., Azemar, N. and Solans, C., Formation and stability of nanoemulsions prepared using the phase inversion temperature method, Langmuir, **18**, (1), 26–30.8 (2002).
[8] Shinoda K. and Saito, H., J. Colloid Interface Sci., **26** , 70 (1968).
[9] Shinoda, K. and Saito, H., The stability of O/W type emulsions as functions of temperature and the HLB of emulsifiers: The emulsification by PIT-method, J. Colloid Interface Sci., 30, 258–263 (1969).
[10] Brooks, B. W. Richmond, H. N. and Zerfa, M. in "Modern Aspects of Emulsion Science", B. P. Binks (ed.), Royal Society of Chemistry Publication, Cambridge (1998).
[11] Sottman, T. and Strey, R., J. Chem. Phys., **108**, 8606 (1997).
[12] Kabalnov A. S. and Shchukin, E. D., Adv. Colloid Interface Sci., **38**, 69 (1992).
[13] Kabalnov, A. S., Langmuir, **10**, 680 (1994).
[14] Walstra, P., Chem. Eng. Sci., **48**, 333 (1993).
[15] Izquierdo, P., Thesis "Studies on Nano-Emulsion Formation and Stability", University of Barcelona, Spain (2002).
[16] Nakajima, H., Tomomossa, S. and Okabe, M., First Emulsion Conference, Paris (1993).
[17] Nakajima, H., in "Industrial Applications of Microemulsions", C. Solans and H. Konieda (eds.), Marcel Dekker, New York (1997).
[18] Kuneida, H., Fukuhi, Y., Uchiyama, H. and Solans, C., Langmuir, **12**, 2136 (1996).
[19] Tadros Th. F. (ed.), "Colloids in Cosmetics and Personal Care", Wiley-VCH, Germany (2008).
[20] Stevens, C. V., Meriggi, A., Peristerpoulou, M., Christov, P. P., Booten, K., Levecke, B., Vandamme, A., Pittevils, N. and Tadros, Th. F., Biomacromolecules, **2**, 1256 (2001).
[21] Tadros, Th. F., Vandamme, A., Levecke, B., Booten, K., Stevens, C. V., Adv. Colloid Interface Sci., **108–109**, 207 (2004).

10.1 Introduction

Swollen micelles or microemulsions are a special class of nanodispersions (transparent or translucent) which actually have little in common with emulsions. The term microemulsion was first introduced by Hoar and Schulman [1, 2] who discovered that by titration of a milky emulsion (stabilized by soap such as potassium oleate) with a medium chain alcohol such as pentanol or hexanol, a transparent or translucent system was produced. A schematic representation of the titration method adopted by Schulman and co-workers is given below.

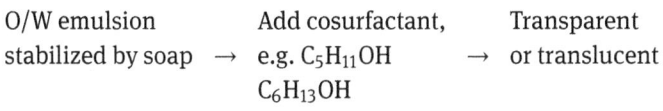

$$
\begin{array}{ccc}
\text{O/W emulsion} & \text{Add cosurfactant,} & \text{Transparent} \\
\text{stabilized by soap} \quad \rightarrow \quad & \text{e.g. } C_5H_{11}OH & \rightarrow \quad \text{or translucent} \\
& C_6H_{13}OH &
\end{array}
$$

The final transparent or translucent system is a W/O microemulsion.

A convenient way to describe microemulsions is to compare them with micelles. The latter, which are thermodynamically stable, may consist of spherical units with a radius that is usually less than 5 nm. Two types of micelles may be considered: normal micelles with hydrocarbon tails forming the core and the polar head groups in contact with the aqueous medium; and reverse micelles (formed in nonpolar media) with a water core containing the polar head groups and the hydrocarbon tails now in contact with the oil. The normal micelles can solubilize oil in the hydrocarbon core forming O/W microemulsions, whereas the reverse micelles can solubilize water forming a W/O microemulsion.

A schematic representation of these systems is shown in Fig. 10.1.

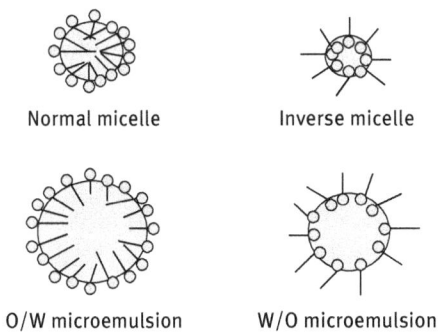

Normal micelle   Inverse micelle

O/W microemulsion   W/O microemulsion

Fig. 10.1: Schematic representation of microemulsions.

A rough guide to the dimensions of micelles, micellar solutions and macroemulsions is as follows: micelles, R < 5 nm (they scatter little light and are transparent); macroemulsions, R > 50 nm (opaque and milky); micellar solutions or microemulsions, 5–50 nm (transparent, 5–10 nm, translucent 10–50 nm).

The classification of microemulsions based on size is not adequate. Whether a system is transparent or translucent depends not only on the size but also on the difference in refractive index between the oil and the water phases. A microemulsion with small size (in the region of 10 nm) may appear translucent if the difference in refractive index between the oil and the water is large (note that the intensity of light scattered depends on the size and an optical constant that is given by the difference in refractive index between oil and water). A relatively large size microemulsion droplet (in the region of 50 nm) may appear transparent if the refractive index difference is very small. The best definition of microemulsions is based on the application of thermodynamics as discussed below.

10.2 Thermodynamic definition of microemulsions

A thermodynamic definition of microemulsions can be obtained from a consideration of the energy and entropy terms for the formation of microemulsions. This is schematically represented in Fig. 10.2 which shows the process of forming a microemulsion from a bulk oil phase (for an O/W microemulsion) or a bulk water phase (for a W/O microemulsion).

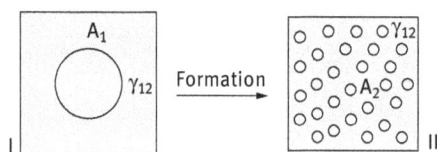

Fig. 10.2: Schematic representation of microemulsion formation.

A_1 is the surface area of the bulk oil phase and A_2 is the total surface area of all the microemulsion droplets. γ_{12} is the O/W interfacial tension.

The increase in surface area when going from state I to state II is $\Delta A \, (= A_2 - A_1)$ and the surface energy increase is equal to $\Delta A \gamma_{12}$. The increase in entropy when going from state I to state II is $T \Delta S^{conf}$ (note that state II has higher entropy since a large number of droplets can arrange themselves in several ways, whereas state I with one oil drop has much lower entropy).

According to the second law of thermodynamics, the free energy of formation of microemulsions, ΔG_m, is given by the following expression:

$$\Delta G_m = \Delta A \gamma_{12} - T \Delta S^{conf} \tag{10.1}$$

With macroemulsions, $\Delta A \gamma_{12} \gg T\Delta S^{conf}$ and $\Delta G_m > 0$. The system is nonspontaneous (it requires energy for formation of the emulsion drops) and it is thermodynamically unstable. With microemulsions, $\Delta A \gamma_{12} \leq T\Delta S^{conf}$ (this is due to the ultra-low interfacial tension accompanying microemulsion formation) and $\Delta G_m \leq 0$. The system is produced spontaneously and it is thermodynamically stable.

The above analysis shows the contrast between emulsions and microemulsions: With emulsions, increasing the mechanical energy and increasing surfactant concentration usually result in the formation of smaller droplets which become kinetically more stable. With microemulsions, neither mechanical energy nor increasing surfactant concentration can result in its formation. The latter is based on a specific combination of surfactants and specific interaction with the oil and the water phases and the system is produced at optimum composition.

Thus, microemulsions have nothing in common with macroemulsions and in many cases it is better to describe the system as "swollen micelles". The best definition of microemulsions is as follows [3]: "System of Water + Oil + Amphiphile that is a single Optically Isotropic and Thermodynamically Stable Liquid Solution". Amphiphile refers to any molecule that consists of hydrophobic and hydrophilic portions, e.g. surfactants, alcohols, etc.

The driving force for microemulsion formation is the low interfacial energy which is overcompensated by the negative entropy of dispersion term. The low (ultra-low) interfacial tension is produced in most cases by combining two molecules, referred to as the surfactant and cosurfactant (e.g. medium chain alcohol).

10.3 Mixed film and solubilization theories of microemulsions

10.3.1 Mixed film theories [4]

The film (which may consist of surfactant and cosurfactant molecules) is considered a liquid "two-dimensional" third phase in equilibrium with both oil and water. Such a monolayer could be a duplex film, i.e. giving different properties on the water side and the oil side. The initial "flat" duplex film (see Fig. 10.3) has different tensions at the oil and water sides. This is due to the different packing of the hydrophobic and hydrophilic groups (these groups have different sizes and cross-sectional areas)

It is convenient to define a two-dimensional surface pressure π,

$$\pi = \gamma_0 - \gamma. \tag{10.2}$$

γ_0 is the interfacial tension of the clean interface, whereas γ is the interfacial tension with adsorbed surfactant.

One can define two values for π at the oil and water phases, π_o and π_w, which for a flat film are not equal, i.e. $\pi'_o \neq \pi'_w$. As a result of the difference in tensions, the film will bend until $\pi_o = \pi_w$. If $\pi'_o > \pi'_w$, the area at the oil side has to expand (resulting

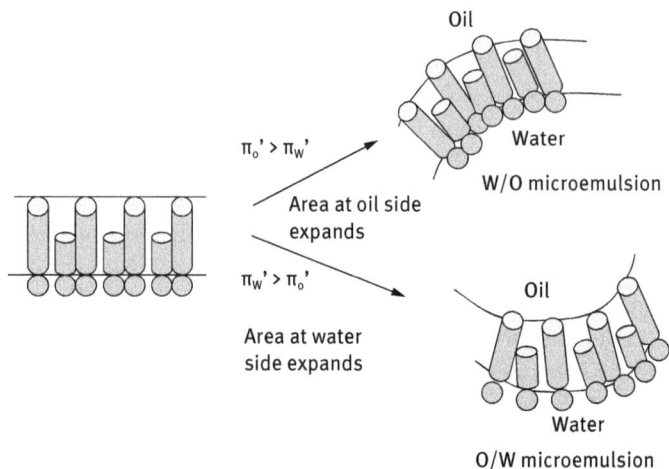

Fig. 10.3: Schematic representation of film bending.

in reduction of π_o') until $\pi_o = \pi_w$. In this case a W/O microemulsion is produced. If $\pi_w' > \pi_o'$, the area at the water side expands until $\pi_w = \pi_o$. In this case an O/W microemulsion is produced. A schematic representation of film bending for production of W/O or W/O microemulsions is illustrated in Fig. 10.3.

According to the duplex film theory, the interfacial tension γ_T is given by the following expression [5]:

$$\gamma_T = \gamma_{(O/W)} - \pi, \tag{10.3}$$

where $(\gamma_{O/W})_a$ is the interfacial tension that is reduced by the presence of the alcohol.

The value of $(\gamma_{O/W})_a$ is significantly lower than $\gamma_{O/W}$ in the absence of the alcohol. For example, for hydrocarbon/water, $\gamma_{O/W}$ is reduced from 50 to 15–20 mN m^{-1} on the addition of a significant amount of a medium chain alcohol like pentanol or hexanol.

Contributions to π are considered to be due to crowding of the surfactant and co-surfactant molecules and penetration of the oil phase into the hydrocarbon chains of the interface.

According to equation (10.3) if $\pi > (\gamma_{O/W})_a$, γ_T becomes negative and this leads to expansion of the interface until γ_T reaches a small positive value. Since $(\gamma_{O/W})_a$ is of the order of 15–20 mN m^{-1}, surface pressures of this order are required for γ_T to approach a value of zero.

The above duplex film theory can explain the nature of the microemulsion: The surface pressures at the oil and water sides of the interface depend on the interactions of the hydrophobic and hydrophilic potions of the surfactant molecule at both sides respectively. If the hydrophobic groups are bulky in nature relative to the hydrophilic groups, then for a flat film such hydrophobic groups tend to crowd forming a higher surface pressure at the oil side of the interface; this results in bending and expansion at the oil side forming a W/O microemulsion. An example for a surfactant with bulky hydrophobic groups is Aerosol OT (dioctyl sulphosuccinate). If the hydrophilic groups are bulky, such as is the case with ethoxylated surfactants containing more than 5 ethylene oxide units, crowding occurs at the water side of the interface. This produces an O/W microemulsion.

10.3.2 Solubilization theories

These concepts were introduced by Shinoda and co-workers [6] who considered microemulsions to be swollen micelles that are directly related to the phase diagram of their components.

Consider the phase diagram of a three-component system of water, ionic surfactant and medium chain alcohol as described in Fig. 10.4. At the water corner and at low alcohol concentration, normal micelles (L_1) are formed since in this case there are more surfactant than alcohol molecules. At the alcohol (cosurfactant corner), inverse micelles (L_2) are formed, since in this region there are more alcohol than surfactant molecules.

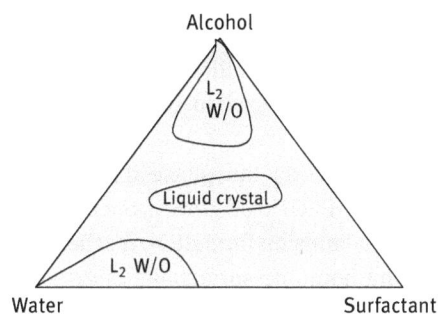

Fig. 10.4: Schematic representation of three-component phase diagram.

These L_1 and L_2 are not in equilibrium but are separated by a liquid crystalline region (lamellar structure with equal number of surfactant and alcohol molecules). The L_1 region may be considered as an O/W microemulsion, whereas the L_2 region may be considered as a W/O microemulsion.

Addition of a small amount of oil miscible with the cosurfactant, but not with the surfactant and water, changes the phase diagram only slightly. The oil may be simply solubilized in the hydrocarbon core of the micelles. Addition of more oil leads to fundamental changes of the phase diagram as is illustrated in Fig. 10.5 when 50 : 50 of W : O are used. To simplify the phase diagram, the ^{50}W/^{50}O are presented on one corner.

Near the cosurfactant (co) corner the changes are small compared to the three phase diagram (Fig. 10.5). The O/W microemulsion near the water-surfactant (sa) axis is not in equilibrium with the lamellar phase, but with a noncolloidal oil + cosurfac-

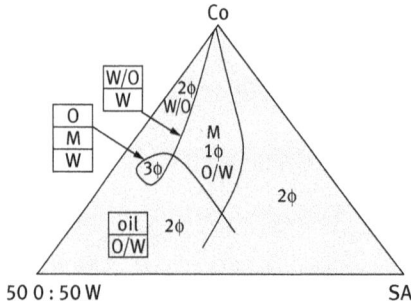

Fig. 10.5: Schematic representation of the pseudoternary phase diagram of oil/water/surfactant/cosurfactant.

tant phase. If co is added to such a two-phase equilibrium at fairly high surfactant concentration all oil is taken up and a one-phase microemulsion appears. Addition of co at low sa concentration may lead to separation of an excess aqueous phase before all oil is taken up in the microemulsion. A three phase system is formed, containing a microemulsion that cannot be clearly identified as W/O or W/O and that is presumably similar to the lamellar phase swollen with oil or to a more irregular intertwining of aqueous and oily regions (bicontinuous or middle phase microemulsion). The interfacial tensions between the three phases are very low ($0.1–10^{-4}\,\mathrm{mN\,m^{-1}}$). Further addition of co to the three phase system makes the oil phase disappear and leaves a W/O microemulsion in equilibrium with a dilute aqueous sa solution. In the large one phase region continuous transitions from O/W to middle phase to W/O microemulsions are found.

Solubilization can also be illustrated by considering the phase diagrams of nonionic surfactants containing poly(ethylene oxide) (PEO) head groups. Such surfactants do not generally need a cosurfactant for microemulsion formation. A schematic representation of oil and water solubilization by nonionic surfactants is given in Fig. 10.6.

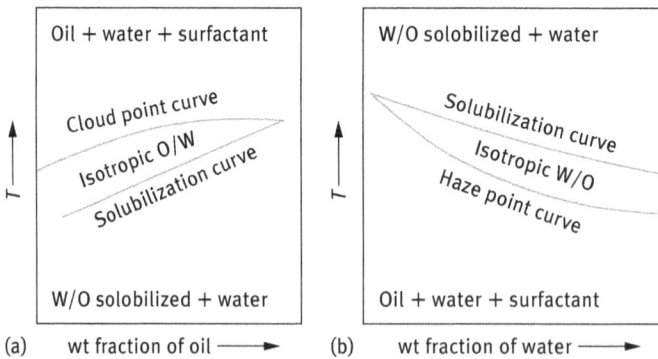

Fig. 10.6: Schematic representation of solubilization; (a) oil solubilized in a nonionic surfactant solution; (b) water solubilized in an oil solution of a nonionic surfactant.

At low temperatures, the ethoxylated surfactant is soluble in water and at a given concentration is capable of solubilizing a given amount of oil. Oil solubilization increases rapidly with increasing temperature near the cloud point of the surfactant. This is illustrated in Fig. 10.6 which shows the solubilization and cloud point curves of the surfactant. Between these two curves, an isotropic region of O/W solubilized system exists. At any given temperature, any increase in the oil weight fraction above the solubilization limit results in oil separation (oil solubilized + oil). At any given surfactant concentration, any increase in temperature above the cloud point results in separation into oil, water and surfactant.

If one starts from the oil phase with dissolved surfactant and adds water, solubilization of the latter takes place and solubilization increases with reduction of temperature near the haze point. Between the solubilization and haze point curves, an isotropic region of W/O solubilized system exists. At any given temperature, any increase in water weight fraction above the solubilization limit results in water separation (W/O solubilized + water). At any given surfactant concentration, any decrease in temperature below the haze point results in separation to water, oil and surfactant.

With nonionic surfactants, both types of microemulsions can be formed depending on the conditions. With such systems, temperature is the most crucial factor since the solubility of surfactant in water or oil depends on temperature. Microemulsions prepared using nonionic surfactants have a limited temperature range.

10.4 Thermodynamic theory of microemulsion formation

The spontaneous formation of a microemulsion with a decrease in free energy can only be expected if the interfacial tension is so low that the remaining free energy of the interface is overcompensated for by the entropy of dispersion of the droplets in the medium [7, 8]. This concept forms the basis of the thermodynamic theory proposed by Ruckenstein and Chi, and Overbeek [7, 8].

10.4.1 Reason for combining two surfactants

Single surfactants do lower the interfacial tension γ, but in most cases the critical micelle concentration (cmc) is reached before γ is close to zero. Addition of a second surfactant of a completely different nature (i.e. predominantly oil soluble such as an alcohol) then lowers γ further and very small, even transiently negative values may be reached [9]. This is illustrated in Fig. 10.7 which shows the effect of addition of the cosurfactant on the γ–log c_{sa} curve. It can be seen that addition of cosurfactant shifts the whole curve to low γ values and the cmc is shifted to lower values.

For a multicomponent system i, each with an adsorption Γ_i (mol m^{-2}, referred to as the surface excess), the reduction in γ, i.e. dγ, is given by the following expression:

$$d\gamma = -\sum \Gamma_i \, d\mu_i = -\sum \Gamma_i RT \, d\ln C_i , \qquad (10.4)$$

where μ_i is the chemical potential of component i, R is the gas constant, T is the absolute temperature and C_i is the concentration (mol dm^{-3}) of each surfactant component.

The reason for the lowering of γ when using two surfactant molecules can be understood by considering the Gibbs adsorption equation for multicomponent systems [9]. For two components sa (surfactant) and co (cosurfactant), equation (10.4) becomes

$$d\gamma = -\Gamma_{sa}RT \, d\ln C_{sa} - \Gamma_{co}RT \, d\ln C_{co} . \qquad (10.5)$$

Integrating equation (10.5) gives,

$$\gamma = \gamma_0 - \int_0^{C_{sa}} \Gamma_{sa}RT \, d\ln C_{sa} - \int_0^{C_{co}} \Gamma_{co}RT \, d\ln C_{co} \qquad (10.6)$$

which clearly shows that γ_0 is lowered by two terms, from both surfactant and cosurfactant.

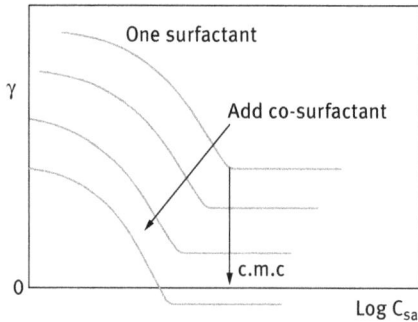

Fig. 10.7: γ-log C_{sa} curves for surfactant + co-surfactant.

The two surfactant molecules should adsorb simultaneously and they should not interact with each other, otherwise they lower their respective activities. Thus, the surfactant and cosurfactant molecules should vary in nature, one predominantly water soluble (such as an anionic surfactant) and the other predominantly oil soluble (such as a medium chain alcohol)

In some cases a single surfactant may be sufficient for lowering γ far enough for microemulsion formation to become possible, e.g. Aerosol OT (sodium diethyl hexyl sulphosuccinate) and many nonionic surfactants.

10.4.2 Free energy of formation of a microemulsion

A simple model was used by Overbeek [9] to calculate the free energy of formation of a model W/O microemulsion: The droplets were assumed of equal size. The droplets are large enough to consider the adsorbed surfactant layer to have constant composition.

The microemulsion is prepared in a number of steps and for each step one calculates the Helmholtz free energy F. This was chosen since the pressure inside the drop is higher by the Laplace pressure $2\gamma/a$ (where a is the droplet radius) than the pressure in the medium.

A summary of the four steps involved in the preparation of a model W/O microemulsion is given below:

(i) Prepare the oil phase in its final concentration,

$$F_1 = \sum n_i' \mu_i' - p_1 V_1 , \tag{10.7}$$

where n_i' and μ_i' are the amount and chemical potential of oil and cosurfactant in the continuous phase, without droplets being mixed in. p_1 is the atmospheric pressure and V_1 is the volume of the oil phase.

(ii) Prepare the aqueous phase in its final concentration,

$$F_2 = \sum n_i' \mu_i' - p_1 V_2 , \tag{10.8}$$

where i are now water, surfactant and salt, and V_2 is the volume of the water phase.

(iii) Form the water phase into droplets close packed in the oil phase (i.e. with a packing fraction $\phi = 0.74$) and add all the adsorbed material,

$$F_3 = \gamma A + \Gamma_{sa} A \left[\mu_{sa}' + \left(2\frac{\gamma}{a} \right) \overline{V}_{sa} \right] + \Gamma_i A \mu_i , \tag{10.9}$$

where i refers to cosurfactant and oil.

The oil must be negatively adsorbed in order to keep the volume of the adsorption layer zero (in accordance with the Gibbs dividing surface). It is assumed in equation (10.9) that the Gibbs plane (the surface of tension in this case) lies close to the surface (where $\Gamma_{water} = 0$).

(iv) Allow the close packed emulsion to expand to its final concentration (volume fraction ϕ),

$$F_4 = n_{dr} RTf(\phi) . \tag{10.10}$$

n_{dr} is the amount of drops (in mol) and $f(\phi)$ is a function of ϕ. $f(\phi)$ may be simply written as

$$f(\phi) = \ln \phi - \ln 0.74 . \tag{10.11}$$

More accurately $f(\phi)$ may be calculated using a hard-sphere model [10],

$$f(\phi) = \ln \phi + \phi \left[\frac{4 - 3\phi}{(1 - \phi)^2} \right] - 19.25 . \tag{10.12}$$

Combining equations (10.7)–(10.12) gives the Helmholtz free energy of the complete emulsion. The free energy is minimized with respect to a change in the interfacial area A. This involves transfer of adsorbed components to or from the interface, thereby changing the bulk concentration and thus γ; the result is

$$\gamma = -\text{const.} \cdot \frac{1}{a^2} g(\phi),$$
(10.13)

where $g(\phi)$ is similar but not identical to $f(\phi)$.

The droplet radius a can be calculated from a knowledge of the total interfacial area A,

$$A = \frac{n_{sa}}{\Gamma_{sa}} - n_{sa} N_{av} \text{ (area/molecule)}.$$
(10.14)

The area per molecule of an anionic surfactant such as sodium dodecyl sulphate (SDS) varies between 0.7 to 1.1 nm^2, depending on the concentration of cosurfactant (pentanol) and salt concentration. The area per pentanol molecule is about 0.3 nm^2. This means that the average area per surfactant molecule is about 0.9 nm^2.

The radius of the droplet can be calculated from the ratio of the volume of the drop to its area,

$$a = \frac{3 \cdot (4/3)\pi a^3}{4\pi a^2} = \frac{3V}{A},$$
(10.15)

where V is the total volume of the droplets and A is the total interfacial area.

The radius a of the microemulsion droplet has to fit both equations (10.13) and (10.15); γ is the most easily varied quantity in these equations. The correct value of γ is obtained by adaptation of C_{sa}.

According to equation (10.13), any value of a is allowed in the accessible range of γ. If γ is close to zero very large radii can be obtained, i.e. very large water/sa ratios are allowed. However, the phase diagram shows that at such high ratios demixing occurs. This analysis shows the inadequacy of the above simple model and it is necessary to add an explicit influence of the radius of curvature on the interfacial tension. The curvature effect is manifested in the packing of the tails and head groups at the O/W interface. With W/O microemulsions the packing of the short chains and the packing of the head groups will favour W/O curvature with a ratio of 3 or more for co/sa. With O/W microemulsions a ratio of sa/co of 2 or less is required. Thus O/W microemulsions need less cosurfactant than W/O microemulsions.

10.4.3 Factors determining W/O versus O/W microemulsions

The duplex film theory predicts that the nature of the microemulsion formed depends on the relative packing of the hydrophobic and hydrophilic portions of the surfactant molecule, which determine the bending of the interface. For example, a surfactant molecule such as Aerosol OT,

```
    O    O    C₂H₅
     \\ / \    |
       C    CH₂–CH–CH₂–CH₂–CH₂–CH₃
      /
    CH₂
    /
Na⁺–O₃–CH
     \
       C    CH₂–CH–CH₂–CH₂–CH₂–CH₃
      // \ /    |
     O    O    C₂H₅
```

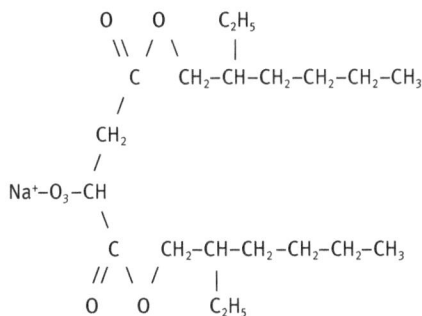

favours the formation of W/O microemulsion, without the need of a cosurfactant. As a result of the presence of a stumpy head group and large volume to length (V/l) ratio of the nonpolar group, the interface tends to bend with the head groups facing onwards, thus forming a W/O microemulsion.

The molecule has V/l > 0.7 which is considered necessary for formation of a W/O microemulsion. For ionic surfactants such as SDS for which V/l < 0.7, microemulsion formation needs the presence of a cosurfactant (the latter has the effect of increasing V without changing l).

The importance of geometric packing was considered in detail by Mitchell and Ninham [11] who introduced the concept of the packing ratio P,

$$P = \frac{V}{l_c a_o}, \tag{10.16}$$

where a_o is the head group area and l_c is the maximum chain length.

P gives a measure of the hydrophilic-lipophilic balance. For values of P < 1 (usually $P \sim \frac{1}{3}$), normal or convex aggregates are produced (normal micelles). For values of P > 1, inverse micelles are produced. P is influenced by many factors: hydrophilicity of the head group, ionic strength and pH of the medium and temperature.

P also explains the nature of the microemulsion produced using nonionic surfactants of the ethoxylate type: P increases with increasing temperature (as a result of the dehydration of the PEO chain). A critical temperature (PIT) is reached at which P reaches 1 and above this temperature inversion occurs to a W/O system.

The influence of the surfactant structure on the nature of the microemulsion can also be predicted from thermodynamic theory. The most stable microemulsion would be that in which the phase with the smaller volume fraction forms the droplets (the osmotic pressure increases with increasing ϕ). For a W/O microemulsion prepared using an ionic surfactant such as Aerosol OT, the effective volume (hard-sphere volume) is only slightly larger than the water core volume, since the hydrocarbon tails may penetrate to a certain extent when two droplets come together. For an O/W microemulsion, the double layers may expand to a considerable extent, depending on the electrolyte concentration (the double layer thickness is of the order of 100 nm in 10^{-5} mol dm^{-3} 1:1 electrolyte and 10 nm in 10^{-3} mol dm^{-3} electrolyte). Thus the effective volume of

O/W microemulsion droplets can be significantly higher than the core oil droplet volume and this explains the difficulty of preparation of O/W microemulsions at high ϕ values when using ionic surfactants.

A schematic representation of the effective volume for W/O and O/W microemulsions is shown in Fig. 10.8.

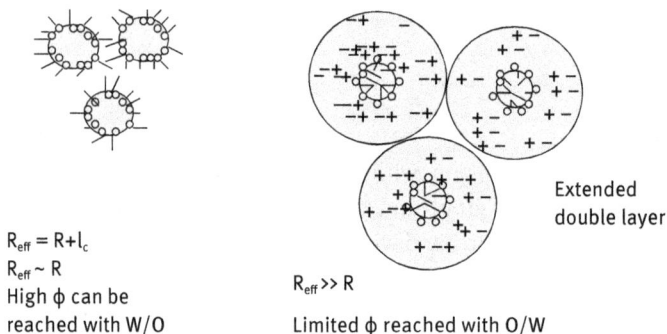

$R_{eff} = R+l_c$
$R_{eff} \sim R$
High ϕ can be
reached with W/O

$R_{eff} \gg R$

Extended
double layer

Limited ϕ reached with O/W

Fig. 10.8: Schematic representation of W/O and O/W microemulsion droplets.

10.5 Characterization of microemulsions using scattering techniques

Scattering techniques provide the most obvious methods for obtaining information on the size, shape and structure of microemulsions. The scattering of radiation, e.g. light, neutrons, X-ray, etc. by particles has been successfully applied for the investigation of many systems such as polymer solutions, micelles and colloidal particles. In all these methods, measurements can be made at sufficiently low concentrations to avoid complications arising from particle-particle interactions. The results obtained are extrapolated to infinite dilution to obtain the desirable property such as the molecular weight and radius of gyration of a polymer coil, the size and shape of micelles, etc. Unfortunately, this dilution method cannot be applied for microemulsions, which depend on a specific composition of oil, water and surfactants. The microemulsions cannot de diluted by the continuous phase since this dilution results in breakdown of the microemulsion. Thus, when applying scattering techniques to microemulsions, measurements have to be made at finite concentrations and the results obtained have to be analyzed using theoretical treatments to take into account droplet-droplet interactions.

Three scattering methods will be discussed below: Time-average (static) light scattering, dynamic (quasi-elastic) light scattering referred to as photon correlation spectroscopy and neutron scattering.

10.5.1 Time average (static) light scattering

The intensity of scattered light I(Q) is measured as a function of scattering vector Q,

$$Q = \left(\frac{4\pi n}{\lambda}\right)\sin\left(\frac{\theta}{2}\right) \tag{10.17}$$

where n is the refractive index of the medium, λ is the wavelength of light and θ is the angle at which the scattered light is measured.

For a fairly dilute system, I(Q) is proportional to the number of particles N, the square of the individual scattering units V_p and some property of the system (material constant) such as its refractive index,

$$I(Q) = [(\text{Material const..})(\text{Instrument const..})]NV_p^2 . \tag{10.18}$$

The instrument constant depends on the geometry of the apparatus (the light path length and the scattering cell constant).

For more concentrated systems, I(Q) also depends on the interference effects arising from particle-particle interaction,

$$I(Q) = [(\text{Instrument const..})(\text{Material const..})]NV_p^2 P(Q)S(Q) , \tag{10.19}$$

where P(Q) is the particle form factor which allows the scattering from a single particle of known size and shape to be predicted as a function of Q. For a spherical particle of radius R,

$$P(Q) = \left[\frac{(3\sin QR - QR\cos QR)}{(QR)^3}\right]^2 . \tag{10.20}$$

S(Q) is the so-called "structure factor" which takes into account particle-particle interaction. S(Q) is related to the radial distribution function g(r) (which gives the number of particles in shells surrounding a central particle) [13],

$$S(Q) = 1 - \frac{4\pi N}{Q}\int_0^\infty [g(r) - 1]r\sin QR\,dr . \tag{10.21}$$

For a hard-sphere dispersion with radius R_{HS} (which is equal to R + t, where t is the thickness of the adsorbed layer),

$$S(Q) = \frac{1}{[1 - NC(2QR_{HS})]} , \tag{10.22}$$

where C is a constant.

One usually measures I(Q) at various scattering angles θ and then plots the intensity at some chosen angle (usually 90°), i_{90} as a function of the volume fraction ϕ of the dispersion. Alternatively, the results may be expressed in terms of the Rayleigh ratio R_{90},

$$R_{90} = \left(\frac{i_{90}}{I_0}\right)r_s^2 . \tag{10.23}$$

I_0 is the intensity of the incident beam and r_s is the distance from the detector.

$$R_{90} = K_0 MCP(90)S(90) \tag{10.24}$$

K_o is an optical constant (related to the refractive index difference between the particles and the medium). M is the molecular mass of scattering units with weight fraction C.

For small particles (as is the case with microemulsions) $P(90) \sim 1$ and,

$$M = \frac{4}{3}\pi R_c^3 N_A , \qquad (10.25)$$

where N_A is Avogadro's constant.

$$C = \phi_c \rho_c , \qquad (10.26)$$

where ϕ_c is the volume fraction of the particle core and ρ_c is their density.

Equation (10.24) can be written in the simple form,

$$R_{90} = K_1 \phi_c R_c^3 S(90) , \qquad (10.27)$$

where $K_1 = K_0 (4/3) N_A \rho_c^2$.

Equation (10.27) shows that to calculate R_c from R_{90} one needs to know $S(90)$. The latter can be calculated using equations (10.21) and (10.22).

The above calculations were obtained using a W/O microemulsion of water/xylene/sodium dodecyl benzene sulphonate (NaDBS)/hexanol [12]. The microemulsion region was established using the quaternary phase diagram. W/O microemulsions were produced at various water volume fractions using increasing amounts of NaDBS: 5, 10.9, 15 and 20 %.

The results for the variation of R_{90} with the volume fraction of the water core droplets at various NaDBS concentrations are shown in Fig. 10.9. With the exception of the 5 % NaDBS results, all the others showed an initial increase in R_{90} with increasing ϕ, reaching a maximum at a given ϕ, after which R_{90} decreases with a further increase in ϕ.

The above results were used to calculate R as a function of ϕ using the hard-sphere model discussed above (equation (10.27)). This is also shown in Fig. 10.9.

It can be seen that with increasing ϕ, at constant surfactant concentration, R increases (the ratio of surfactant to water decreases with increasing ϕ). At any volume fraction of water, increasing surfactant concentration results in decreasing microemulsion droplet size (the ratio of surfactant to water increases).

10.5.2 Calculation of droplet size from interfacial area

If one assumes that all surfactant and cosurfactant molecules are adsorbed at the interface, it is possible to calculate the total interfacial area of the microemulsion from a knowledge of the area occupied by surfactant and cosurfactant molecules.

Total interfacial area = Total number of surfactant molecule$_s$ × area per surfactant molecule A_s + total number of cosurfactant molecules × area per cosurfactant molecule A_{co}.

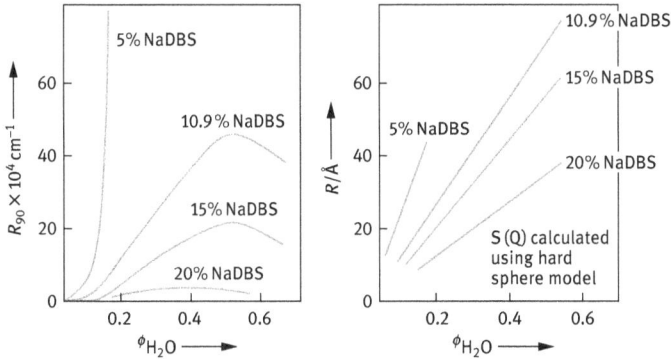

Fig. 10.9: Variation of R_{90} and R with the volume fraction of water for a W/O microemulsion based on xylene-water-NaDBS-hexanol.

The total interfacial area A per kg of microemulsion is given by the expression,

$$A = \frac{(n_s N_A A_s + n_{co} N_A A_{co})}{\phi}.$$ (10.28)

n_s and n_{co} are the number of moles of surfactant and cosurfactant.

A is related to the droplet radius R (assuming all the droplets are of the same size) by

$$A = \frac{3}{R\rho}.$$ (10.29)

Using reasonable values for A_s and A_{co} ($30A^2$ for NaDBS and $20A^2$ for hexanol) R was calculated and the results were compared with those obtained using light scattering results. Two conditions were considered: (a) All hexanol molecules were adsorbed 1A1; (b) part of the hexanol adsorbed to give a molar ratio of hexanol to NaDBS of 2 : 1 (1A2). Good agreement is obtained between the light scattering data and R calculated from interfacial area particularly for 1A2.

10.5.3 Dynamic light scattering (photon correlation spectroscopy, PCS)

In this technique one measures the intensity fluctuation of scattered light by the droplets as they undergo Brownian motion [14]. When a light beam passes through a colloidal dispersion, an oscillating dipole movement is induced in the particles, thereby radiating the light. Due to the random position of the particles, the intensity of scattered light, at any instant, appear as random diffraction ("speckle" pattern). As the particles undergo Brownian motion, the random configuration of the pattern will fluctuate, such that the time taken for an intensity maximum to become a minimum (the coherence time), corresponds approximately to the time required for a particle to move one wavelength λ. Using a photomultiplier of active area about the diffraction maximum (i.e. one coherent area) this intensity fluctuation can be measured. The analogue output is digitized using a digital correlator that measures the photocount (or intensity) correlation function of scattered light.

The photocount correlation function $g^{(2)}(\tau)$ is given by

$$g^{(2)} = B[1 + \gamma^2 g^{(1)}(\tau)]^2 \,, \tag{10.30}$$

where τ is the correlation delay time.

The correlator compares $g^{(2)}(\tau)$ for many values of τ.

B is the background value to which $g^{(2)}(\tau)$ decays at long delay times. $g^{(1)}(\tau)$ is the normalized correlation function of the scattered electric field and γ is a constant (~ 1).

For monodispersed noninteracting particles,

$$g^{(1)}(\tau) = \exp(-\Gamma\gamma) \,. \tag{10.31}$$

Γ is the decay rate or inverse coherence time that is related to the translational diffusion coefficient D,

$$\Gamma = DK^2 \,, \tag{10.32}$$

where K is the scattering vector,

$$K = \left(\frac{4\pi n}{\lambda_0}\right) \sin\left(\frac{\theta}{2}\right) \,. \tag{10.33}$$

The particle radius R can be calculated from D using the Stokes–Einstein equation,

$$D = \frac{kT}{6\pi\eta_0 R} \,, \tag{10.34}$$

where η_0 is the viscosity of the medium.

The above analysis only applies for very dilute dispersions. With microemulsions which are concentrated dispersions, corrections are needed to take interdroplet interaction into account. This is reflected in plots of $\ln g^{(1)}(\tau)$ versus τ which become nonlinear, implying that the observed correlation functions are not single exponentials.

As with time-average light scattering, one needs to introduce a structure factor in calculating the average diffusion coefficient. For comparative purposes, one calculate the collective diffusion coefficient D which can be related to its value at infinite dilution D_0 by [15]

$$D = D_0(1 + \alpha\phi) \,, \tag{10.35}$$

where α is a constant that is equal to 1.5 for hard spheres with repulsive interaction.

10.5.4 Neutron scattering

Neutron scattering offers a valuable technique for determining the dimensions and structure of microemulsion droplets. The scattering intensity I(Q) is given by

$$I(Q) = (\text{Instrument const.})(\rho - \rho_0)NV_p^2 P(Q)S(Q) \,, \tag{10.36}$$

where ρ is the mean scattering length density of the particles and ρ_0 is the corresponding value for the solvent.

One of the main advantages of neutron scattering over light scattering is the Q range at which one operates: With light scattering, the range of Q is small (~ 0.0005–$0.0015\,\text{A}^{-1}$) while for small angle neutron scattering the Q range is large (0.02–$0.18\,\text{A}^{-1}$). In addition, neutron scattering can give information on the structure of the droplets.

As an illustration Fig. 10.10 shows plots of $I(Q)$ versus Q for W/O microemulsions (xylene/water/NaDBS/hexanol) [16].

The Q values at the maximum can be used to calculate the lattice spacing using Bragg's equation. Alternatively, one can use a hard-sphere model to calculate $S(Q)$ and then fit the data of $I(Q)$ versus Q to obtain the droplet radius R.

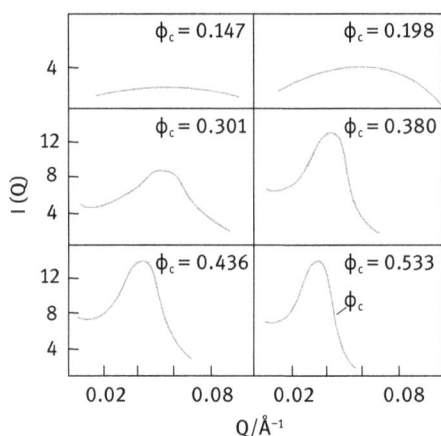

Fig. 10.10: I(Q) versus Q for W/O microemulsions at various water volume fractions.

10.5.5 Contrast matching for determining the structure of microemulsions

By changing the isotopic composition of the components (e.g. using deuterated oil and H_2O–D_2O) one can match the scattering length density of the various components: By matching the scattering length density of the water core with that of the oil, one can investigate the scattering from the surfactant "shell". By matching the scattering length density of the surfactant "shell" and the oil, one can investigate the scattering from the water core.

10.6 Characterization of microemulsions using conductivity

Conductivity measurements may provide valuable information on the structural behaviour of microemulsions. In the early applications of conductivity measurements,
the technique was used to determine the nature of the continuous phase. O/W microemulsions should give fairly high conductivity (that is determined by that of the
continuous aqueous phase) whereas W/O microemulsions should give fairly low conductivity (that is determined by that of the continuous oil phase).

As an illustration Fig. 10.11 shows the change in electrical resistance (reciprocal
of conductivity) with the ratio of water to oil (V_w/V_o) for a microemulsion system
prepared using the inversion method [17]. Figure 10.11 indicates the change in optical
clarity and birefringence with the ratio of water to oil.

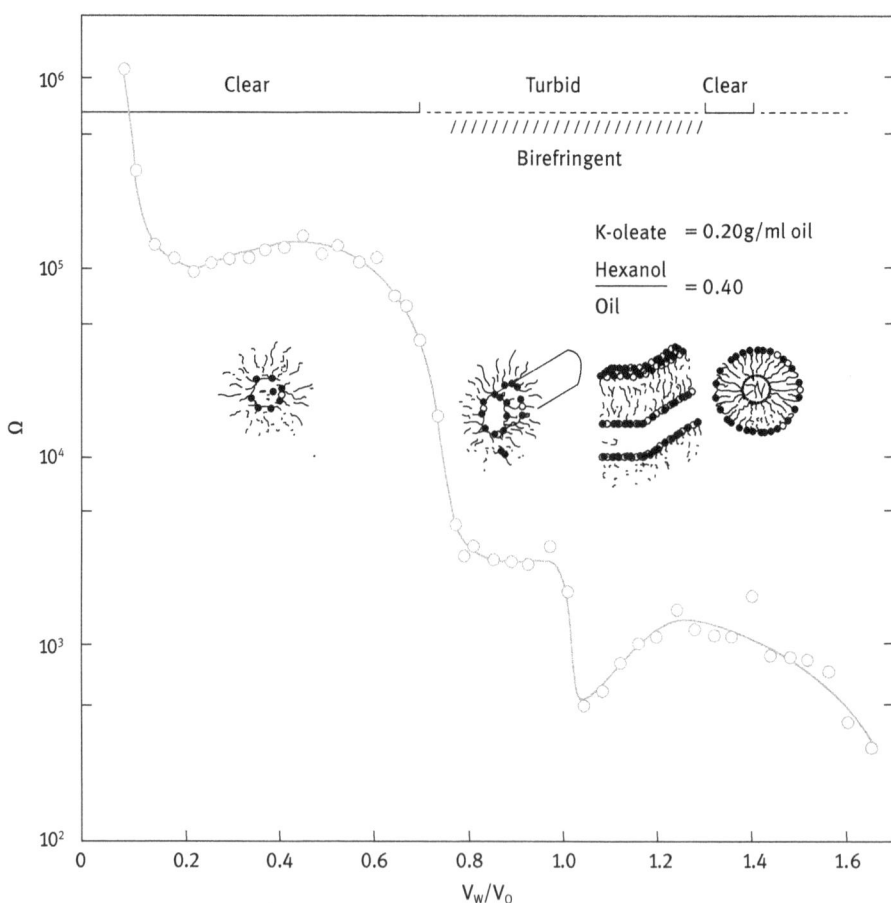

Fig. 10.11: Electrical resistance versus V_w/V_o.

At low V_w/V_o, a clear W/O microemulsion is produced with a high resistance (oil continuous). As V_w/V_o increases, the resistance decreases, and in the turbid region, hexanol and lamellar micelles are produced. Above a critical ratio, inversion occurs and the resistance decreases producing O/W microemulsion.

Conductivity measurements were also used to study the structure of the microemulsion, which is influenced by the nature of the cosurfactant. This is illustrated in Fig. 10.12 for two systems based on water/toluene/potassium oleate/butanol and water/hexadecane/potassium oleate/hexanol [18]. The difference between the two systems is in the nature of the cosurfactant, namely butanol (C_4 alcohol) and hexanol (C_6 alcohol). The first system based on butanol shows a rapid increase in κ above a critical water volume fraction value, whereas the second system based on hexanol shows much lower conductivity values with a maximum and minimum at two water volume fractions values ϕ_w' and ϕ_w''.

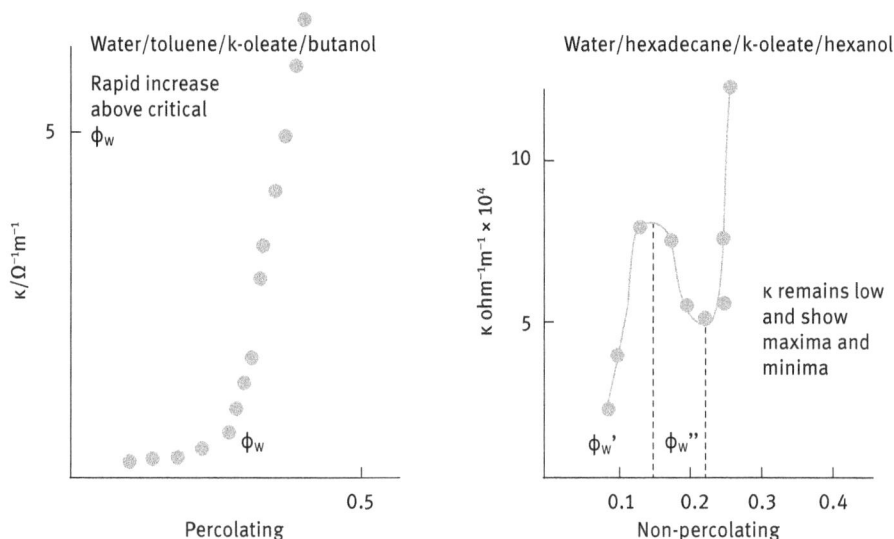

Fig. 10.12: Conductivity versus water volume fraction for two W/O microemulsion systems.

In the first case (when using butanol), the κ-ϕ_w curve can be analyzed using the percolation theory of conductivity [19]. In this model, the effective conductivity is practically zero as long as the volume fraction of the conductor (water) is below a critical value ϕ_w^p (the percolation threshold). Beyond this value, κ suddenly takes a nonzero value and it increases rapidly with a further increase in ϕ_w.

In the above case (percolating microemulsions), the following equation were theoretically derived.

$$\kappa = (\phi_w - \phi_w^p)^{8/5} \quad \text{when } \phi_w > \phi_w^p, \tag{10.37}$$

$$\kappa = (\phi_w^p - \phi_w)^{-0.7} \quad \text{when } \phi_w < \phi_w^p. \tag{10.38}$$

By fitting the conductivity data to equations (10.37) and (10.38), ϕ_w^p was found to be 0.176 ± 0.005 in agreement with the theoretical value.

The second system based on hexanol does not fit the percolation theory (nonpercolating microemulsion). The trend of the variation of κ with water volume fraction is due to more subtle changes in the system on changing ϕ_w. The initial increase in κ with increasing ϕ_w can be ascribed to enhanced surfactant solubilization with added water. Alternatively, it could be due to increasing surfactant dissociation on the addition of water. Beyond the maximum, addition of water mainly causes micelle swelling, i.e. a definite water core (microemulsion droplets) begins to be formed, which may be considered a dilution process leading to decreasing conductivity (the decrease in κ beyond the maximum may be due to the replacement of the hydrated surfactant-cosurfactant aggregates with microemulsion droplets). The sharp increase in κ beyond the minimum must be associated with a facilitated path for ion transport (formation of nonspherical droplets) resulting from swollen micelle clustering and subsequent cluster interlinking.

A systematic study of the effect of cosurfactant chain length on the conductive behaviour of W/O microemulsions was carried out by Clausse and his co-workers [20]. The cosurfactant chain length was gradually increased from C_2 (ethanol) to C_7 (heptanol). The results for the variation of κ with ϕ_w are shown in Fig. 10.13. With the short chain alcohols (C < 5), the conductivity shows a rapid increase above a critical ϕ value. With longer chain alcohols, namely hexanol and heptanol, conductivity

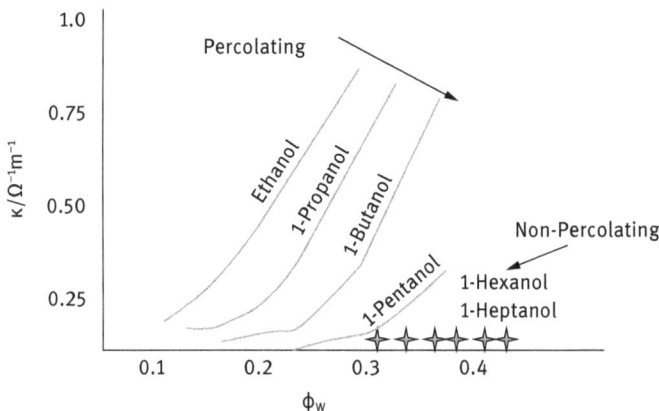

Fig. 10.13: Variation of conductivity with water volume fraction for various cosurfactants.

remains very low up to a high water volume fraction. With the short chain alcohols, the system shows percolation above a critical water volume fraction. Under these conditions the microemulsion is "bicontinuous". With the longer chain alcohols, the system is nonpercolating and one can define definite water cores. This is sometimes referred to as a "true" microemulsion.

10.7 NMR measurements

Lindman and co-workers [21–23] demonstrated that the organization and structure of microemulsions can be elucidated from self-diffusion measurements of all the components (using pulse gradient or spin echo NMR techniques). Within a micelle, the molecular motion of the hydrocarbon tails (translational, reorientation and chain flexibility) is almost as rapid as in a liquid hydrocarbon. In a reverse micelle, water molecules and counterions are also highly mobile. For many surfactant-water systems, there is a distinct spatial separation between hydrophobic and hydrophilic domains. The passage of species between different regions is an improbable event and this occurs very slowly.

Thus, self-diffusion, if studied over macroscopic distances, should reveal whether the process is rapid or slow depending on the geometrical properties of the inner structure. For example, a phase that is water continuous and oil discontinuous should exhibit rapid diffusion of hydrophilic components, while the hydrophobic components should diffuse slowly. An oil continuous but water discontinuous system should exhibit rapid diffusion of the hydrophobic components. One would expect that a bicontinuous structure should give rapid diffusion of all components.

Using the above principle, Lindman and co-workers [21–23] measured the self-diffusion coefficients of all components consisting of various components, with particular emphasis on the role of the cosurfactant. For microemulsions consisting of water, hydrocarbon, an anionic surfactant and a short chain alcohol (C_4 and C_5), the self-diffusion coefficient of water, hydrocarbon and cosurfactant was quite high, of the order of 10^{-9} m^2 s^{-1}, i.e. two orders of magnitude higher than the value expected for a discontinuous medium (10^{-11} m^2 s^{-1}). This high diffusion coefficient was attributed to three main effects: bicontinuous solutions, easily deformable and flexible interface and absence of any large aggregates. With microemulsions based on long chain alcohols (e.g. decanol), the self-diffusion coefficient for water was low, indicating the presence of definite (closed) water droplets surrounded by surfactant anions in the hydrocarbon medium. Thus, NMR measurements could clearly distinguish between the two types of microemulsion systems.

10.8 Formulation of microemulsions

The formulation of microemulsions or micellar solutions, like that of conventional macroemulsions, is still an art. In spite of exact theories that explain the formation of microemulsions and their thermodynamic stability, the science of microemulsion formulation has not advanced to the point where one can predict with accuracy what happens when the various components are mixed. The very much higher ratio of emulsifier to disperse phase, which differentiates microemulsions from macroemulsions, appears at first sight to make the application of various techniques for formulation less critical. However, in the final stages of the formulation one immediately realizes that the requirements are very critical due to the greater number of parameters involved.

The mechanics of forming microemulsions differ from those used in making macroemulsions. The most important difference lies in the fact that putting more work into a macroemulsion or increasing emulsifier usually improves its stability. This is not so for microemulsions. Formation of a microemulsion depends on specific interactions of the molecules of oil, water and emulsifiers. These interactions are not exactly known. If such specific interactions are not realized, no amount of work nor excess emulsifier can produce the microemulsion. If the chemistry is right, microemulsification occurs spontaneously.

One should remember that for microemulsions the ratio of emulsifier to oil is much higher than that used for macroemulsions. The emulsifier used is at least 10 % depending on the oil and in most cases it can be as high as 20–30 %. W/O systems are made by blending the oil and emulsifier with some heating if necessary. Water is the added to the oil-emulsifier blend to produce the microemulsion droplets and the resulting system should appear transparent or translucent. If the maximum amount of water that can be microemulsified is not high enough for the particular application, one should try other emulsifiers to reach the required composition.

The most convenient way of producing O/W microemulsion is to blend the oil and emulsifier and then pour the mixture into water with mild stirring. In the case of waxes, both oil/emulsifier blend and the water must be at higher temperature (above the melting point of the wax). If the melting point of the wax is above the boiling temperature of water, the process can be carried out at high pressure. Another technique to mix the ingredients is to make a crude macroemulsion of the oil and one of the emulsifiers. By using low volumes of water, a gel is formed and the system can then be titrated with the co-emulsifier until a transparent system is produced. This system may be further diluted with water to produce a transparent or translucent microemulsion.

Four different emulsifier selection methods can be applied for formulation of microemulsions: (i) The hydrophilic-lipophilic balance (HLB) system. (ii) The phase inversion temperature (PIT) method. (iii) The cohesive energy ratio (CER) concept. (iv) Partitioning of cosurfactant between the oil and water phases. The first three methods are essentially the same as those used for the selection of emulsifiers for macroemulsions, described in detail in Chapter 8. However, with microemulsions one should try to match the chemical type of the emulsifier with that of the oil. Cosurfactant portioning plays a major role in microemulsion formation. According to the thermodynamic theory of microemulsion formation, the total interfacial tension of the mixed film of surfactant and cosurfactant must approach zero. The total interfacial tension is given by the following equation:

$$\gamma_T = \gamma_{(O/W)_a} - \pi, \tag{10.39}$$

where $\gamma_{(O/W)_a}$ is the interfacial tension of the oil in the presence of alcohol cosurfactant and π is the surface pressure. $\gamma_{(O/W)_a}$ seems to reach a value of 15 mN m^{-1} irrespective of the original value of $\gamma_{O/W}$. It seems that the cosurfactant which is predominantly oil soluble distributes itself between the oil and the interface and this causes a change in the composition of the oil which now is reduced to 15 mN m^{-1}.

Measuring the partition of the cosurfactant between the oil and the interface is not easy. A simple procedure to select the most efficient cosurfactant is to measure the oil/water interfacial tension $\gamma_{O/W}$ as a function of cosurfactant concentration. The lower the percentage of cosurfactant required to lower $\gamma_{O/W}$ to 15 mN m^{-1}, the better the candidate.

10.9 Industrial applications of microemulsions

The first marketed microemulsions were dispersions of carnauba wax in water [2]. They were prepared by adding potassium oleate to melted wax followed by incorporation of boiling water in small aliquots. The resulting opalescent formulations were used as a floor polish and formed a glossy surface on drying. The effectiveness and stability of the liquid wax formulations stimulated the development of many other formulations consisting of either O/W or W/O microemulsions [2]. Later, microemulsions were used in several industrial applications as described below.

10.9.1 Microemulsions in pharmaceuticals

Microemulsions act as supersolvents of drugs [24] (including drugs that are relatively insoluble in both aqueous and hydrophobic solvents), probably as a consequence of the presence of surfactant and cosurfactant. The O/W microemulsion can behave as a reservoir of lipophilic drugs, whereas the W/O microemulsion can behave as a reservoir of hydrophilic drugs. The drug will be partitioned between the dispersed and continuous phases, and when the system comes in contact with a semipermeable membrane (skin or mucous membrane), the drug can be transported through the barrier. Drug release with pseudo-zero-order kinetics can be obtained, depending on the volume of the dispersed phase, the partition of the drug among interphase and continuous and dispersed phases, and the transport rate of the drug. Since the mean diameter of the microemulsion droplets is below 100 nm, a large interfacial area is produced, from which the drug can be quickly released into the external phase when in vitro or in vivo absorption takes place, maintaining the concentration in the external phase close to the initial level.

Microemulsions can be sterilized by filtration since the droplet diameter is below 0.22 μm. Autoxidation of lipids in O/W microemulsions is lower than in emulsions or micellar solutions. Solid colloidal therapeutic systems can be obtained with both O/W and W/O microemulsions. In addition, the microemulsion has a low viscosity thus facilitating its administration. The use of microemulsions as delivery systems can improve the efficacy of the drug, allowing the total dose to be reduced and thus minimizing side effects.

The main limitations on the use of microemulsions in drug delivery arise chiefly from the need for all the components to be approved by the Food and Drug Administration (FDA), particularly surfactants and cosurfactants. For this reason, microemulsions based on lecithin (naturally occurring surfactant that can be produced from egg yolk or soybean) and a short chain alcohol like butanol have been introduced.

In predicting the release of a drug from a microemulsion, one of the most important problems is to evaluate its partition among interphase and continuous and dispersed phases. In order to know how a drug is partitioned, its partition coefficient P_{cos} among oil, water and cosurfactant, in the ratios present in the microemulsion can be determined. To evaluate the significance of P_{cos} five drugs with various lipophilicity, namely phenylbutazone, betamethasone, nitrofurazone, menadione and prednisone, were dissolved in an O/W microemulsion (consisting of isopropyl myristate, buffer pH 7, dioctylsulphosuccinate and butanol). Table 10.1 shows the octanol/water (pH 7) partition coefficient P_{oct}, isopropyl myristate/water (pH 7) partition coefficient P_{IPM} and P_{cos}. This table shows how the partition coefficient of the drugs varies in the presence and absence of the cosurfactant butanol. In the presence of the cosurfactant, the partition coefficient of the drug increases above its value in the absence of the alcohol. In addition, the logarithms of the coefficient of permeation of the drugs through a hydrophilic membrane are inversely proportional to the logarithms of P_{cos}; the higher the concentration of the drug in the internal phase (reservoir), the lower the amount released over time from the systems.

Modulation of drug release can be advantageous. This may be achieved in a microemulsion by keeping the amounts of the other components fixed and changing only the amount of cosurfactant. As a consequence, P_{cos} and permeation coefficients of the drug vary.

Ziegenmeyer and Fuhrer [25] compared the release rate of tetracycline hydrochloride microemulsion with that of the gel and cream formulation. This is illustrated in Fig. 10.14.

One of the most useful applications of microemulsions in pharmacy is for percutaneous administration. Drug transport from microemulsions is usually better than that from ointments, gels and creams. Systemic medication has also been achieved. The

Tab. 10.1: Partition coefficient of drugs.

Drug	P_{oct}	P_{IPM}	P_{cos}
Nitrofurazone	1.8 ± 0.1	0.12 ± 0.02	1.5 ± 0.1
Phenylbutazone	5.3 ± 0.2	2.1 ± 0.2	4.5 ± 0.2
Prednisone	29 ± 2	0.3 ± 0.05	6.4 ± 0.2
Betamethasone	95 ± 5	2.3 ± 0.3	30 ± 2
Menadione	160 ± 10	100 ± 8	180 ± 12

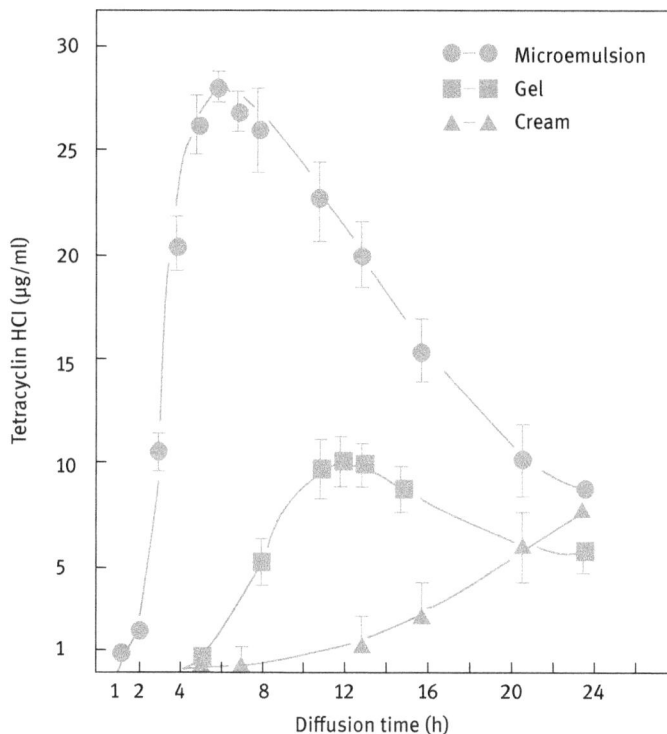

Fig. 10.14: In vitro permeation of tetracycline hydrochloride through skin membranes from a microemulsion, gel and cream.

facilitated transport is due to the complete dissolution of the drug in the microemulsions, reaching relatively high concentrations as a consequence of the supersolvent properties of microemulsions, and the dispersed phase can act as a reservoir, making it possible to maintain an almost constant concentration in the continuous phase; thus, pseudo-zero-order kinetics can be achieved. Moreover, some of the components of the microemulsion can operate as enhancers. The release rate of the formulation can also be controlled when using a microemulsion.

Ziegenmeyer and Fuhrer [25] compared the release rate through skin membranes of tetracycline hydrochloride microemulsion with that of a gel and cream. The results are shown in Fig. 10.14.

The results of Fig. 10.14 show that the transport of the drug is significantly better for the microemulsion. Moreover, the time lag is notably lower due to the microemulsion components which act as enhancers.

Another application of microemulsions in pharmacy is in the area of ocular administration of drugs that are delivered topically to treat eye disease. These drugs are essentially delivered as O/W microemulsions to dissolve poorly soluble drugs, to increase absorption and to prolong the release time. Siebenbrodt and Keipert [26] developed and characterized lecithin-Tween 80 based microemulsions, which dissolve some poorly soluble drugs such as atropine, chloramphinicol and indomethacin. These drugs were solubilized in therapeutically relevant concentrations (0.5%) in microemulsions. With the aim of enhancing the amounts of drug transport through the conjunctiva, timolol, a β-blocker used in glaucoma therapy, was dissolved in lecithin-based O/W microemulsion [27]. Aliquots of timolol in solution, as an ion pair in solution, and as an ion pair in microemulsion was administered in rabbits. The aqueous humour concentration was measured as a function of time for timolol alone, timolol as an ion pair in solution and timolol as an ion pair in microemulsion. The results are shown in Fig. 10.15.

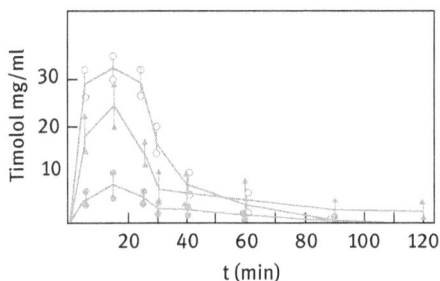

Fig. 10.15: Aqueous humour concentration time profiles following multiple instillations in rabbits' eyes: (black circles) timolol alone; (white circles) timolol as an ion pair in solution; (triangle) timolol as an ion pair in microemulsion.

The results of Fig. 10.15 show that the areas under the curves for timolol in aqueous humour after administration of the microemulsion and the ion pair solution, are 3.5 and 4.2 higher, respectively, than that observed for timolol alone. The absorption times lengthened for timolol in microemulsion. It is probably that the tiny nanodroplets remained on the cornea for some time and acted as a microemulsion reservoir for timolol, appreciably prolonging the absorption time.

A further application of microemulsions is in the area of peroral administration, which can be applied for protection of biodegradable drugs from the biological environment. Drew et al. [28] enhanced the oral absorption of cyclosporine by administering hard gelatine capsules containing two O/W microemulsions (slow and fast release) and a solid micellar solution of cyclosporine to healthy male volunteers. Absorption increased on average by 45 % for the solid solution and 49 % for the faster releasing microemulsion, compared to the reference soft gelatine capsule.

Another application of microemulsions is in parenteral administration. O/W microemulsions are used mainly as carriers of lipophilic drugs in order to attain prolonged release and to administer parenterally lipophilic drugs that are not soluble in water. They can be administered intravenously, intramuscularly or subcutaneously. W/O microemulsions can be used for subcutaneous and intramuscular administration of hydrophilic drugs with the aim of prolonged release.

A very important application of microemulsions is the preparation of solid colloidal therapeutic systems. Colloidal drug carriers are potential tools for achieving site-specific drug delivery. Polyalkycyanoacrylate nanoparticles have been investigated as potential lisosomotropic carriers of drugs. They are prepared by microemulsion polymerization, in which droplets of the insoluble monomers are microemulsified in the aqueous phase [29]. Samples of the nanoparticles showed that the inner structure was a highly porous matrix. The acute toxicity of the polycyanoacrylate nanoparticles was rather low and the toxicity of doxorubicin adsorbed onto such particles was markedly reduced [30]. Nanocapsules are prepared by dissolving the monomer alkylcyanoacrylate and a lipophilic drug in a lipid phase that is slowly injected into an aqueous solution of a nonionic surfactant [31]. The main advantage of using polyalkylcyanoacrylate is its biodegradability.

10.9.2 Applications of microemulsions in cosmetics

One important area of application of microemulsions in cosmetics is in the solubilization of fragrance and flavour oils [32]. Dartnell et al. [33] reported a microemulsion containing a perfuming concentrate, i.e. a nonalcoholic perfuming product. This microemulsion contains approximately 5 to 50 % of the concentrate of odoriferous substances, without using ethanol; polyethylene glycol derivatives are used as the primary surfactant and polyglycerol and ether phosphate derivatives are used as co-surfactants. Chiu and Yang [34] showed that a vitamin E microemulsion has high resistance to oxidation in air. The microemulsions are obtained by solubilizing vitamin E in aqueous solution with pure nonionic surfactants, polyethylene glycol monoalkyl ethers. Parra et al. [35] reported microemulsions as vehicles for nucleophilic reagents in cosmetic formulations. The modifications of chemical reactivity induced in human hair during its treatment with a reactive agent ($NaHSO_3$) or oxidative agent (H_2O_2) with a micellar or microemulsion system as the vehicle were investigated. The W/O type of microemulsion consists of sodium dodecyl sulphate, pentanol (heptane) and water. The treatment of human hair with $NaHSO_3$ carried in a micellar solution system favours reductive attack, and this effect is enhanced when the reagent is in a microemulsion vehicle. In oxidization of human hair with H_2O_2 the order of reactivity is the following: aqueous medium < micellar solution < microemulsion. The reactivity becomes higher as the water content decreases and the hydrocarbon content increases. Solans et al. [36] also reported the activity of thioglycolic acid, incorporated in microemulsions, towards cysteine residues. The influence of microemulsion structure on cysteine reactivity with keratin fibres was investigated using an appropriate model system. The realm of hydrocarbon-continuous microemulsion-type media was found to induce the highest activity. This is due to the existence of water pools containing most of the reagent in the inverse microemulsion and this appears to enhance cysteine activity. Another factor may be that the reactive medium preferentially wets the hydrophobic fractions of keratin and spreads onto and penetrates into the individual fibres.

10.9.3 Applications in agrochemicals

A very attractive alternative for formulation of agrochemicals is to use microemulsion systems. As mentioned above, these are single optically isotropic and thermodynamically stable dispersions consisting of oil, water and amphiphile (one or more surfactants). These systems offer a number of advantages over O/W emulsions for the following reasons [37]. Once the composition of the microemulsion is identified, the system is prepared by simple mixing all the components without the need of any appreciable shear. Due to their thermodynamic stability, these formulations undergo no separation or breakdown on storage (within a certain temperature range depending on the system). The low viscosity of the microemulsion systems ensures their ease of pourability, dispersion on dilution and they leave little residue in the container. Another main attraction of microemulsions is their possible enhancement of biological efficacy of many agrochemicals. This is due to the solubilization of the pesticide by the microemulsion droplets.

The role of microemulsions in enhancing biological efficiency can be described in terms of the interactions at various interfaces and their effect on transfer and performance of the agrochemical. The application of an agrochemical as a spray involves a number of interfaces, where interaction with the formulation plays a vital role. The first interface during application is that between the spray solution and the atmosphere (air) which governs the droplet spectrum, rate of evaporation, drift, etc. In this respect the rate of adsorption of the surfactant at the air/liquid interface is of vital importance. Since microemulsions contain high concentrations of surfactant and mostly more than one surfactant molecule is used for their formulation, then on diluting a microemulsion on application, the surfactant concentration in the spray solution will be sufficiently high to cause efficient lowering of the surface tension γ. As discussed above, two surfactant molecules are more efficient in lowering γ than either of the two components. Thus, the net effect will be production of small spray droplets, which as we will see later, adhere better to the leaf surface. In addition, the presence of surfactants in sufficient amounts will ensure that the rate of adsorption (which is the situation under dynamic conditions) is fast enough to ensure coverage of the freshly formed spray by surfactant molecules [37].

The second interaction is between the spray droplets and the leaf surface, whereby the droplets impinging on the surface undergo a number of processes that determine their adhesion and retention and further spreading on the target surface. The most important parameters that determine these processes are: the volume of the droplets and their velocity, the difference between the surface energy of the droplets in flight, E_0, and their surface energy after impact, E_s. As mentioned above, microemulsions which are effective in lowering the surface tension of the spray solution ensure the formation of small droplets which do not usually undergo reflection if they are able to reach the leaf surface. Clearly, the droplets need not to be too small otherwise drift may occur. One usually aims at a droplet spectrum in the region of 100–400 μm. The adhesion of droplets is governed by the relative magnitude of the kinetic energy of the droplet in flight and its surface energy as it lands on the leaf surface. Since the kinetic energy is proportional to the third power of the radius (at constant droplet velocity), whereas the surface energy is proportional to the second power, one would expect that sufficiently small droplets will always adhere. For a droplet to adhere, the difference in surface energy between free and attached drop $(E_0 - E_s)$ should exceed the kinetic energy of the drop, otherwise bouncing will occur. Since E_s depends on the contact angle, θ, of the drop on the leaf surface, it is clear that low values of θ are required to ensure adhesion, particularly with large drops that have high velocity. Microemulsions when diluted in a spray solution usually give low contact angles of spray drops on leaf surfaces as a result of lowering the surface tension and their interaction with the leaf surface.

Another factor which can affect biological efficacy of foliar spray application of agrochemicals is the extent to which a liquid wets and covers the foliage surface. This, in turn, governs the final distribution of the agrochemical over the areas to be protected. Several indices may be used to describe the wetting of a surface by the spray liquid, of which the spread factor and spreading coefficient are probably the most useful. The spread factor is simply the ratio between the diameter of the area wetted on the leaf, D, and the diameter of the drop, d. This ratio is determined by the contact angle of the drop on the leaf surface. The lower the value of θ, the higher the spread factor. As mentioned above, microemulsions usually give a low contact angle for the drops produced from the spray. The spreading coefficient is determined by the surface tension of the spray solution as well as the value of θ. Again, with microemulsions diluted in a spray both γ and θ are sufficiently reduced and this results in a positive spreading coefficient. This ensures rapid spreading of the spray liquid on the leaf surface [37].

Another important factor for controlling biological efficacy is the formation of "deposits" after evaporation of the spray droplets, which ensure the tenacity of the particles or droplets of the agrochemical. This will prevent removal of the agrochemical from the leaf surface by the falling rain. Many microemulsion systems form liquid crystalline structures after evaporation, which have high viscosity (hexagonal or lamellar

liquid crystalline phases). These structures will incorporate the agrochemical particles or droplets and ensure their "stickiness" to the leaf surface [37].

One of the most important roles of microemulsions in enhancing biological efficacy is their effect on penetration of the agrochemical through the leaf. Two effects
may be considered which are complimentary. The first effect is due to enhanced penetration of the chemical as a result of the low surface tension. For penetration to occur
through fine pores, a very low surface tension is required to overcome the capillary
(surface) forces. These forces produce a high pressure gradient that is proportional to
the surface tension of the liquid. The lower the surface tension, the lower the pressure
gradient and the higher the rate of penetration. The second effect is due to solubilization of the agrochemical within the microemulsion droplet. Solubilization results in
an increase in the concentration gradient, thus enhancing the flux due to diffusion.
This can be understood from by considering Fick's first law,

$$J_D = D \left(\frac{\partial C}{\partial x} \right) , \qquad (10.40)$$

where J_D is the flux of the solute (amount of solute crossing a unit cross-section in
unit time), D is the diffusion coefficient, and $(\partial C/\partial x)$ is the concentration gradient.
The presence of the chemical in a swollen micellar system will lower the diffusion coefficient. However, the presence of the solubilizing agent (the microemulsion droplet)
increases the concentration gradient in direct proportionality to the increase in solubility. This is because Fick's law involves the absolute gradient of concentration which
is necessarily small as long as the solubility is small, but not its relative rate. If saturation is denoted by S, Fick's law may be written as,

$$J_D = D100S \left(\frac{\partial \%S}{\partial x} \right) , \qquad (10.41)$$

where $(\partial \%S/\partial)$ is the gradient in relative value of S. Equation (10.41) shows that for
the same gradient of relative saturation, the flux caused by diffusion is directly proportional to saturation. Hence, solubilization will in general increase transport by
diffusion, since it can increase the saturation value by many orders of magnitude
(that outweighs any reduction in D). In addition, the solubilization enhances the rate
of dissolution of insoluble compounds and this will have the effect of increasing the
availability of the molecules for diffusion through membranes.

10.9.4 Applications in the food industry

Microemulsions (with the size range 5–50 nm) offer distinct advantages over macroemulsions (with the size range 0.1–10 μm) for formulation of many food products
[38]. In the first place, microemulsions are obtained spontaneously by mixing all the
ingredients and they are thermodynamically stable. Unlike macroemulsions, which

are only kinetically stable, microemulsions show no creaming, flocculation or coalescence. They can also be produced with narrow droplet size distribution. Hence microemulsions can potentially bring unique advantages to foods, providing systems that are reproducible and well characterized. In addition, microemulsions facilitate the addition of food gradients such as flavours, preservatives and nutrients (that are poorly soluble in water) by solubilizing them within the surfactant aggregates. These solubilized ingredients become protected from unwanted degradative reactions.

Unfortunately, the thermodynamic stability of the microemulsion phases means that they can only be produced in specific temperature, pressure and composition ranges. Knowing what conditions will allow a microemulsion to be established requires information on the phase diagram of the mixture of interest. In general, the ingredients of the mixture must be specifically chosen in order to achieve a microemulsion with the desired characteristics.

A major thrust of research in the area of food microemulsions is to investigate the aggregates formed by edible surfactants with oil/water systems. A key determinant of the type of assembly amphiphilic molecules will form at equilibrium is the "geometry" of the surfactant [39]. Within a micellar aggregate one observes that the hydrophobic tail groups of the surfactant molecule must fit within a relatively confined core of the aggregate, whereas the distance between the head groups is significantly less constrained. Thus, surfactant molecules with bulky head groups and nonbulky tails will have the easiest time forming a micelle. For an O/W microemulsion droplet, the presence of a second solvent in the droplet core means that the tails of the surfactant are less constrained than they are in the micellar aggregate. Thus, surfactants with a more sterically hindered tail group will be likely to form such microemulsions. For a W/O microemulsion, the situation is reversed; the tails need to be widely spaced and the heads are constrained. Hence surfactants with small head groups and large tail groups favour this form of microemulsion.

In practice, there are two ways in which the surfactant geometry required for microemulsion formation can be achieved. For a W/O microemulsion, a surfactant with bulky tail group can be selected for the system, often by using surfactants with two tail groups such as phospholipid or Aerosol OT (see below). Alternatively, the added bulk needed in the tail region can be achieved by adding to the system a second long-chain molecule (cosurfactant), typically alcohols, which can reside between the tails of the surfactant within the microemulsion droplet.

Based on the above concepts, two food-grade surfactants, namely glycerophospholipid and Aerosol OT (Fig. 10.16), could be used to prepare W/O microemulsions. The phospholipids can have a variety of structures, resulting from variation in the R, R', and R'' groups indicated in Fig. 10.16. Of these, phosphatidylcholine (lecithin) is the most widely investigated. Dilinolyly phosphatidylcholine has been shown using phosphorous[31] nuclear magnetic resonance (^{31}P-NMR) to form W/O microemulsion with chlorobenzene as a solvent. ^{31}P-NMR and proton NMR (^{1}H-NMR) have been used to demonstrate that dipalmitoyl phosphatidylcholine for W/O microemulsions at 52 °C in benzene, chlorobenzene and o-dichlorobenzene. Both these surfactants appear to form droplets containing up to 20 molecules of water for every phosphatidylcholine molecule. This molar ratio is referred to as w_0.

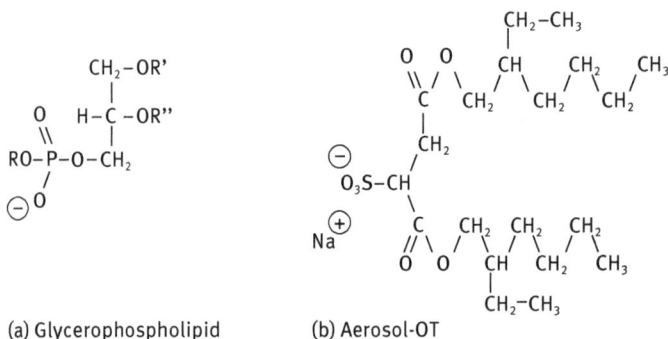

(a) Glycerophospholipid   (b) Aerosol-OT

Fig. 10.16: Double-tailed surfactant molecules.

Phospholipids extracted from foods have been investigated as well; these extracts contain a mixture of different phospholipid types. Phosphatidylcholine from egg forms W/O microemulsions in benzene, o-dichlorobenzene, diethyl ether, carbon tetrachloride, cyclohexane, octane and dodecane as shown by light scattering, electron microscopy, ^{31}P-NMR and ^{1}H-NMR. The water content of these microemulsions ranges from w_0 0 to 46. Soybean phosphatidylcholine forms a W/O microemulsion with a w_0 value of 16 at 4 °C.

The above investigations showed that W/O microemulsions formed from lecithin contain spherical droplets with ~ 20 nm radius. However, the droplet radius did not show the expected increase with increasing w_0. At high water contents the droplets begin to change shape and a clear viscous gel phase is observed in many systems. This gel contains arrays of water tubules within the external organic phase. The amount of water that can be incorporated per mole of surfactant within these phosphatidylcholine microemulsions depends on the type of phospholipid, the type of organic solvent and temperature. The presence of electrolyte in the aqueous phase also affects the value of w_0. Also, w_0 for egg lecithin microemulsions increases with increasing temperature above ~ 10 °C; below that temperature the microemulsion does not form.

The effect of addition of a cosurfactant (e.g. propanol) on the lecithin W/O microemulsions showed an increase in water solubilization when the lecithin concentration was 3 %. However, at 1 % lecithin addition of propanol had little effect on w_0. Addition of cholesterol enabled the system to incorporate more water without forming a gel.

Aerosol OT (sodium bis(2-ethylhexyl) sulphosuccinate), a double-tailed surfactant with a small head group, forms W/O microemulsions without the need for adding a cosurfactant. The microemulsion contains approximately monodisperse and spherical water droplets, whose size is linearly related to the amount of water per mole of surfactant incorporated into the aqueous phase. These microemulsions can solubilize large quantities of water, yielding w_0 values up to 60 (corresponding to 2.5 gram of water per gram of surfactant). The amount of water that can be incorporated decreases with increasing electrolyte concentration in the aqueous phase. The amount of water increases with increasing temperature. AOT can be combined with food-grade surfactants, such as sorbitan monostearate, to yield a microemulsion with considerable capacity for water solubilization.

Investigations on O/W microemulsions from food-grade materials are scarce. In contrast to the lecithin-based W/O microemulsion, a cosurfactant is needed to form oil-continuous phases from these molecules. Shinoda et al. [40] found that at least 14 wt % propanol in water is needed to form a hexadecane/water microemulsion with lecithin as the surfactant. In addition to these studies of lecithins, some success has been achieved with solubilizing edible oils in other food-grade surfactants such as Tween 20. A wide range of O/W microemulsions could be produced from monoglycerides, sodium palmitate or Tween 60. These microemulsions required the addition of a cosurfactant such as propylene glycol.

10.9.5 Microemulsions in biotechnology

There has been considerable interest in the use of microemulsions, particularly W/O (L2 type) for various applications in biotechnology [41]. This stems from the fact that many proteins can be solubilized in W/O microemulsions based on nonpolar aliphatic hydrocarbons. One of the main applications of microemulsions is in the field of enzymatic reactions. The major potential advantages of employing enzymes in media of low water content are the increased solubility of nonpolar components, the possibility of shifting the thermodynamic equilibria in favour of condensation and improved thermal stability of the enzymes thus enabling carrying out the reaction at higher temperatures. Enzymatic catalysis in microemulsions has been used for a variety of reactions, such as synthesis of esters, peptides and sugar acetals, transesterification and steroid transformation. The enzymes used include lipases, phospholipases, trypsin, lysozyme, peptidases, glucosidases and oxidases.

The catalytic activity of enzymes confined in microemulsion water droplets varies with overall water content (w_0). For many enzymes, the activity seems to increase with increasing w_0 reaching a maximum at an optimum maximum w_0 value, after which it decreases with a further increase in w_0. It seems that the maximum activity occurs around a w_0 at which the water droplets are somewhat larger than those of the enzyme.

The hydrophobicity of the solvent plays a key factor in the catalytic activity of the enzyme. A good correlation is between the hydrophobicity of the organic solvent (as measured by its log P value, where P is the partition coefficient of the solvent in the water/octanol system) and biocatalytic activity. The highest activity is obtained for solvents with log P > 4 and the lowest activity is obtained for solvents with log P < 2.

The choice of surfactant also plays a major role for the rate of enzymatic reactions in W/O microemulsions. For example, the lipase-catalyzed hydrolysis of triglycerides is rapid in microemulsions based on Aerosol OT and is extremely sluggish in microemulsions based on nonionic surfactants. This is due to the access of the enzyme to the interface between water and oil. This access is best when using Aerosol OT, whereas with the nonionic surfactant system, the stretching of the poly(ethylene glycol) (PEG) chain out from the interface prevents the enzyme from entering the interface.

10.9.6 Microemulsions in enhanced oil recovery (EOR)

In an oil well, oil is accumulated together with water and gas in reservoirs consisting of porous and permeable rocks [42]. The amount of oil spontaneously produced due to natural pressure that exists in the reservoir (primary oil recovery) and that produced by water or gas injection (secondary oil recovery) seldom exceeds 30–40 % of the original oil in place. This results from the unfavourable reservoir characteristics. Firstly, high oil viscosity and rock heterogeneity are responsible for areas unswept by the injected fluid and delay oil production. Secondly, capillary forces (see below) highly active in such porous media result in poor microscopic displacement efficiency. The oil remaining trapped after primary and secondary oil recovery represents a considerable amount and the main target is to recover some of this trapped oil, a phenomena referred to as tertiary oil recovery. The utilization of microemulsions for enhanced oil recovery (EOR) is performed within the framework of the chemical process, which concerns the use of polymers, surfactants and possibly alkaline agents, generally in combination.

In well-swept zones, capillary forces cause large amounts of oil to be left behind, as in any immiscible displacement. In these zones, the oil phase becomes discontinuous, i.e. producing droplets. If the rock is water wet, each droplet can then flow until it encounters a pore narrower than its own diameter, where it is trapped (residual oil). To enter such narrow pores, the surface curvature of the droplet should increase. The capillary pressure P_c is given by Laplace's equation,

$$P_c = P_o - P_w = \frac{2\gamma}{r},\qquad(10.42)$$

where P_o and P_w are the pressures in the oil phase and the water phase respectively, γ is the oil/water interfacial tension and r is the curvature of the supposedly spherical interface. Thus, a pressure gradient is required to exceed the capillary force retaining the droplet to make it flow. To keep this gradient within a range of values compatible with field situations, the interfacial tension has to be reduced to values $< 10^{-2}$ mN m^{-1} which can be achieved using microemulsions.

The displacement of residual oil depends on the competition between the viscous and capillary forces, as expressed by the dimensionless capillary number N_c,

$$N_c = \frac{(\Delta P/L)k_d}{\gamma \cos \theta}, \tag{10.43}$$

where ΔP is the pressure drop over the distance L, k_d is the displacing fluid permeability of the reservoir rock and θ is the contact angle.

N_c can also be expressed as the ratio between viscous and capillary forces,

$$N_c = \frac{\eta v}{\gamma \cos \theta}, \tag{10.44}$$

where η and v are the displacing fluid viscosity and Darcy velocity (flow rate per unit area), respectively.

No residual oil can be displaced until a minimum value of N_c is reached [43]. For typical water flood conditions $(N_c)_{crit} \approx 10^{-6}$ and to displace the residual oil $(N_c)_{crit}$ has to be increased by three or four orders of magnitude and this can be achieved by lowering γ by the same order. Values of $\gamma < 10^{-3}$ mN m^{-1} can be achieved by using micellar or microemulsion flooding.

The choice of surfactant(s) in the process of microemulsion flooding is crucial and one must consider the degree of interfacial tension lowering by considering the phase behaviour of the oil/brine/surfactant mixture. Phase behaviour depends on the surfactant partition coefficient between oil and brine, resulting from the preferential surfactant solubility into one of the phases.

As an illustration, Fig. 10.17 shows the (pseudo-)ternary phase diagrams of a typical oil/brine/anionic surfactant system. The top apex more often represents a mixture of amphiphilic compounds.

A hydrophilic surfactant, a mixture whose overall composition representative is located below the bimodal curve, will separate into two phases: an O/W microemulsion containing brine, surfactant and oil solubilized in the micelles and excess oil phase. This is denoted Winsor I. With an oleophilic surfactant, phase splitting preferably yields W/O microemulsion containing oil, surfactant solubilized in inverted micelles and excess brine phase (Winsor II). For intermediate surfactant solubility, a three-phase region forms, where a middle-phase microemulsion with a bicontinuous structure is in equilibrium with excess oil and brine phases (Winsor III).

Any change of a variable that favours the surfactant partitioning in the oil tends to promote the I–III–II transition. This variable can be increasing salinity, a decrease in oil molecular weight or a lengthening of the lipophilic part of the surfactant. The partitioning of anionic surfactant in brine is favoured by temperature rise, and that of nonionics is favoured by temperature lowering.

Anionic surfactant

O
Me

IV

I

PP

Brine

Oil

O
Me
W

IV

PP M PP
II III I

Brine

Oil

Me
W

IV

PP

II

Brine

Oil

Salinity
Surfactant chain lenght

Temperature
Hydrocarbon chain lenght

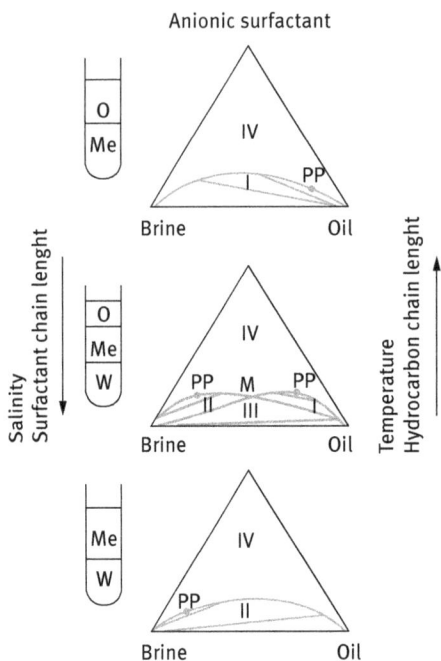

Fig. 10.17: Phase behaviour of a typical oil/brine/anionic surfactant system: O, excess oil phase; Me, microemulsion phase; W, excess water phase; PP, plait point; I and II, two-phase zones (with O and W as excess phases); III, three-phase zone; IV, single phase zone; M, representative point of the microemulsion (middle phase) composition.

As an illustration, Fig. 10.18 shows a schematic representation of the phase behaviour of multiphase microemulsion, where the changes from Winsor I–Winsor III–Winsor II are produced by increasing salinity [44].

With anionic surfactants, phase behaviour may depend on surfactant concentration through fractionation. NaCl is partly excluded from the surfactant rich phase, whereas the opposite trend occurs with $CaCl_2$ due to the strong association of the surfactant anion with the divalent salt. This association makes the transition I–III–II easier.

The lowest interfacial tensions are reached when a Winsor III environment occurs as is illustrated in Fig. 10.18. In such an environment, there are two types of interfaces. Near a phase transition, the larger of the two interfacial tensions between microemulsion and excess phases is associated with a high surface pressure in the surfactant layer, whereas the lower tension is associated with the vicinity of the critical end points in the phase diagram. Thus, the three-phase configuration is considered as an optimum, and any parameter value that leads to it is said to be optimal. Thus, the middle-phase microemulsion (Winsor III) appears to be the central element of the chemical process.

There is some correlation between the amounts of oil and brine solubilized in the micellar phase and the interfacial tension. This is shown in Fig. 10.18 where the solubilization expressed in terms of the solubilized parameter, SP, is plotted versus wt %

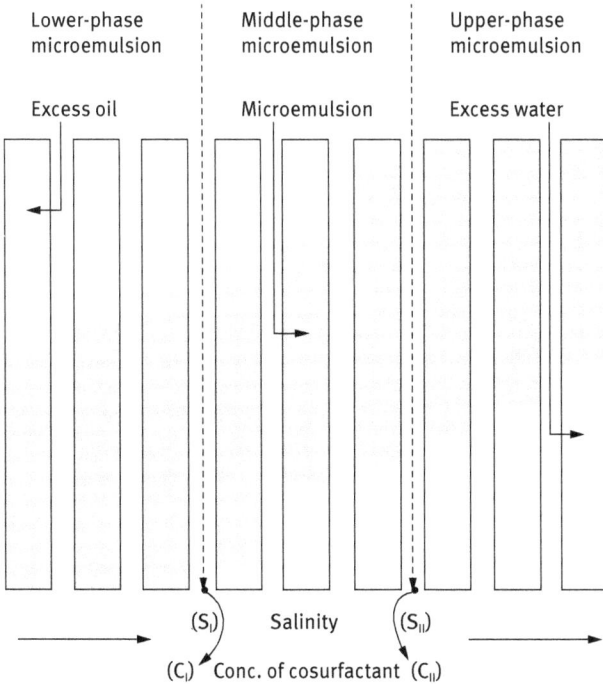

Fig. 10.18: Phase behaviour of multiphase microemulsions with increasing salinity: Lower-phase microemulsion and excess oil (Winsor I); middle-phase microemulsion with excess oil and water (Winsor III); upper-phase microemulsion and excess water (Winsor II).

salinity. SP is defined as

$$(SP)^2 = \frac{\alpha}{\gamma}, \tag{10.45}$$

where α is a constant that depends on the surfactant.

It can be seen from Fig. 10.19 that the interfacial tension decreases as solubilization increases. Through solubilization, interfacial tension is related to the molecular attractive interaction between surfactant, oil and water molecules. Winsor expressed the overall effect resulting from these interactions in terms of the cohesive energy ratio. R, defined as the ratio of solvent attraction between amphiphile (surfactant), C, and oil, O, i.e. A_{CO}, and between amphiphile and water, A_{CW},

$$R = \frac{A_{CO}}{A_{CW}}. \tag{10.46}$$

The stronger the solvent attraction, the higher the solubilization. If the surfactant is preferentially water soluble, $R < 1$ and the interfacial region will be convex toward water, giving a Winsor I system. In contrast, if $R > 1$, the interfacial region will be convex toward oil giving a Winsor II system. When $R \sim 1$, a Winsor III system forms and this leads to optimum conditions for EOR.

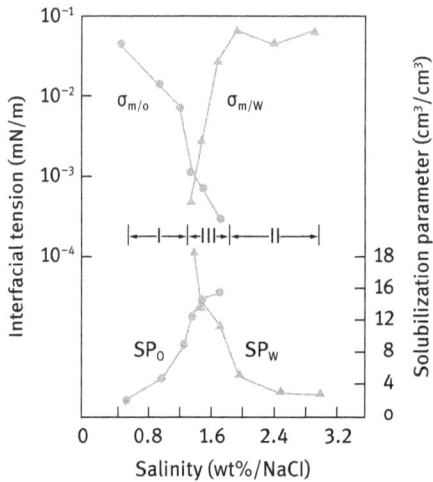

Fig. 10.19: Interfacial tension and solubilization parameter versus salinity. $\sigma_{m/o}$ and $\sigma_{m/w}$ interfacial tension between oil and microemulsion phase and between water and microemulsion phase respectively; SP_o and SP_w solubilization parameters of oil and water respectively.

The surfactants used for EOR must be chemically stable, brine soluble and compatible with the other usual components, especially polymers and alkaline agents. Cationics and amphoterics are rejected because of their adsorption on negatively charged clay surfaces. With anionics (usually sodium salts) and nonionics, the conditions must be established to minimize rock interactions. Petroleum sulphonates and alkylxylene sulphonates are the most commonly used surfactants in EOR. However, their performances drop as salinity and divalent cation concentrations increase, resulting in low brine solubility and poor interfacial efficiency. Other surfactant families which display higher salt tolerance have been considered such as sulphated or sulphonated ethoxylated alcohols or alkylphenols, α-olefin sulphonates and nonionics. The latter are often used as cosurfactants. Cosolvents such as C_3–C_5 alcohols are sometimes combined with the primary surfactant to improve solubility and to prevent the formation of highly viscous liquid crystalline phases.

10.9.7 Microemulsions as nanosize reactors for synthesis of nanoparticles

Microemulsions are suitable reactors for the preparation of nanoparticles which may contain crystalline, quasi-crystalline or amorphous phases. The particles can be metals, ceramics or composites with rather unique and improved mechanical, electronic and optical properties compared with normal coarse-grained polycrystalline materials [45]. W/O microemulsions with aqueous droplets in the size range 5–25 nm, stabilized by an interfacial film of surfactants in the continuous hydrocarbon phase are ideal microreactors. These aqueous droplets continuously collide, coalesce and break apart resulting in continuous exchange of solute content. When the chemical reaction is fast, the overall reaction rate is controlled by coalescence of the droplets. A relatively

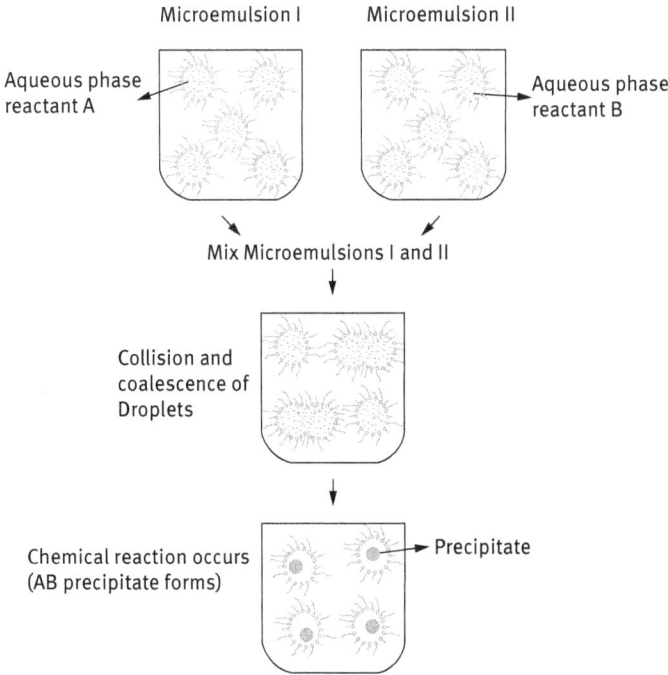

Fig. 10.20: Schematic representation of synthesis of nanoparticles in microemulsions.

rigid interface decreases the rate of coalescence and leads to a low precipitation rate. In contrast, a fluid interface in the microemulsion enhances the rate of precipitation.

Conceptually, one takes two identical W/O microemulsions and dissolves reactants A and B in the aqueous phase of these two microemulsions. Upon mixing the two microemulsions, the two reactants A and B come into contact due to collision and coalescence of the two microemulsions, thus resulting in the formation of precipitate AB. This is schematically represented in Fig. 10.20.

The above method was applied for the synthesis of silver halides, barium ferrite, and Y–Ba–Cu–O and Bi–Pb–Sr–Ca–Cu–O superconductors [45]. All these materials exhibit unique properties unlike those of materials synthesized using conventional techniques.

References

[1]   Hoar, T. P. and Schulman, J. H., Nature (London) **152**, 102 (1943).
[2]   Prince, L. M., "Microemulsion Theory and Practice", Academic Press, New York (1977).
[3]   Danielsson, I. and Lindman, B., Colloids and Surfaces, **3**, 391 (1983).
[4]   Schulman, J. H., Stoeckenius, W. and Prince, L. M., J. Phys. Chem., **63**, 1677 (1959).
[5]   Prince, L. M., Adv. Cosmet. Chem., **27**, 193 (1970).

[6]   Shinoda, K. and Friberg, S., Adv. Colloid Interface Sci., **4**, 281 (1975).
[7]   Ruckenstein, E. and Chi, J. C., J. Chem. Soc. Faraday Trans. II, **71**, 1690 (1975).
[8]   Overbeek, J. Th. G., Faraday Disc. Chem. Soc., **65**, 7 (1978).
[9]   Overbeek, J. Th. G., de Bruyn, P. L. and Verhoeckx, F., in "Surfactants, Th. F. Tadros (ed.), Academic Press, London (1984) pp. 111–132.
[10] Carnahan, N. F. and Starling, K. E., J. Chem. Phys., **51**, 635 (1969)
[11] Mitchell, D. J. and Ninham, B. W., J. Chem. Soc. Faraday Trans. II, **77**, 601 (1981).
[12] Baker, R. C., Florence, A. T., Ottewill, R. H. and Tadros, Th. F., J. Colloid Interface Sci., **100**, 332 (1984).
[13] Ashcroft, N. W. and Lekner, J., Phys. Rev., **45**, 33 (1966).
[14] Pusey, P. N., in "Industrial Polymers: Characterisation by Molecular Weights, J. H. S. Green and R. Dietz (eds.), Transcripta Books, London (1973).
[15] Cazabat, A. N. and Langevin, D., J. Chem. Phys., **74**, 3148 (1981).
[16] Cebula, D. J., Ottewill, R. H., Ralston, J. and Pusey, P., J. Chem. Soc. Faraday Trans. I, **77**, 2585 (1981).
[17] Prince, L. M., "Microemulsions", Academic Press, New York (1977).
[18] Lagourette, B., Peyerlasse, J., Boned, C. and Clausse, M., Nature, **281**, 60 (1969).
[19] Kilpatrick, S., Mod. Phys., **45**, 574 (1973).
[20] Clausse, M., Peyerlasse, J., Boned, C., Heil, J., Nicolas-Margantine, L. and Zrabda, A., in "Solution Properties of Surfactants", K. L. Mittal and B. Lindman (eds.), Plenum Press (1984), Vol. 3, p. 1583.
[21] Lindman, B. and Winnerstrom, H., in "Topics in Current Chemistry", F. L. Borschke (ed.), Springer-Verlag, Heidelberg, (1980), 1–83.
[22] Winnerstrom, H. and Lindman, B., Phys. Rep., **52**, 1 (1970).
[23] Lindman, B., Stilb, S. P. and Moseley, M. E., J. Colloid Interface Sci., **83**, 569 (1981).
[24] Gasco, M. R., "Microemulsions in the Pharmaceutical Field: Perspectives and Applications", in "Industrial Applications of Microemulsions", C. Solans and H. Kuneida (eds.), Marcel Dekker, New York (1997), Ch. 5.
[25] Ziegenmeyer, J. and Fuhrer, C., Acta Pharm. Technol., **26**, 273 (1980).
[26] Siebenbrodt, I. and Keipert, S., Pharmazie, **46**, 435 (1991).
[27] Gasco, M. R., Gallarate, M., Trotta, M., Bauchiro, L., Gremmo, E. and Chipparo, O., J. Pharm. Biomed. Anal., **7**, 433 (1989).
[28] Drew, J., Meier, R., Vonderscher, J., Kiss, D., Posanski, U., Kissel, T. and Gyr, K., Br. J. Clin. Pharmacol., **34**, 60 (1992).
[29] Couvreur, P., Roland, M. and Speiser, P., US Patent 4,329,332 (1982).
[30] Couvreur, P., Kante, B., Grislain, L., Roland, M. and Speiser, P., J. Pharm. Sci., **72**, 790 (1982).
[31] Al-Khouri, N., Roblot-Treupel, L., Fessi, H., Devissaguet, J. P. and Puisieux, F., Int. J. Pharm., **28**, 125 (1986).
[32] Nakajima, H., "Microemulsions in Cosmetics", in "Industrial Applications of Microemulsions", C. Solans and H. Kuneida (eds.), Marcel Dekker, New York (1997), Chapter 8.
[33] Dartnell, N. and Breda, B., Microemulsion containing a perfuming concentrate and corresponding product, US Patent 5,252,555 (1993).
[34] Chiu, Y. C. and Yang, W. L., Colloids and Surfaces, **63**, 311 (1992).
[35] Parra, J. L., Garcia Dominguez, J. J., Comelles, J., Sanchez, C., Solans, C. and Balaguer, F., Int. J. Soc. Cosmet. Sci., **7**, 127 (1985).
[36] Solans, C., Parra, J. L., Erra, P., Azemar, N., Clausse, M. and Touraud, D., Int. J. Soc. Cosmet. Sci., **9**, 215 (1987).
[37] Tadros, Th. F., "Colloids in Agrochemicals", Wiley-VCH, Germany (2009), Ch. 10.

[38] Dungan, S. R., "Microemulsions in Foods: Properties and Applications", in "Industrial Applications of Microemulsions", C. Solans and H. Kuneida (eds.), Marcel Dekker, New York (1997), Ch. 7.

[39] Israelachvili, N., Mitchell, D. J. and Ninham, B. W., J. Chem. Soc. Faraday Trans II, **72**, 1525 (1976).

[40] Shinoda, K., Araki, M., Sadaghiani, A., Khan, A. and Lindman, B., J. Phys. Chem., **95**, 989 (1991).

[41] Holmberg, K., "Microemulsions in Biotechnology", in "Industrial Applications of Microemulsions", C. Solans and H. Kuneida (eds.), Marcel Dekker, New York (1997), Ch. 4.

[42] Baviere, M. and Canselier, J. P., "Microemulsions in the Chemical EOR Process", in "Industrial Applications of Microemulsions", C. Solans and H. Kuneida (eds.), Marcel Dekker, New York (1997), Ch. 16.

[43] Taber, J. J., Soc. Petrol. Eng. J., **9**, 3 (1969).

[44] Abe, M., "Microemulsions in Enhanced Oil Recovery", in "Industrial Applications of Microemulsions", C. Solans and H. Kuneida (eds.), Marcel Dekker, New York (1997), Ch. 14.

[45] Pillai, V. and Shah, D. O., "Microemulsions as Nanosize Reactors for the Synthesis of Nanoparticles of Advanced Materials", in "Industrial Applications of Microemulsions", C. Solans and H. Kuneida (eds.), Marcel Dekker, New York (1997), Ch. 11.

Index

www.ingramcontent.com/pod-product-compliance
Lightning Source LLC
Chambersburg PA
CBHW081055220326
41598CB00038B/7104